Recent Developments in Agriculture

Recent Developments in Agriculture

Edited by Adriana Winkler

SYRAWOOD
PUBLISHING HOUSE

New York

Published by Syrawood Publishing House,
750 Third Avenue, 9th Floor,
New York, NY 10017, USA
www.syrawoodpublishinghouse.com

Recent Developments in Agriculture
Edited by Adriana Winkler

International Standard Book Number: 978-1-68286-716-7 (Hardback)

Cataloging-in-Publication Data

Recent developments in agriculture / edited by Adriana Winkler.
    p. cm.
Includes bibliographical references and index.
ISBN 978-1-68286-716-7
1. Agriculture. I. Winkler, Adriana.
S493 .R43 2019
630--dc23

TABLE OF CONTENTS

Preface VII

Chapter 1 **Isolation and Screening of Multifunctional Rhizobacteria from the Selected Sites of Madhupur, Narshingdi and Mymensingh, Bangladesh** 1
Moonmoon Nahar Asha, Atiqur Rahman, Quazi Forhad Quadir and
Md. Shahinur Islam

Chapter 2 **Comparative Effects on Storage Period of Varieties Pineapple Fruits** 9
Shah Md. Yusuf Ali, Md. Ahiduzzaman, Sharmin Akhter, M Abdul Matin Biswas,
Nafis Iqbal, Jakaria Chowdhury Onik and M Hafizur Rahman

Chapter 3 **Biological Control of Anthracnose of Soybean** 25
Akida Jahan, Nushrat Jahan, Farjana Yeasmin, Mohammad Delwar Hossain and
Muhammed Ali Hossain

Chapter 4 **Corruption and the Agricultural Production Efficiency of the European Countries during the Recent Economic Crisis** 33
Mohammad Monirul Hasan and József Tóth

Chapter 5 **Genetic Divergence of Indigenous Pummelo Genotypes** 44
Md. Sarowar Alam, Md. Sultan Mia, Md. Salim, Jubair Al Rashid and
Md. Saidur Rahman

Chapter 6 **Effects of Organic and Inorganic Fertilizers on the Growth, Yield and Nitrogen Uptake by BRRI dhan28** 51
Md. Rafiqul Islam, Aurunima Kanchi Suprova Shawon,
Most. Lutfun Nesa Begum and Azmul Huda

Chapter 7 **Climate Change and Crop Production Challenges: An Overview** 57
Sushan Chowhan, Shapla Rani Ghosh, Tushar Chowhan, Md. Mahmudul Hasan
and Md. Shyduzzaman Roni

Chapter 8 **Effects of Different Levels of Urea and Magic Growth Spray Solution on the Yield and Yield Attributes of BRRI dhan29** 76
Monika Nasrin, Md. Abdus Salam, Md. Akhter Hossain Chowdhury,
Md. Arif Hossain Khan and Md. Muzammel Hoque

Chapter 9 **Post-Harvest Factors Affecting Quality and Shelf Life of Mango Cv. Amropali** 84
Sherajum Monira, M. Abdur Rahim, MAB Khalil Rahad and M. Ashraful Islam

Chapter 10 **Variation in Morphological Attributes and Yield of Tomato Cultivars** 92
Amit Malaker, AKM Zakir Hossain, Tahmina Akter and Md. Shariful Hasan Khan

Chapter 11 **Screening of Potato Genotypes based on Glucose and Asparagines Content to Minimize Acrylamide Formation in Potato Chips and French Fries** 100
Fatema Zahan, Md. Masudul Karim, Tahmina Akter and Md. Alamgir Hossain

Chapter 12   **Evaluation of Productive Performance of Broiler in Response to Koroch**
              (*Pongamia pinnata*) **Cake Feeding**...**110**
              Masuma Habib, Abu Jafur Md. Ferdaus, Md. Touhidul Islam,
              Begum Mansura Hassin and Md. Shawkat Ali

Chapter 13   **Performance of Broiler Fed on Diet Containing Deoiled Koroch** (*Pongamia*
              Pinnata) **Seed Cake Treated with NaOH and HCl** **117**
              Masuma Habib, Abu Jafur Md. Ferdaus, Md. Touhidul Islam,
              Begum Mansura Hassin and Md. Shawkat Ali

Chapter 14   **Climate Change Effects and Adaptation Measures for Crop Production in**
              South West Coast of Bangladesh...**124**
              Rajib Jodder, Mohammad Asadul Haque, Tapan Kumar, M Jahiruddin,
              M. Zulfikar Rahman and Derek Clarke

Chapter 15   **Agricultural Waste Management Practices in Trishal Upazilla, Mymensingh**...........**134**
              Tangina Akhter, Md. Ali Ashraf, Md. Monirul Hassan, Farzana Akhter and
              Azmira Nasrin Riza

Chapter 16   **Control of Seed Borne Fungi on Tomato Seeds and their Management by**
              Botanical Extracts...**142**
              Imam Mehedi, Afia Sultana and Md. Amanut Ullah Raju

Chapter 17   **Comparative Study on Host Preference and Damage Potentiality of Red**
              Pumpkin Beetle, *Aulacophora foveicollis* **and Epilachna Beetles,** *Epilachna*
              dodecastigma..**150**
              Md. Mahbubur Rahman and Mohammad Mahir Uddin

Chapter 18   **Pathological Conditions of Avian Coccidiosis in the Small Scale Commercial**
              Broiler Farms in Dinajpur District..**156**
              Md. Manik Hossain, Md. Shahadat Hossain, Md. Tareq Mussa,
              SM Harunur Rashid and Md. Nazrul Islam

Chapter 19   **Empowerment of Resource Poor Women through Income Generation**
              Activities (IGAs): A Case of Slum Area in Dhaka City Corporation of Bangladesh **163**
              Sadia Jahan Moon and Md. Abdul Momen Miah

Chapter 20   **Effect of Different Doses of Ipil-Ipil** (*Leucaena leucocephala*) **(LAM.) De Wit.**
              Tree Green Leaf Biomass on Rice Yield and Soil Chemical Properties.........................**170**
              Niloy Paul, Mohammad Kamrul Hasan and Md. Nasir Uddin Khan

Chapter 21   **Postharvest Behavior and Keeping Quality of Potted Poinsettia**.....................................**180**
              M. Ashraful Islam and Daryl C. Joyce

Chapter 22   **Abundance and Composition of Zooplankton at Sitakunda Coast of**
              Chittagong, Bangladesh..**192**
              Md. Shahzad Kuli Khan, Sheikh Aftab Uddin and Mohammed Ashraful Haque

              Permissions

              List of Contributors

              Index

PREFACE

The purpose of the book is to provide a glimpse into the dynamics and to present opinions and studies of some of the scientists engaged in the development of new ideas in the field from very different standpoints. This book will prove useful to students and researchers owing to its high content quality.

Agriculture refers to the breeding and cultivation of plants and animals to provide economically important products like food, fur, fuels, medicine, etc. Crop production has significantly increased due to advances in modern agronomy, plant breeding and use of agrochemicals. Animal husbandry has seen similar increase in outputs due to advanced breeding techniques like selective breeding. Some of the common agricultural practices include shifting cultivation, subsistence farming, intensive farming, etc. Currently, new techniques are being explored to make agriculture more sustainable such as organic farming, integrated pest management, regenerative agriculture and no-till farming among many others. This book elucidates the concepts and innovative models around prospective developments with respect to agriculture. It studies, analyzes and upholds the pillars of agriculture science and its utmost significance in modern times. This book, with its detailed analyses and data, will prove immensely beneficial to professionals and students involved in this area at various levels.

At the end, I would like to appreciate all the efforts made by the authors in completing their chapters professionally. I express my deepest gratitude to all of them for contributing to this book by sharing their valuable works. A special thanks to my family and friends for their constant support in this journey.

Editor

ISOLATION AND SCREENING OF MULTIFUNCTIONAL RHIZOBACTERIA FROM THE SELECTED SITES OF MADHUPUR, NARSHINGDI AND MYMENSINGH, BANGLADESH

Moonmoon Nahar Asha, Atiqur Rahman, Quazi Forhad Quadir* and Md. Shahinur Islam

Department of Agricultural Chemistry, Faculty of Agriculture, Bangladesh Agricultural University, Mymensingh 2202, Bangladesh

*Corresponding author: Quazi Forhad Quadir, E-mail: qfq@bau.edu.bd

---

ARTICLE INFO	ABSTRACT

Key words

Rhizobacteria
Bioinoculants
Phosphorus
Solubilization
N-fixation
Plant growth

A laboratory experiment was performed to isolate some native rhizobacteria that could be used as bioinoculants for sustainable crop production. A total of 43 rhizobacteria were isolated from undisturbed plant rhizosphere soils of three different locations of Bangladesh and evaluated their plant growth promoting traits, both direct and indirect. The study has screened out isolates on the basis of their phosphorous solubilization and nitrogen (N) fixation. The phosphate solubilization assay in National Botanical Research Institute of Phosphate (NBRIP) medium revealed that 12 bacterial isolates were able to solubilize tricalcium phosphate and the rhizobacteria M25 showed best performance with a PSI of 3.33 at 5 day. Exactly 47% (20 isolates) of the isolated rhizobacteria were able to grow in N-free Winogradsky's medium, which is an indication of potential N_2-fixers. Among the 20 potential N-fixers, 15 were able to grow within 24 hours of incubation indicating that they are more efficient in N-fixation. The present study successfully isolated and characterized 43 rhizobacteria. Some of these isolated rhizobacteria have potential plant growth promoting traits and are potential plant growth promoting rhizobacteria (PGPR) candidate. Considering all plant growth promoting traits, the isolate F37 was the best followed by M6. However, further experiments are needed to determine the effectiveness of these isolates under *in vitro* and different field conditions to understand the nature of interaction with the plant and environment.

INTRUDUCTION

In recent years, focus has been on the use of plant growth promoting rhizobacteria (PGPR) as an alternative, environmentally friendly and effective strategy for plant control (Babalola and Glick, 2012; Patel *et al.*, 2012). Research on plants associated with microorganisms is currently expanding quite rapidly with the identification of new bacterial strains, which are more effective in promoting plant growth (Trivedi and Pandey, 2008). PGPR are among the most complex, diverse, and important assemblages in the biosphere (Khan, 2005). They are considered as a group of beneficial free living soil bacteria for sustainable agriculture and environment (Babalola, 2010).

PGPR are characterized by a number of activities, which include the capacity to colonize plant roots surfaces closely adhering to soil interface, increase mineral nutrient solubilization (i.e. P) and N- fixation (Khan, 2005; Shanab *et al.*, 2003) promote plant growth and yield, suppress plant diseases and soil borne pathogens by the production of hydrogen cyanide (HCN), siderophores, antibiotics, and/or competition for nutrients (Kamnev and Lelie, 2000; Shanab *et al.*, 2003; Idris *et al.*, 2007). Furthermore, PGPR improve plant stress tolerance especially to drought, salinity, metal toxicity and production of phytohormones such as indole-3-acetic acid (IAA) (Khan *et al.*, 2009; Verma *et al.*, 2010; Figueiredo *et al.*, 2010).

Research and development on beneficial rhizobacteriaas well as information on the exploitation of other plant growth promotional activities in Bangladesh is scanty. Though Atiqur *et al.*, (2010) have contributions relating those activities. The undisturbed forest flora of Madhupur and Narshingdi may harbour diverse rhizobacteria which may be rhizosphere competent, i.e. able to compete well with other rhizosphere microbes for nutrients secreted by the root. The present study was aimed to isolate and purify the rhizobacteria from the forest flora of Madhupur, Narshingdi and Mymensigh and to screen for P-solubilizing and N-fixing bacteria.

MATERIALS AND METHODS

Collection of rhizosphere plant samples

Fern (*Pteris* spp.) was collected from acidic soils of Madhupur forest, *Cyperus* spp. was collected from Rasulpur during the month of September 2013 and *Melastroma* samples were collected from red soils of Shivpur Upazilla, Narshingdi and Botanical Garden of Bangladesh Agricultural University, Mymensingh during January 2013. Plant samples were collected along with their root and carefully kept in plastic bags, labeled and sealed (in order to minimize the evaporation loss) and stored in a refrigerator at 4°C. A total of four plant samples were collected for the isolation of rhizosphere bacterial isolates.

Isolation and purification of bacterial strains

All the glassware including petridishes were sterilized for 20 minutes at 120°C by autoclave (Model: JSAC-80 JSR). A small part of roots from each sample were separately taken in a test tube. After pouring 10 mL of sterilized distilled water, vigourous shaking was done in bio-safety cabinet (Model: JSCB-900SB JSR). At the end of shaking, the samples were serially diluted upto 10^{-1} and 10^{-2} with sterilized distilled water. Then, 2-3 drops from dilutions and originals were placed on nutrient broth agar (NBA) by Spread Plate Technique and incubated at 28°C for 48h in a microbial incubator (Model: EN-120, Nuve) (Atiqur, 2010). Several screenings were done to isolate pure cultures. Pure colonies were isolated and maintained on NBA plates at 4°C for further studies. For long term storage, the pure isolates were preserved in eppendorf tubes and stored in 10% glycerol solution at -20°C.

Characterization of bacterial strains

Morphological characterization of bacterial strains

The bacterial isolates were grown in NBA medium (containing nutrient broth 1%, sucrose (1%, agar 1.5% and pH=6.5) and incubated in microbial incubator at 28°C for 24 hours to study their colony structure, shape, color, elevation and pigmentation. All the glassware and media were sterilized before incubation.

Biochemical characterization of bacterial strains

Gram reaction test

To study the biochemical characteristics, all the bacterial isolates were grown in NBA medium and incubated in microbial incubator at 28°C for 48 hours. On glass slide a loopful of bacteria from a well grown colony was mixed with a drop of 3% KOH aqueous solution. Mixing was continued for less than l0 seconds. A toothpick was used for picking as well as mixing bacteria from a colony. The toothpick was raised a few centimeters from the glass slide. Strands of viscid material confirmed that the bacterium was gram-negative. Producing thread with a toothpick indicates that it was gram (+ve) bacteria (Ahmed, 2011).

Catalase test

A small amount of bacterial isolate was placed from culture onto a clean microscope slide. A few drops of H_2O_2 were added onto the smear. A positive result is the rapid evolution of O_2 as evidenced by bubbling. A negative result is no bubbles or only a few scattered bubbles (Wheelis, 2008).

Screening for Phosphate Solubilizing Bacteria

All bacterial strains were tested by an agar assay using National Botanical Research Institute Phosphate (NBRIP) containing sucrose = 10 gL^{-1}, $Ca_3(PO_4)_2$ =5 g L^{-1}, $MgCl_2.6H_2O$ = 5 g L^{-1}, $MgSO_4.7H_2O$ = 0.25 g L^{-1}, KCl = 0.2 g L^{-1}, $(NH_4)_2SO_4$ = 0.1 g L^{-1}, pH=7.0 medium supplemented with 1.5% agar (Nautiyal, 1999). Six strains per plate were stabbed in triplicate using sterile toothpicks. The halo and colony diameters were measured after 5 day of incubation of the plates at 25°C. The ability was described by the following equation (Edi *et al.,* 1996).

$$\text{Solubilization index} = \frac{\text{Colony diameter} + \text{Clearing zone}}{\text{Colony Diameter}}$$

Screening for N-fixing bacteria

To study nitrogen fixation ability of the bacteria, isolates were grown in modified Winogradsky's medium (Hashidoko *et al.,* 2002) which is an N-free medium. (Winogradsky's medium without tryptophan and yeast extract) and kept for 48 hours in microbial incubator at 28°C.

RESULTS

Collection and isolation of bacteria

A total of 43 rhizobacterial strains as shown in Table 1 were successfully isolated from the rhizosplane of four plant root samples collected from red soils of Rasulpur area of Madhupur forest, Sonaimuri Upazila of Narshingdi and Botanical Garden of BAU, Mymensingh.

Morphological characteristics

The morphological characteristics of PGPR isolates varied widely. All the isolates produced different size, shape and elevation and pigment production capacity. Some of the rhizobacterial isolates showed clear pigment producing ability, while others are creamy, colorless or white in colour. The morphological characteristics of the isolated bacteria have been presented in Table 2.

Biochemical characteristics

All the isolates were studied for their biochemical characteristics. Both gram reaction and catalase tests were done. Among the 43 rhizobacteria, 17 isolates were gram negative and the rests were gram positive (Table 3). In catalase test, only 3 isolates were catalase negative (*viz.* M10, M11 and M31). Catalase positive bacteria was indicated by the production of bubble with 3% H_2O_2.

Screening phosphate solubilizing bacteria

The isolated rhizobacteria were screened for P solubilizing capability using tricalcium phosphate. After five days of incubation in NBRIP agar medium, only 12 out of 43 isolates were able to grow and showed phosphate solubilizing capacity in the form of $Ca_3(PO_4)_2$ with various phosphate solubilizing indexes. Out of 43 isolates, 12 isolates grew and solubilize phosphate on NBRIP medium as evidenced by hallow zone around the colony (Figure 1). The phosphate solubilizing capacity of the isolated rhizobacteria have been shown in Table 4.

The remaining isolates were unable to grow in NBRIP medium. Among the solubilizers, 3 isolates showed weak solubilization (PSI<1.5, 5 showed moderate solubilization (PSI upto 2.5) and 5 isolates have high phosphate solubilizing capacity (PSI >2.5). Maximum phosphate solubilizing activity was shown by M25 (PSI 3.33) followed by F39 (PSI 3.0). Least P solubilization was exhibited by M35 (PSI 1.11).

Screening N-fixing Bacteria

To screen nitrogen fixing bacteria, N-free modified Winogradsky's medium was used. The bacteria able to grow in N-free medium were identified as nitrogen fixing bacteria. Among 43 isolates, 20 grew in N-free Winogradsky's medium with 15 isolates showing high growth only after 24 hours (Table 5).

Table 1. List of isolated Rhizobacteria from collected plant roots

Source	Sampling location	Isolate code
Melastroma spp.	Narshingdi	M1, M2, M3, M4, M5, M6, M7, M8, M9, M10, M11, M12, M13, M14, M15, M16, M17, M18
*Melastroma*spp.	Botanical garden	M19, M20, M21, M22, M23, M24, M25, M26, M27, M28, M29, M30, M31, M32, M33, M34, M35
*Pteris*spp.	Madhupur	F36, F37, F38, F39
*Cyperus*spp.	Rasulpur	FR40, FR41, FR42, FR43

Table 2. Morphological characteristics of isolated Rhizobacteria

Morphological properties	Isolate No.
Shape	
Round	M1, M2, M3, M4, M5, M7, M8, M9, M12, M16, M17, M19, M23, M25, M26, M27, M28, M31, M32, M34, M35, F36
Irregular	M6, M11, M13, M14, M18, M20, M21, M22, M24, M29, M30, M33, F37, F38, F39, FR40, FR41, FR42, FR43
Oval	M10, M15
Colour	
Deep red	M9, M12
Grey	M5
Greyish	M1, M2, M7, M11, M40
Pink	M26
Reddish	M23
Red	M28, M30, M34
Slight green	M17, F37, F38
Slight red	M27
Transparent	M32
Whitish	M10, M13, M18, M20, M21, M25, FR41
White	M3, M4, M8, M14, M15, M19, M22, M31, M33, M35, F39, FR42
Yellowish	M16, M24
Yellow	M29, F36, FR43
Elevation	
Raised	All except M18, M22, M26
Depressed	M18, M22, M26

Table 3. Biochemical characteristics of isolated Rhizobacteria

Biochemical Characterization	Isolate No.
Gram test	
Gram (-)	M1, M2, M11, M17, M19, M21, M22, M24, M27, M28, M34, M35, F37, F38, F39, FR40
Gram (+)	M3, M4, M5, M6, M7, M8, M9, M10, M12, M13, M14, M15, M16, M18, M20, M23, M25, M26, M29, M30, M31, M32, M33, FR41, FR42, FR43
Catalase test	
Catalase (+)	M1, M2, M3, M4, M5, M6, M7, M8, M9, M12, M13, M14, M15, M16, M17, M18, M19, M20, M21, M22, M23, M24, M25, M26, M27, M28, M29, M30, M32, M33, M34, M35, F36, F37, F38, F39, FR40, FR41, FR42, FR43
Catalase (-)	M10, M11, M31

Table 4. Phosphate solubilization capacity of isolated Rhizobacteria with PSI

PSI value	Isolate No.
0.0 – 1.0	M3, M5, M7, M8, M9, M10, M12, M13, M14, M15, M18, M19, M20, M21, M22, M23, M24, M26, M27, M28, M29, M30, M31, M32, M33, M34, , F36, FR40, FR41, FR42, FR43
1.0 – 2.0	M4, M11, M16, M35
2.0 – 3.0	M1, M2, M6, M17, F37, F38, F39
3.0 – 4.0	M25

Table 5. Bacterial isolates response in N free media

Growth in N free Medium*	Isolate No.
-	M3, M4, M5, M6, M8, M9, M12, M19, M21, M24, M26, M27, M28, M29, M30, M32, M33, M34, M35, F38, F39, FR41, FR42, FR43
+	M7, M22, M25, FR40
++	M1, M2, M10, M11, M13, M14, M15, M16, M17, M18, M20, M23, M31, F36, F37

*(-), (+) and (++) designate low, medium and high bacterial growth, respectively in N-free Winogradsky's medium after 24 hrs.

DISCUSSION

Plants provide a nutrient rich habitat for the growth and development of various groups of microorganisms, especially bacteria. Bacteria profit from plants because of the enhanced availability of nutrients and plants. In turn, plant benefit from the bacterial associations by growth enhancement, stress reduction or protection from pathogens. In present study, 43 rhizobacteria were isolated from 4 root samples collected from undisturbed red soils of Madhupur, forest soil of Narshingdi and one samples collected from Botanical Garden of Bangladesh Agricultural University, Mymensingh. The morphological and biochemical characteristics of the isolates were studied which revealed that the undisturbed rhizosphere was a rich source of diverse group of rhizobacteria. The isolated rhizobacteria varied in shape, colour, pigment production and other morphological features studied.

Identification and characterization of plant growth promoting traits of the rhizobacteria is an important step in determining their potential utilization in sustainable soil management and crop production system. The beneficial effect of rhizobacteria on plant growth stimulation may be due to a number of mechanisms. To determine the potential of plant growth stimulation, all the 43 isolated rhizobacteria were studied for their biofunctionalities. The isolates were tested for their mineral phosphate solubilization and growth under N-free conditions and the results were discussed in following sections.

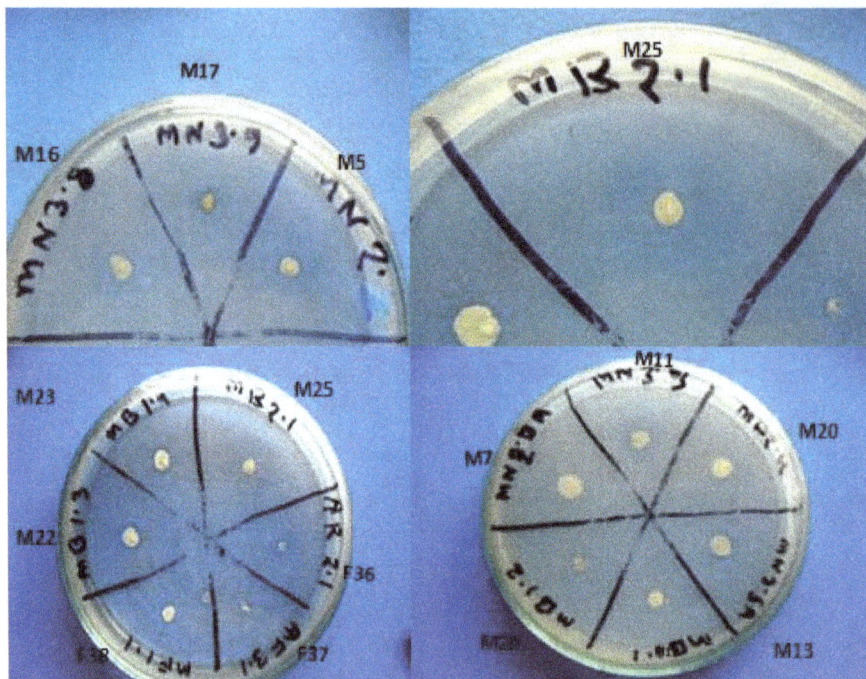

Figure 1. Bacterial isolates showing P-solubilization in NBRIP media supplemented with tricalcium phosphate. Hallow zones around the bacterial colony indicates phosphate solubilization.

Phosphate solubilization

Phosphorus is a major essential macronutrient for biological growth and development. Plant growth is often limited by insufficient phosphate availability. The low solubility of common phosphates such as $Ca_3(PO_4)_2$, hydroxyapatite and aluminium phosphate may cause low phosphate availability in agricultural soils. Microorganisms offer a biological rescue system capable of solubilizing the insoluble inorganic P of soil and make it available to the plants. The ability of some microorganisms to convert insoluble P to an accessible form, like orthophosphate, an important trait in a PGPB for increasing plant yields (Chen *et al.*, 2006). High proportion of phosphate solubilizing microbes is concentrated in the rhizosphere and they are metabolically more active than other sources (Vazquez *et al.*, 2000).

Qualitative P-solubilization potential estimated by observing the large clear/halo zones on agar media revealed that out of 43 bacterial isolates tested, 12 isolates had P-solublizing ability (Table 4) in agar medium supplemented with insoluble $Ca_3(PO_4)_2$. The isolates exhibited different degrees of phosphate solublizing capacity as revealed by PS index ranging from 1.11 to 3.33 (at day 5). The isolated rhizobacteria showed variation in their phosphate solubilization capacity. This variation in P-solubilization by the rhizobacteria might be due to difference in their organic acids production capacity both in terms of amount and type of acids. Similar results were reported by Rashid *et al.* (2004). The rhizobacterial isolates M25 and F39 showed the highest P solubilization (PSI >3) in plate culture which is generally a reliable method for preliminary screening and characterization of P-solubilizing microorganisms. In order to precisely characterize the isolates for their P-solubilization ability, liquid culture and field trial of the P-solubilizers identified through plate assay. The cause of P solubilization could probably be due to secretion of organic acids, such as gluconic, 2-ketogluconic, oxalic, citric, acetic, malic, and succinic acid (Zaidi *et al.* 2009).

The ability of several isolates to solubilize tricalcium phosphate *in vitro* suggested the application of those isolates in crop fields. Rodriguez and Fraga (1999) studied that *Pseudomonas* and other PSB like *Bacillus* and *Rhizobium* were capable of increasing the availability of P in soil. Specifically, all those isolates might be potential inoculants for alkaline soil based on the ability to solubilize phosphate bounded by calcium which mostly exists in alkaline soils (Goldstein, 1995). Soil inoculation with PSB has been shown to improve solubilization of fixed soil P and applied phosphates resulting in higher crop yields (Nautiyal and Mehta, 2001).

Several PSB could also promote plant growth by rendering phosphate into solution more than they need for their metabolism and the surplus can be absorbed by plant (Kloepper et al., 1980).

Nitrogen fixation

Nitrogen is the most significant yield-limiting element in many agricultural production systems. The nitrogen-fixing bacteria have stimulating effect on the plant; they are able to fix the nitrogen in symbiosis with leguminous plants using the nitrogenase enzyme. Biological N_2 fixation is gaining importance in rice ecosystem because of current concern on the environmental and soil health that are caused by the continuous use of nitrogenous fertilizers and the need for improved sustainable rice productivity.

Therefore, the isolated rhizobacteria were tested for their response in N-free medium. Of the 43 isolates, 19 grew in the medium indicating that these bacteria were able to utilize the atmospheric nitrogen. Fifteen rhizobacterial isolates showed high growth response only after 24 hours, while others grew after 48 hours of incubation. The remaining isolated bacteria (24) could not grow in N-free medium indicating that these bacteria were unable to utilize atmospheric N_2 and therefore, these bacteria are non N_2-fixing bacteria. However, to confirm their N-fixing ability, acetylene reduction assay and NifH gene identification is necessary before their intended utilization as PGPR.

CONCLUSION

Further study is needed to evaluate microbe-microbe and plant-microbe-environment interactions and their plant growth-promoting activities to ensure their use in a sustainable manner in crop production system.

COMPETING INTEREST

The authors declare that they have no competing interests.

ACKNOWLEDGEMENTS

The research work was funded by Ministry of Science and Technology (MOST), Bangladesh and HEQEP-AIF Sub-project (CP#2013). Authors would like to thank Mr. Istiaq Ahmed for his kind assistance during the conduction of the research work.

REFERENCES

1. Ahmed I, 2011. Isolation and characterization of As resistant bacteria from the contaminated soil and their effects on seed germination of rice, MS Thesis, Department of Agricultural Chemistry, Bangladesh Agricultural University.
2. Atiqur R, IR Sitepu, S Tang, and Y Hashidoko, 2010. Salkowski's reagent test as a primary screening index for functionalities of rhizobacteria isolated from wild dipterocarp samplings growing naturally on medium-strongly acidic tropical peat soil. Journal of Bioscience, Biotechnology and Biochemistry, 74: 2202-2208.
3. Babalola OO, 2010. Beneficial bacteria of agriculture importance. Biotechnological Letters, 32: 1559-1570.
4. Babalola OO and BR Glick, 2012. Indigenous African agriculture and plant associated microbes: Current practice and future transgenic prospects. Scientific Research Essays, 7: 2431-2439.
5. Chen YP, PD Rekha, AB Arun, FT Shen, WA Lai, and CC Young, 2006. Phosphate solubilizing bacteria from subtropical soil and their tricalcium phosphate solubilizing abilities. Applied Soil Ecology, 34: 33-41.
6. Edi PM, AM Moawad, and PLG Vlek, 1996. Effect of phosphate-solubilizing Pseudomonas putida on the growth of maize and its survival in the rhizosphere. Indonesia Journal of Crop Science, 11: 13-23.

7. Figueiredo MVB, L Seldin, FF Araujo and RLM Mariano, 2010. Plant growth promoting rhizobacteria: Fundamentals and applications. Plant Growth and Health Promoting Bacteria, Microbiology Monographs, 18: 21-43.

8. Goldstein AH, 1995. Recent progress in understanding the molecular genetics and biochemistry of calcium phosphate sulubilization by gram negative bacteria. Biological, Agriculture and Horticulture, 12: 185-193.

9. Hashidoko Y, M Tada, M Osaki and S Tahara, 2002. Soft gel medium solidified with gellab gum for preliminary screening for root-associating, free-living nitrogen-fixing bacteria inhabiting the rhizoplane of plants. Bioscience, Biotechnology and Biochemistry, 66: 2259-2263.

10. Idris SE, DJ Iglesias, M Talon and R Borriss, 2007. Tryptophan-dependent production of Indole-3-Acetic Acid (IAA) affects level of plant growth promotion by Bacillus amyloliquefaciens FZB42. Molecular Plant-Microbe Interactions, 20: 619-626.

11. Kamnev AA and D Lelie, 2000. Chemical and biological parameters as tools to evaluate and improve heavy metal phytoremediation. Bioscience Report, 20: 239-258.

12. Khan AG, 2005. Role of soil microbes in the rhizosphere of plants growing on trace metal contaminated soils in phytoremediation. Journal of Trace Elements in Medicine and Biology, 18: 355-364.

13. Khan MS, A Zaidi, PA Wani and M Oves, 2009. Role of plant growth promoting rhizobacteria in the remediation of metal contaminated soils. Environmental Chemistry Letters, 7: 1-19.

14. Kloepper JW, J Leong, M Teintze and MN Schroth, 1980. Enhanced plant growth by siderophores produced by plant growthpromotingrhizobacteria. Nature, 286: 885- 886.

15. Nautiyal CS, 1999. An efficient microbiological growth medium for screening phosphate solubilizing microorganisms. FEMS Microbiology Letters, 170: 265-270

16. Nautiyal CS and S Mehta, 2001. An Efficient Method for Qualitative Screening of Phosphate-Solubilizing Bacteria. Microbiology, 43: 51-56.

17. Patel HA, RK Patel, SM Khristi, K Parikh and G Rajendran, 2012. Isolation and characterization of bacterial endophytes from Lycopersiconesculentum plant and their plant growth promoting characteristics. Nepal Journal of Biotechnology, 2: 37-52.

18. Rashid M, S Khalil, N Ayub, S Alam and F Latif, 2004. Organic acids production and phosphate solubilization by phosphate solubilizing microorganisms (PSM) under in vitro conditions. Pakistan Journal of Biological Science, 7: 187-196.

19. Rodriguez H and R Fraga, 1999. Phosphate solubilizing bacteria and their role in plant growth promotion, Biotechnology Advances, 17: 319-339.

20. Shanab RI, JS Angle, TA Delorme, RLChaney, P Berkum, H Moawad, K Ghanem and HA Ghozlan, 2003. Rhizobacterial effects on nickel extraction from soil and uptake by Alyssum murale. New Phytologist, 158: 219-224.

21. Trivedi P and A Pandey, 2008. Recovery of plant growth-promoting rhizobacteria from sodium alginate beads after 3 years following storage at 4°C. Journal of Industrial Microbiology and Biotechnology, 35: 205-208.

22. Vazquez P, G Holguin, M Puente, AE Cortes and Y Bashan, 2000. Phosphate solubilizing microorganisms associated with the rhizosphere of mangroves in a semi-arid coastal lagoon. Biology and Fertility of Soils, 30: 460-468.

23. Verma JP, J Yadav, KN Tiwari, Lavakush and V Singh, 2010. Impact of plant growth promoting rhizobacteria on crop production. International Journal of Agricultural Research, 5: 954-983.

24. Wheelis M, 2008. Principles of Modern Microbiology, Jones and Bartlett Publishers, Inc., Sudbury, MA.

25. Zaidi A, MS Khan, M Ahemad and M Oves, 2009. Plant growth promotion by phosphate solubilizing bacteria. Acta Microbiologica et Immunologica Hungarica, 56: 263-284.

COMPARATIVE EFFECTS ON STORAGE PERIOD OF VARIETIES PINEAPPLE FRUITS

*Shah Md. Yusuf Ali[1], Md. Ahiduzzaman[2], Sharmin Akhter[1], M Abdul Matin Biswas[1], Nafis Iqbal[2], Jakaria Chowdhury Onik[2] and M Hafizur Rahman[2]

[1]Department of Agricultural Extension, Gazipur Sadar-1701, Bangladesh; [2]Department of Agro-processing, Faculty of Agriculture, Bangabandhu Sheikh Mujibar Rahman Agricultural University, Salna, Gazipur-1706, Bangladesh

*Corresponding author: Shah Md. Yusuf Ali; E-mail: s.m.yusuf1968@gmail.com

ARTICLE INFO

Key words

Pineapple
p^H meter
Refract meter
Weighing balance
Sstorage period

ABSTRACT

Pineapple is considered as one of the most wanted tropical fruits and it is widely taken for fresh consumption as well as their flesh and juice are used for preparation of different product in Agro-processing industries. For such industrial processes, it is important to know the information of characteristics changes of pineapple during day after storage. Four varieties of pineapple were collected from different areas of Bangladesh named Honey Queen (H.Q), Giant Kew (G.K), Asshini and Ghorasal. Some Physico-chemical properties (weight loss, moisture content, ash and edible portion, pH, TSS, titrable acidity (TA), total sugar, reducing sugar) biochemical properties (ascorbic acid) and sensorial attributes (color, odor, firmness, appearances, sweetness and overall acceptability) of pineapple juice were studied during day after storage. This study examined the Comparison of different varieties of pineapple fruit characteristics and sensory quality of the pineapple fruits during storage. It was shown that there was a significant changes between the storage periods in relation to different varieties of fruits. The firmness of pineapple fruits were in outside and inside to be 0.21 to 0.27 N/m^2 and 0.06 to 0.10 N/m^2, respectively. The pH values of different varieties were found to be in the range of 4.30 to 4.36. The highest and lowest sweetness index were estimated to be 36.30 and 22.15 for Honey Queen and Asshini respectively. The highest and lowest magnitude of sugar contents of four pineapple varieties were found to be in the range of 14.16 to 15.8 mg/100g.The average TSS values were found to be 15.12%, 12.33%, 13.14% and 12.95% for H.Q., G.K., Asshini and Ghorashal, respectively. The comparative study indicated the characteristics of different varieties of pineapple changes during after storage.

INTRODUCTION

Pineapple (*Ananus comosus* L. Merr.) has long been one of the most appreciated fruit tropical and sub-tropical area, because of its attractive flavor and refreshing sugar-acid balance. Pineapple has long been an important cash crop. The climate and the soils of many parts of Bangladesh are suitable for pineapple production. It is widely cultivated in Bandarban, Khagrachari, Moulvibazar, Sylhet, Chittagong, Rangamati Hill tract, Tangail (The daily Star, 2014). At least ninety varieties of pineapple are cultivated in the world. In Bangladesh, however, three varieties of pineapple are mostly grown. The cultivated varieties are Giant Kew (locally Kalandar), Honey Queen (Jaldubi) and Red Spanish (Ghorashal) etc. Apart from this variety, one local variety named "Asshini" (late variety) is grown by a few farmers of Madhupur, Tangail district. About 45,685 ha of land are now 2 under pineapple cultivation with a total production of about 2, 34,865 metric tons. Another local variety named "Ghorashal" is cultivated at Palash upazila in Narsingdi district. In respect of the total production of fruits which grown in Bangladesh it ranks 4th among the major fruits (BBS, 2014). But Bangladeshi pineapples are more juicy and tasty than any other countries (Banglapedia, 2006). The H.Q variety is largely produced in Sylhet, Habigang, Moulviganj, Kamalgonj and Satgaon with coverage of 4454.00 ha (The Daily Star, 2014). However, the Honey Queen is the small fruit compared to other varieties which has a necessary demand. Rabon and Borab areas of Palash, Narsingdi district are found that farmers of the area are involved themselves in pineapple cultivation and they raise a new orchard with their own initiative (Banglapedia, 2006).

Pineapple is used for the preparation of alcohol, calcium citrate, citric acid, vinegar, oxalic acid, pineapple gum and flavor. It is a good source of vitamin A, B & C, carotene, ascorbic acid and is rich in calcium, phosphorus, magnesium, potassium and iron (USDA Nutrient data base 2008 and Rashid et.al 1987). There are some of the many health benefits of eating ripens pineapples such as packed with vitamins and minerals, prevents cough and colds, strengthens bones, keeps gums healthy, lowers risk of macular degeneration, alleviates arthritis, improves digestion etc. (USDA nutrient database 2008). Besides, it is also a source of bromelin, a digestive enzyme (Lodh et al., 1973). Various food items like squash, syrup, jelly, etc. are prepared from pineapple. Its juice is helpful for healing fever, jaundice, influenza and cold, among other ailments (USDA nutrient database 2008). Diversified uses of pineapple have also led to develop many fruit processing industries both in developed and developing countries including Bangladesh. However, the processing characteristics of pineapple such as different physio- chemical properties are not properly studied for the different varieties grown in Bangladesh. That's why it is necessary to know the different properties of different varieties of pineapple. This study was conducted to find out the physio- chemical characteristics and sensory quality of the different pineapple varieties during days after storage. The objective of the research was to study the effect of storage period of pineapple varieties on physico-chemical properties and sensorial attributes and to make a comparison of different properties among the varieties of pineapple.

MATERIALS AND METHODS

Experimental materials

The experiment was carried out in the laboratories of the Department of Agro-Processing, Bangabandhu Sheikh Mujibur Rahman Agricultural University, Gazipur, during the period from July, 2014 to June 2015. The laboratory of Agro-Processing department was used for chemical analysis of fruits and the laboratory of Post-Harvest Technology division, BARI was used for preliminary studies.

Materials used for this experiment

The materials used for the experiment were the freshly harvested pineapple fruits of the variety Giant Kew, Honey Queen, Asshini and Ghorashal. The fruits Giant Kew and Asshini were collected from field of the growers of Gachabari village and Jalchittra market, Madhupur Upazila in Tangail district, Honey queen variety was collected from Sylhet and Ghorashal variety was collected from Palash upazilla, Narsinghdi district.The fruits of optimum maturity stage were identified and harvested in the morning hours and immediately transferred with careful handling and placed to the Department of Agro-Processing laboratory, Bangabandhu

Sheikh Mujibur Rahman Agricultural University, Gazipur. The fruit was kept for twelve days for storage to analyze the different properties of fruits at different days after storage. The experimental pineapples were loaded immediately in a recognized wooden box with utmost care and covered with a polythene paper to protect the fruits from direct sunlight. For transport fruits were placed on leaves with crowns and layer to layer to decrease mechanical damage. Then the fruits were immediately transferred to the Agro-processing Laboratory.

Pineapple Storage for experimental use

The different varieties of pineapple were placed in the Agro-Processing laboratory at the room temperature located at the floor. The floor was covered with newspaper. Very little information available in Bangladesh on physic- chemical changes, shelf life changes during storage and ripening. Storing at room temperature will increase the acidity level of the pineapple, but will not improve sweetness.

Sample preparation

Pineapple were weighed using a top loading balance (Salter Model, Japan), the weight of the samples were determined and juice extraction from samples were performed according to the method described by Lim (1985). The crown and stem portions were removed and the skin were peeled by using knife. After that the fruits were sliced and the fruit slices were pressed using a blender (Jensons Lab, Japan) into paste for one to two minutes to get the pineapple juice and then the juice was collected. The mixture was then filtered through a muslin cloth Juice was collected is similar way for other varieties and the extracts were then stored in a normal room temperature.

The extracted juice was used for present study. Standard procedure was used for determining the different characters. The data obtained from different pineapples samples were recorded. The experimental treatments were as follows:

Table1. Treatments of the study

Variety	Days after storage (DAS)											
	D_0			D_4			D_8			D_{12}		
H.Q.	T1	T2	T3	T1	T2	T3	T1	T2	T3	T1	T2	T3
G.K.	T1	T2	T3	T1	T2	T3	T1	T2	T3	T1	T2	T3
Asshini	T1	T2	T3	T1	T2	T3	T1	T2	T3	T1	T2	T3
Ghorasal	T1	T2	T3	T1	T2	T3	T1	T2	T3	T1	T2	T3

N.B. D0: 0 DAS, D_4: 4 DAS, D_8: 8DAS, D12: 12DAS and T1, T2 and T3 are recap

For experiment, fifty six numbers of different varieties of pineapple were taken. For experiment three number of each variety fruit were taken. Table 1 show that there were forty eight number of different varieties of pineapple were taken for physio- chemical parameters and total eight number of varieties were kept for shelf life observation after 0, 4, 8 and 12 DAS.

Analyses of physical properties

The effect of different postharvest treatments on shelf life and the quality changes of pineapples storage were studied here. The data were collected on different characteristics of sensory evaluation and analyzed. The following parameters were studied.

Percent weight loss of fruit

Per weight loss was calculated by using the following equation:

$$\text{Percent weight loss (\% WL)} = \frac{IW - FW}{IW} \times 100$$

Where,
WL = Percent total weight loss, IW = Initial weight of fruits (kg), FW = Final weight of fruits (kg)

Moisture content

Moisture content was determined by using the Standard Official methods of Analysis (AOAC 1990). This involved drying to a constant weight at 105^0 C at calculated moisture as the loss in weight of the dried samples. The crucible was thoroughly washed and dried in an oven at 100^0C for 30 min and allowed to cool inside desiccators. After cooling they were weighed using a weighing balance and recorded as W_1, then 2.0 gm of the finely ground samples were put into crucibles and weighed to determine the value of W_2. Thereafter, the sample and crucible were placed inside the oven and dried at 105^0 C for 4 hour, then cooled & weighed at the same temperature for 30 min until constant weights were obtained to get W_3.

The percent moisture content was calculated using the following equation:

$$\% \text{ moisture content} = \frac{W_1 - W_2}{W_1 - W_3} \times 100$$

Where,

W_1= Initial pineapple sample weight with crucible

W_2= Final pineapple sample weight with crucible

W_3= Initial weight of empty crucible

Determination of total ash

Total ash content of the sample was determined by incineration in a muffle furnace, as described by AOAC based on the vaporization of water and volatile with being organic substances in the presence of Oxygen in the air to CO_2 at a Temperature of 550^0 C (Dry ash) . About 1.0 gm of finely ground dried sample was placed in a porcelain crucible and incinerated at 550^0 C for 6 hour in a muffle in furnace. The ash was cooled in a desiccators and weighed. The percentage of ash content in the sample was calculated.

$$\text{Ash content (\%)} = \frac{wt. of ash}{Total\ wt. of\ sample} \times 100$$

Edible portion (%)

Initially the total weight of fruit without crown was weighed by using a balance. Then the fruit was peeled by using sharp knife and the central core was removed and the remaining fruit pulp was weighed. Finally the percentage of edible portion of the fruit was measured with the following equation (Ranganna 1994).

$$\text{Edible portion of fruit (\%)} = \frac{wt\ of\ edible\ portion}{Total\ wt. of\ fruit} \times 100$$

Shelf life

The shelf life was recorded by counting the days required to attain the last stage of ripening, but the fruit remaining still ready for marketing (Ranganna 1994).

Firmness Test

The firmness of pineapple fruit can be determined (outside and inside) by measuring penetration force using an Instron Universal testing machine.

Analyses of Chemical properties

pH measurement of pulp juice

The pH of pineapple juice samples was evaluated with the Sartorius Professional Meter PP-50. The measurement was a mean value of triplicates.

Determination of TSS

The Total soluble solids of pineapple juice samples were measured using digital refractometer Digit at 19-20^0 C. Results were reported as degrees Brix.

Table 3. Effects of physical properties of different varieties of pineapple during day after storage.

Characteristics	Varieties of pineapple															
	Honey Queen				Giant Kew				Asshini				Ghorashal			
	D0	D4	D8	D12	D0	D4	D8	D12	D0	D4	D8	D12	D0	D4	D8	D12
Moisture content %	84.2	83.3	83.1	82.5	86.1	85.2	83.2	82.7	86.3	85.3	84.0	82.7	86.1	84.4	83.8	82.3
Ash content %	0.32	0.34	0.36	0.41	0.25	0.28	0.32	0.36	0.28	0.3	0.34	0.38	0.31	0.32	0.34	0.40
Edible portion %	60.5	62.3	63.2	63.8	62.6	63.9	64.8	65.8	64.5	65.5	66.40	66.8	58.14	60.13	60.90	61.16
Wt loss %	4.26	7.06	9.2	11.7	5.04	7.40	9.50	11.9	-	--	-	-	3.5	5.47	7.5	9.4

Table 3 shows the physical properties such as ash content (%) and the edible portion (%) of different varieties of pineapple are increased during increased storage time. But the moisture content of fruits are decreased during increased storage period.

Vitamin C estimation

Ascorbic acid content was determined to the method of Ranganna (1994) by using 2,6-Dichlorophenol-Indophenol Visual Titration Method. i) 3% metaphosphoric acid (HPO_3): It was prepared by dissolving 50 g HPO3 in distilled water and volume make up to 1000 ml.ii) Ascorbic acid standard: Weigh accurately 100 mg of L-ascorbic acid and make up to 100 ml with 3% HPO3. Dilute 10 ml to 100 ml with 3% HPO3 (1 ml= 0.1 mg of ascorbic acid).iii) Dye Solution: 0.0525 gm of sodium bcarbonate dissolve in hot glass distilled water and add 0.0625 gm of 2,6 dichlorophenol- indophenols cool and make up 250 ml (Lane and Eynon 1923). The following steps were followed for estimation of ascorbic acid:

a) **Standardization of dye solution**: Five ml of standard ascorbic acid solution was taken in a conical flask and 5 ml of metaphosphoric acid (HPO_3) solution was added to it and shaken. A micro burette was filled with the dye solution and the mixed ascorbic acid solution was titrated with dye where appearance of pink color indicated the end point, which persisted at least15 seconds. The milliliters of dye solution required to complete the titrations recorded.
Dye factor was calculated using of the following formula-

$$\text{Dye factor} = \frac{1}{Titre}$$

b) **Preparation of sample**: Ten grams of fresh pulp was homogenized with 50 ml of 3% metaphosphoric acid solution in a blender machine. After blending it was filtered and transferred to a 5000 ml volumetric flask and was made up to the mark with 3% metaphosphoric acid.

c) **Titration**: Ten ml of pulp extracted sample was taken in an aliquot and titrated with dye solution (2, 6-dichlorphenol-Indophenol) till pink color was appeared which persisted at least 15 seconds. The titration was replicated thrice for each time.

The ascorbic acid content of the sample was calculated by sing the following formula-

$$\text{Vitamin C content (mg/100g)} = \frac{T \times D \times V_1}{V_2 \times W} \times 100$$

Where,

T = Titer
D = Dye factor
V_1 = Volume made up
V_2 = Volume of extract taken for estimation
W = Weight of sample taken for estimation

Titrable acidity

The titratable acidity, expressed in % of citric acid was determined by the titrametric method (Ranganna 1994). The following reagents were used for the determination of titrable acidity.

Reagent preparation:

Dissolve 0.40gm of NaOH in water, and volume make up 100 ml. Dissolve 1.00 gm of phenolphthalein indicator in ethanol and volume make up 100 ml.

Extraction of pineapple juice

10g of fruit pulp was taken in 100 ml beaker and homogenized with distilled water in blender. The blended materials were than filtered and transferred to a 250 ml volumetric flask and the volume was made up to the mark with distilled water.

Procedure

10 ml of pulp solution was taken in a 100 ml conical flask. Two or three drops of phenolphthalein indicator were added and then the conical flask was shaken vigorously. It was then filtrated immediately with 0.1N NaOH solution from a burette till a permanent pink color was appeared. The titration was done for three times. The percentage of titrable acidity in fruit pulp was calculated by using following formula-

$$\text{Titrable acidity (\%)} = \frac{T \times N \times V_1 \times E}{V_2 \times W \times 1000} \times 100$$

Where,

T = Titre

N = Normality of NaOH

V1 = Volume made up

E = Equivalent weight of acid

V2 = Volume of extract taken for estimation

W = Weight of sample taken for estimation

Determination of total sugars

Reagents preparation

Fehling"s solution (A) Dissolve 69.28 gm of copper sulphat ($CuSO_4.5H_2O$) in water, make up volume 1000ml. Fehling"s solution (B) Dissolve 346 gm of Potassium Sodium Tartrate and 100 gm of NaOH in water, make up volume 1000ml.

Methylene blue indicator

Dissolve 1 gm of Methylene blue in 100 ml of water. 10% lead acetate solution: Dissolve 10 gm of lead acetate in water and volume make up 100 ml.

10% Potassium Oxalate solution: Dissolve 10 gm of Potassium Oxalate in water and volume make up 100 ml. Standard invert sugar solution: Weight accurately 0.25 gm of glucose in water and volume make up 100 ml. 1N NaOH solution: Take 40 gm of NaOH in water volume make up 1000 ml. The Standardization of Fehling"s solution could be calculated according to following procedure (Ranganna, 1994).

a) Standardization of Fehling's solution: Fifty ml of both Fehling"s solution A and Fehling"s solution B were mixed together in a beaker. Ten millimeter of the mixed solution was pipette into a 250 ml conical flask and 25 ml distilled water was added to it standard sugar solution was taken in a burette. The conical flask containing mixed solution was heated on a hot plate. When the solution began to boil, three drops of ethylene blue indicator solution was added to it without removing the flask from the hot plate. Mixed solution was titrated by standard sugar solution.

The end point was indicated by depolarization of the indicator. Fehling"s Factor was calculated by using the following formula-

$$\text{Factor for Fehling solution} = \frac{Titre \times 2.5}{1000}$$

b) Preparation of sample: Twenty gram of fresh pineapple fruit pulp was taken in a 100 ml beaker an then it was transferred to a blender machine and homogenized with distilled water. After blending it was made up to the mark with distilled water. The pulp solution was filtered. One hundred milliliter of filtrate was taken in a 250ml volumetric flask. Five milliliter of 45% neutral lead acetate solution was added to it and then shaken and waited for 10 minute. Five milliliter of 22% potassium oxalate solution was further added to the flask and the volume was made up to the mark with distilled and filtered.

Reducing sugar content (%)

10 mL of mixed Fehling"s solution was taken in a 250 ml conical flask and 50 mL distilled water was added to it. Filtrated pulp solution was taken in a burette. Conical flask containing the mixed Fehling"s solution was heated on a hot plate. Three to five drops of methylene blue indicator were added to the flask when

boiling started, and titrate with solution taken in the burette. The end point was indicated by decolourization of indicator. Percentage of reducing sugar was calculated according to the following equation:

$$\text{Reducing sugar content (\%)} = \frac{F \times D}{T \times W} \times 100$$

Where,

F= Fehling"s factor, D = Dilution, T = Titre and W = Weight or volume of the sample

Titration of total invert sugar

Fifty milliliter purified solution (filtrate) was taken in a 250 ml conical flask. Five gram citric acid and 50 ml distilled water were added to it. The conical flask containing sugar solution was boiled for inversion of sucrose and finally cooled. Then the solution was transferred to a 250 ml volumetric flask and neutralized by1N NaOH using phenolphthalein indicator. The volume was made up to the mark with distilled water. Then the mixed Fehling"s solution was titrated using similar procedure followed as in case of invert sugar (reducing sugar) mentioned earlier. The percentage of total invert sugar was calculated by using the formula used in case of reducing sugar.

Non-reducing sugar

Non-reducing sugar was estimated by subtracting reducing sugar from total invert sugar as following:
% non-reducing sugar = % total invert sugar - % reducing sugar.

Estimation of total sugar

Total sugar was estimated by adding the reducing sugar and non-reducing sugar as following:
% total sugar = % reducing sugar + % non-reducing sugar

Determination of the sweetness index

The sweetness index can be calculated according to the (Ranganna, 1994) formula

$$SI = \frac{Total\ soluble\ solid}{Acidity}$$

RESULTS AND DISCUSSIONS

Physical characteristics of different varieties of pineapple

Results on physical changes of pineapple as obtained at 4 days interval from initial stage to 12 DAS for the present experiment were presented and discussed below:

General Physical Parameters

Weight loss

Table 2. Effects of weight loss during days after storage

Characteri-stics	Physical parameters at different varieties of pineapple at DAS															
	Honey queen				Giant Kew				Asshini				Ghorashal			
	D_0	D_4	D_8	D_{12}	D_0	D_4	D_8	D_{12}	D_0	D_4	D_8	D_{12}	D_0	D_4	D_8	D_{12}
Fruit weight with crown (g)	820	710	625	520	1.85	1780	1685	1500	2000	1980	1685	1600	790	631	525	425
Diameter (cm)	35	34	34	33	35	34	32	29	36	33	32	29	33.3	32	30	28.9
Length (cm)	11	9	8.6	8.2	13.5	13	14	12	15	14	14	12	10	8.6	8.16	8.9

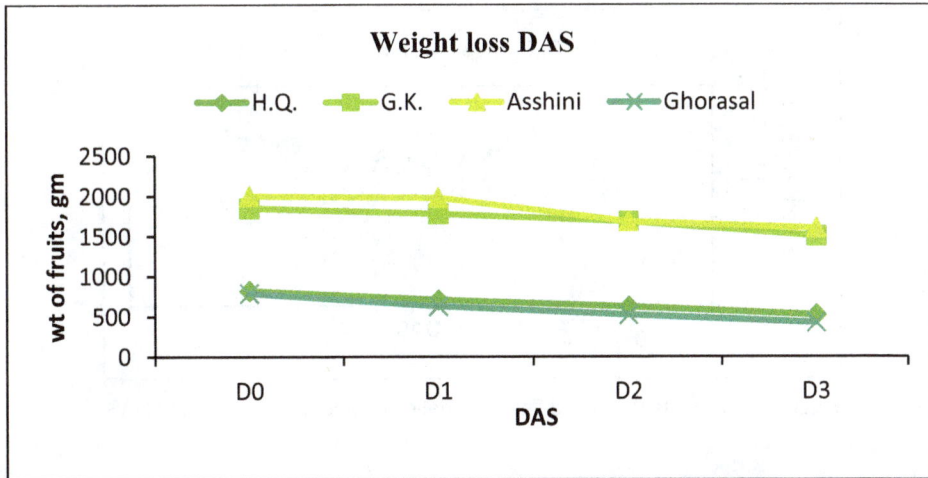

Figure 1. Effects of different varieties of pineapple on weight loss at different DAS

The Figure 1 showed that the weight loss was slightly decreased during the storage period of different verities. It was observed that the higher wt. of Asshini and G.K., so the wt. losses trends above the other H.Q. and Ghorasal variety. It was observed that there was a slightly decreased between D0 to D1 but rapidly decreased between D1 to D2. The other parameters length and diameters are almost same condition.

Moisture content

It was observed that fig.2.the moisture content decreased with the increase during storage period. Moisture content of pineapple pulp decreases from 82.5 to 84.2% for H.Q., 82.7% to 86.1% for G.K., 82.7 to 86.3% for Asshini and 82.30 to 86.19 % for Ghorashal during the storage period of 0 to 12 DAS. The highest decrease in moisture content was found 3.7% in Asshini variety and lowest decrease in moisture content was found 1.4% in H.Q. variety. These values are approximately similar to those values as found by Hasan (1980) who found 87.30% and 84.50% in H.Q. and G.K., respectively.

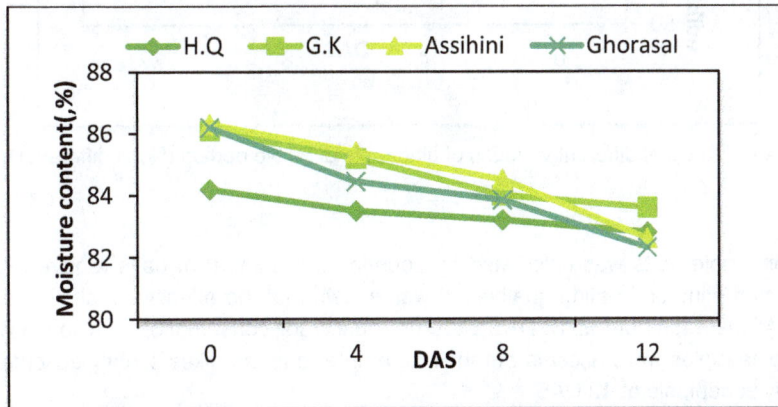

Figure 2. Moisture content (%) of pineapple juice of different varieties at different DAS.

Ash content (%)

It was observed that the ash slightly increased with the increase in storage time. Ash in the pulp of varieties of pineapple were found to be in the range of 0.32 to 0.41% (Honey queen), 0.25 to 0.36% (Giant Kew), 0.28 to 0.38% (Asshini) and 0.31 to 0.40% (Ghorashal) during the storage of 12 days. These values are very similar to those values as found by Das and Medhi (1996) who got found ash content of 0.23% to 0.50% fresh weight (fw) pineapples which is very similar to experiment variety.

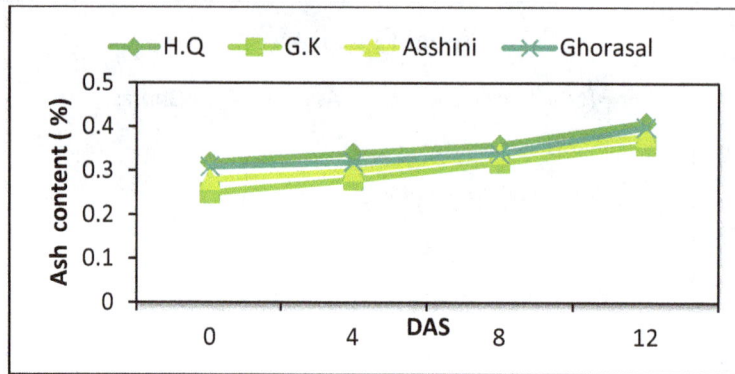

Figure 3. Ash content (%) of different varieties of pineapple at different DAS

Edible portion

The edible portion of fruits was significantly affected with days after storage (Fig.4). The edible portion of H.Q, G.K., Asshini and Ghorashal varieties were 60.5 to 63.86%, 62.60 to 65.80%, 64.56 to 66.80% and 58.14 to 61.16% increased respectively during the storage period of 0 to 12 days. It was observed that at the 12th day of storage, the highest (66.8%) values of Asshini were recorded. The trend of increasing edible portion was similar with the varieties when Asshini was top and Ghorashal was smaller producing the edible portion. Ahmed and Rahman (1974) studied that the edible portion of pineapple fruits contained 67.70% of their whole weight.

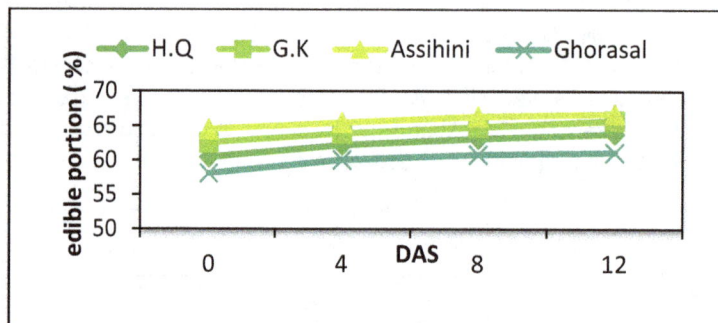

Figure 4. Effects of different varieties of pineapple of edible portion (%) at different DAS

Shelf life

Shelf life of pineapple fruits was calculated by counting the number of days required to ripen fully with retained optimum marketing and eating qualities. It was shown that the effective shelf life of pineapple which acceptable till 8 DAS as eating but at 12 DAS were not quality for consumption. It also showed that the peel and crown of the pineapples were accepts but their core, inter pulp are likes slightly acceptable but the juice color are not quality acceptable at 12 DAS.

Color

It has been shown that the peel color of pineapple was changed from green color to bright yellow which increased as the duration of storage progressed. The change in color during storage might be increase in carotene pigments of the pulp caused by enzymatic oxidation and photo degradation due to series of physical and chemical changes like the breakdown of chlorophyll and increase pigments of the pulp caused by enzymatic oxidation.

Color

Variety	Days after storage			
	0	4	8	12
Whole fruit				
Inside color				
Juice color				

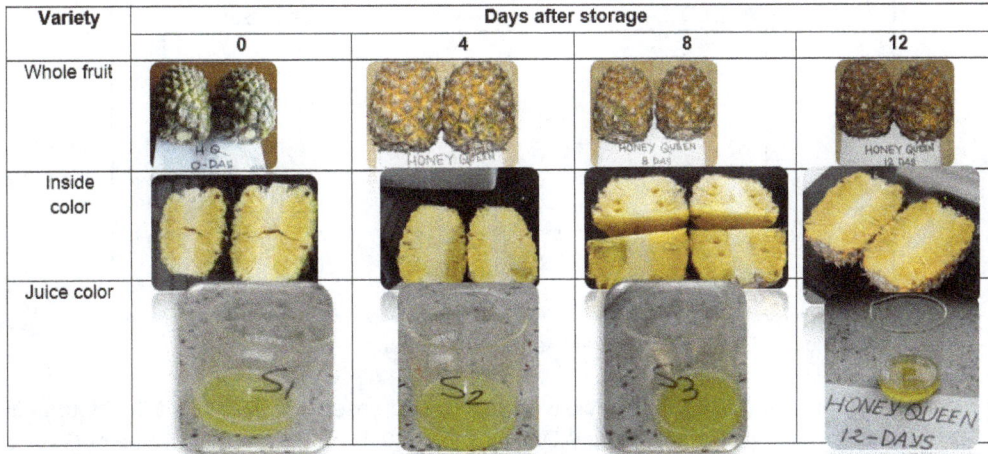

Effects of chemical properties of fruits during storage period

pH

It was observed that pH values show increase in trends with time. The pH values changes in the range of 4.31 to 4.38, 4.28 to 4.33, 4.21 to 4.32 and 4.29 to 4.36 for the varieties of H.Q, J.K, Asshini and Ghorashal, respectively.

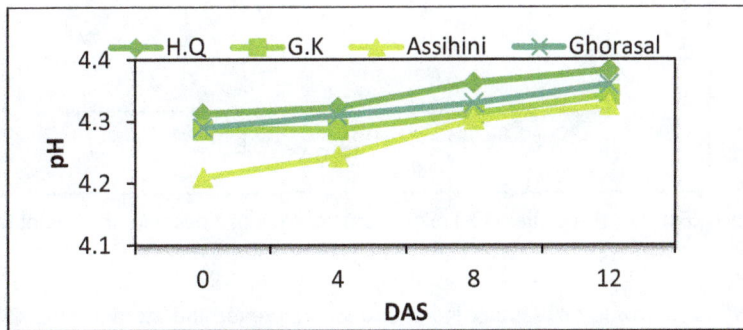

Figure 5. pH of different pineapple varietiesat different DAS

TSS

Figure 6 showed that the TSS of pineapple fruits was increased with the increased 0, 4, 8 and 12 days after storage. Among for variety H.Q. exhibits highest amount of TSS (14.26 to 15.96%) following by Asshini (12.73 to 13.56%), G.K. (11.40 to 13.26%) and Ghorashal (11.55 to 14.36%).TSS is one of the most important quality factors for most of the fruits and for pineapple; a TSS of 13.8 to 17.0% indicates the highest quality of fruits to attain the optimum harvesting stage (Morton, 1987). As a result, the increasing trend of percent total soluble solids contents of fruit during storage could be attributed mainly to the breakdown of starch into simple sugars during ripening along with a proportional increase in TSS and further hydrolysis decreased the TSS during storage.

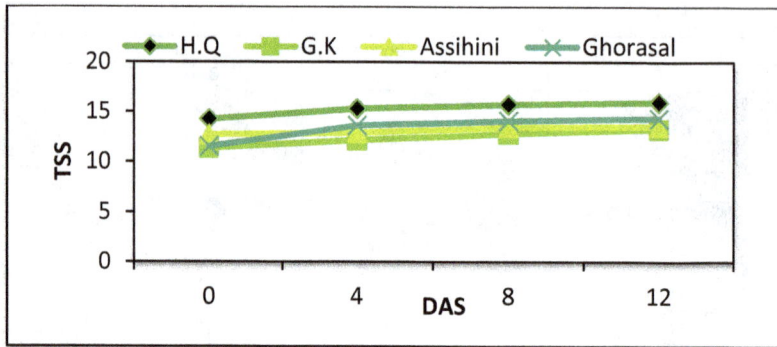

Figure 6. TSS content of different varieties of pineapple at different DAS

Titrable acidity

The Figure 7 showed that the titratable acidity is slightly decreased with increased the storage life. The values of titrable acidity for varieties are: Honey Queen: 0.49 to 0.41 % with a mean value of 0.46 % which was the maximum, Asshini: 0.68 to 0.62%, Giant Kew: 0.62 to 0.54% and Ghorashal : 0.56 to 0.46 % .Singleton (1958) suggested that acid content increase during maturation in warm condition and also said that decreased in the titrable acidity of pineapple in acidity during the ripening of pineapple was due to the loss in the dominant citric acid.

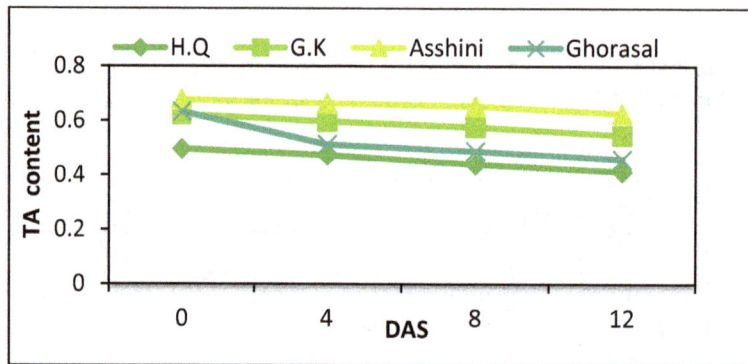

Figure 7. TA content (%) of different varieties of pineapple at different DAS

Sweetness index

Figure 8 shown that the sweetness index increased with increase the storage time. The acidity was the highest in the fruits Asshini refers to Giant Kew while the soluble solids were the highest in the H.Q. variety. These results showed that the fruits H.Q. are sweeter than other varieties.

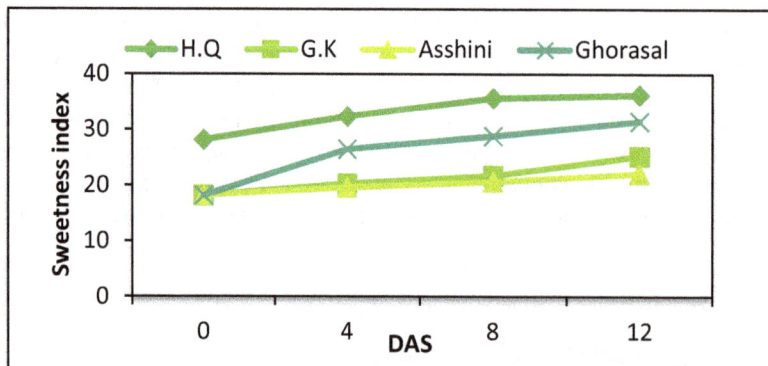

Figure 8. Sweetness index of different varieties of pineapple at different DAS

Vitamin C content

It was observed that among the storage period (0, 4, 8 and 12 DAS), the ascorbic acid content in Honey Queen (14.2 to 8.73), Giant Kew (15.8 to 11.6), Asshini (15.5 to 10.9) and Ghorashal (14.70 to 9.89) were similar for the variety. Vitamin C content in the variety Giant Kew and Asshini are higher than that of Honey Queen and Ghorashal variety, indicating that Giant Kew and Asshini are slightly more acidic than Honey Queen and Ghorashal. Rashid (1987) reported that pineapple contains 10-25 mg/100 gm of vitamin C.

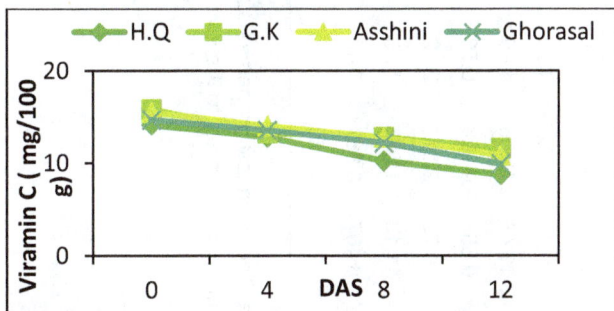

Figure 9. Vitamin C content of the different varieties of pineapple at different DAS

Total sugar content

It was shown that the total sugar gradually increased with increased during storage day (Fig. 10). Of the four varieties the Honey Queen (11.71 to 14.7 mg/100 gm) contains higher amount of total sugar than the Giant Kew (10.2 to 12.0 mg/100 gm),Asshini (9.9 to 11.5 mg/100 gm) and Ghorashal (11.55 to13.20 mg/100 gm) during storage periods. Amankwa (1995), Upadhayay and Tripathi (1985) observed that total sugar in the pineapple juice increased gradually with time which was similar trend as found in the present study.

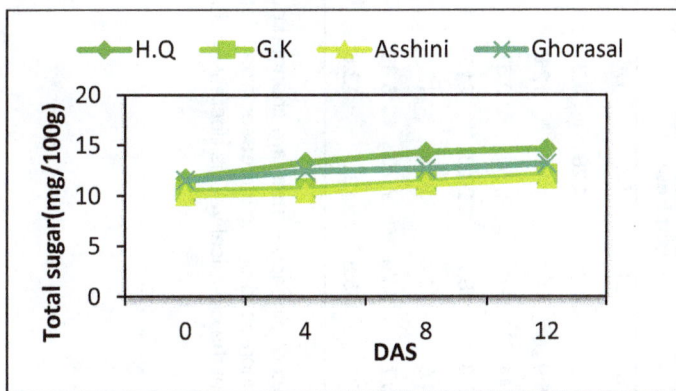

Figure 10. Total sugar content of different varieties of pineapple at different DAS

Sensory evaluation

Table 5. Comparison of different varieties of pineapple under sensory evaluation

Characteristics (observed at12 DAS)	Different varieties of pineapple			
	Honey Queen	**Giant Kew**	**Asshini**	**Ghorashal**
Appearance	fairly good	slightly bad	acceptable	fairly good
Firmness	very firm	fairly firm	slightly firm	firm
Sweetness	very strong	fair	fair	strong
Tartness	strong	fair	fair	strong
Overall acceptability	likes slightly	accepts	accepts	likes slightly

Table 4. Effects of chemical properties of different varieties of pineapple during day after storage.

Characteristics	Varieties of pineapple															
	Honey Queen				Giant Kew				Asshini				Ghorashal			
	D0	D4	D8	D12	D0	D4	D8	D12	D0	D4	D8	D12	D0	D4	D8	D12
pH	4.31	4.32	4.36	4.38	4.28	4.28	4.31	4.34	4.21	4.24	4.30	4.32	4.29	4.31	4.33	4.36
TSS	14.2	15.3	15.7	15.9	11.4	12.1	12.8	13.2	12.7	12.9	13.46	13.56	11.5	13.63	14.10	14.36
TA	0.49	0.47	0.44	0.41	0.62	0.59	0.57	0.54	0.67	0.66	0.653	0.626	0.56	0.512	0.488	0.458
SI	28.2	32.3	35.7	36.3	18.2	20.3	21.7	25.2	18.2	19.6	20.72	22.15	20.2	26.50	28.89	31.50
TS	11.7	13.3	14.3	14.6	10.4	10.7	11.4	12.1	10.1	10.3	11.13	11.73	11.5	12.44	12.68	13.20
RS	8.26	9.73	10.2	10.7	6.76	7.70	8.83	9.16	6.23	7.50	8.467	8.867	7.40	9.234	9.850	10.20
VC	14.2	12.8	10.2	8.73	15.8	13.3	12.8	11.5	15.5	14.0	12.76	10.80	14.7	13.56	12.17	9.89

Table 4 shows the chemical values of different varieties of pineapple fruits are changed during different storage time. The chemical properties such as pH, TSS, SI, TS, RS of H.Q., G.K., Asshini and Ghorashal varieties are increased during increased from starting day to 12 DAS which are almost similar to the level for pineapple fruit. The TA and VC of different varieties of pineapple shows that the chemical values decreased during storage time.

The color, firmness, sweetness, appearance, sweetness and tartness, overall acceptability were recorded by sensory evaluation. Sensory analysis was carried out by a panel of six non- trained assessors, recruited among students, Professors and employees of BSMRAU, Gazipur. Assessors evaluated the pineapple fruit quality using sorting preference tests and hedonic scale. Each assessor evaluated three sample of each cultivar, previously randomized to avoid position bias and presented in recipients with lids, coded with random three symbols. The sensory evaluation test was done to determine the quality and acceptance of the pineapple fruits derived from four varieties of pineapple. The results from the taste panel are shown in Table 5. The average score by the taste panelists showed an acceptability of the fruits appearance.

Table 6. Effects of shelf life of pineapple during storage

Date	Whole fruit, kg	Firmness, N/m^2		Edible portion, kg	% edible portion
		Outside	Inside		
0 DAS	0.915	0.270	0.077	0.560	61.2
4 DAS	0.795	0.210	0.075	0.495	62.3
8 DAS	0.830	0.227	0.075	0.538	64.8
12 DAS	0.860	0.220	0.095	0.547	63.7

Firmness of tissue makes sure the quality of fruits at storage condition. The fruits lost its firmness with increase in storage period. From Table 6, it was shown that the firmness of outside and inside of pineapple ranges are 0.210 to 0.270 N/m2 and 0.075 to 0.095 N/m2 respectively. It was observed that the firmness of pineapple are slightly changed because when their moisture percentages are decreased during storage and then unchanged their inside and outside firmness.

CONCLUSION

The present research was conducted on physico-chemical characteristics changes, shelf life and quality of pineapple through various experimental analyses of pineapple varieties during the different storage period. The data were recorded on color change, firmness, moisture content, TSS, reducing sugar, non-reducing sugar, total sugar, titratable acidity, TSS, ascorbic acid, pH, edible portion and shelf life. From the physical analysis peel color of pineapple was matured green at the preliminary time of experiment. Peel color and firmness changes were significantly influenced during storage. Moisture content of pineapple fruits was significantly affected during storage. From the analysis it is found that the moisture content, titratable acidity, Vitamin C were decreased during storage and the properties such as ash (%), edible portion (%) , pH, TSS, sweetness index, total sugar and reducing sugar were increased during storage.

The firmness of pineapple fruit was measured both in outside and inside to be 0.21 to 0.27N/m^2 and 0.075 to 0.095 N/m^2, respectively. It might be concluded that storage of harvested pineapple could be extended up to twelve days after harvesting for its improved quality except vitamin C which was higher during consumption right after harvest.

RECOMMENDATIONS

- Different processed product should be prepared from different varieties of pineapples and their comparison study should be carried out.
- Imparting short training for the farmers to acquaint with varieties of pineapple, modern technology of cultivation of pineapple and storage facilities.

REFERENCES

1.   Ahmaed K and Rahman MS, 1974. Edible portion and non-edible wastes of some fruits of Bangladesh. Bangladesh Horticulture, 2: 38-41.

2.   Amankwa APA,Martin PP and Hugon R,1995. Effect of time of harvest on the fruit. Quality characteristics of pineapple in two areas of Southern Ghana.Acta Horticulture, 425: 531-538.

3.   AOAC, 1990. Official Methods of Analysis. 15th ed. Association of Official Analytical chemist, Washinton D.C. P.1005.

4.   Das R and Medhi G, 1996. Physico-chemical changes of pineapple fruits under certain post-harvest treatments. South Indian Horticulture Journal, 44: 5-7.

5.   Hassan MK, 1980. Final Report of USAID and EC funded project (jointly implemented by FAO) entitled "Postharvest Loss Assessment: A Study to Formulate Policy for to Loss Reduction of Fruits and Vegetables and Socio Economic Uplift of the Stakeholders". P.50-53.

6.   Lane J H and Eynon L, 1923. Method for determination of reducing and non- reducing sugar. Journal of the Chemical Society, India, 42: 32-37.

7.   Lodh SB, Diwaker NG, Chadha KL, Melanta KR and SelvarajY, 1973. Biochemical Changes associated with growth and development of pineapple fruits var. Kew. III. Indian Journal of Horticulture, 30: 381-383.

8.   Morton JF, 1987. Fruits of Warm Climates. Miami, *USA. pp.* 18-22.

9.   Ranganna S, 1994. Manual of Analysis of Fruit and Vegetable Products. Tata Mc Graw-Hill Publishing Company Limited, *New Delhi*, p. 634.

10. Rohrbach KG and Pall DJ, 1982. Post-harvest diseases of pineapple. Acta horticulture, 269,503-08.

11. Singleton VL, 1958. A test for the degree of bruising of pineapple flesh. Pineapple Res. Inst. Hawaii, Honolulu. Pineapple Res. Inst. News, 6:111-114 (Private document).

12. The Daily Star, 2014.W.W.W. The daily star net accessed on 22.06.2014

13. Upadhyay NP and Tripathi BM, 1985. Postharvest changes during storage and ripening of mango (*Mangiferaindica 1.*) fruit. Progressive Horticulture, 17: 25-27.

14. USDA, 2008.United States Department of Agriculture, Food Security in the United States Measuring Household Food Security.

3

BIOLOGICAL CONTROL OF ANTHRACNOSE OF SOYBEAN

Akida Jahan, Nushrat Jahan, Farjana Yeasmin, Mohammad Delwar Hossain and Muhammed Ali Hossain*

Department of Plant Pathology, Faculty of Agriculture, Bangladesh Agricultural University, Mymensingh-2202, Bangladesh

*Corresponding author: Muhammed Ali Hossain; E-mail: alihossain.bau@gmail.com

---

ARTICLE INFO

Key words

Soybean
Anthracnose
Trichoderma
BAU-Biofungicide
IPM Lab-
Biopesticide

ABSTRACT

Soybean (*Glycine max* L. Meril) is one of the most important and well recognized oil seed and grain legume crops of the world. A field experiment was conducted to investigate the efficacy of BAU-Biofungicide (*Trichoderma harzianum*), *Trichoderma* based IPM Lab bio-pesticide and Bavistin against anthracnose of soybean. Five soybean cultivars viz. Sohag, BARI Soybean-6, BINA Soybean-1, BINA Soybean-2 and BINA Soybean-3 were used in this experiment. The field experiment was carried out following Randomized Complete Block Design in the field laboratory of the Department of Plant Pathology, Bangladesh Agricultural University, Mymensingh. Anthracnose infections were found initially lower at 80 DAS, moderate infections were found at 95 DAS and the highest infections were recorded at 110 DAS in case of all the tested five soybean varieties. The highest percent reduction of anthracnose infected plants/plot over control was observed in Bavistin treated plot (76.25%) that was near to BAU Bio-fungicide treated plots. All the growth parameters of soybean plants such as plant height, number of pods/plant, seed weight/plant, yield/plot and yield/ha were increased significantly in BAU-Biofungicide treated plots that showed best performance in compare to IPM Lab bio-pesticide (2%) and Bavistin @ 0.2% when these treatments were applied two times with 15 days interval. However, anthracnose incidence was reduced significantly when all the treatments applied in this experiment and the effect of BAU-Biofungicide was found almost similar to Bavistin in reduction of anthracnose of soybean.

INTRODUCTION

Soybean is called the "golden bean" or "Miracle bean" or "Protein hope of future" because of its high nutritive value. It is a major food and feed source that mainly cultivated for high-quality oil (20%) and high protein content (40%) (Napoles *et al.*, 2009, Osho, 2003). Two essential fatty acids namely linoleic and linolenic are found in soybeans, aid in the body's absorption of vital nutrients, and regulate smooth muscle contraction, blood pressure and the growth of cells. It can meet up minerals like Ca and P including vitamin A, B, C, D and other different nutritional needs (Rahman, 1982). A variety of soya products as food such as soya dal, soya chatni, soya khichuri, soya milk, soya curd, soya flour, soya meat and roasted soybean snacks are becoming familiar (Osho, 2003). Moreover, Soybean root nodules contain *Rhizobium*, which fixes atmospheric nitrogen and enriches soil fertility. The nodule soybean can fix 94Kg nitrogen in a hectare in one season (Satter, 2001).

Soybean was domesticated in the eleventh century BC around northeast China (Hymowitz and Shurtleff, 2005) with satisfactory yield. But, due to lack of suitable climatic conditions, the yield of soybean is very low in Bangladesh. The average yield of soybean is about 3.0 t ha^{-1} in the world, whereas the yield of soybean is only 2-2.25 t ha^{-1} in Bangladesh (SAIC, 2007). The lack of high yielding as well as the lack of pest and disease resistance varieties are the main causes for the lower yield of soybean at farmers' level in Bangladesh. All parts of soybean plant are susceptible to a number of pathogens which reduce quality and/or quantity of seed yield. Soybean suffers nearly from 150 different diseases (Sinclair, 1994) and 51 out of 150 are identified as seed borne diseases, 26 diseases out of 51 seed borne diseases are known to be transmitted through seed (McGee, 1992). Among the seed born diseases of soybean, anthracnose caused by *Colletotrichum dematium* var *truncatum* is the most serious fungal disease. This *C. dematium* var *truncatum* fungus can cause severe damage of soybean by reducing seed yield and quality in warmer, tropical and sub-tropical regions of the world (Sinclair, 1994).

Since *Colletotrichum dematium* var *truncatum* is increasingly destructive in oil production of soybean, ways of controlling the disease need to be developed resistant varieties are required to stabilize seed production and to promote sustainable agriculture without hazardous chemical control. But it's very difficult, laborious, costly and time consuming to developed a resistant variety against *Colletotrichum dematium* var *truncatum*. Hence, the introgression of biological agent for the control of anthracnose of soybean is the only viable way for the long term control of this disease and save the nature as well as getting balanced the environment from the hazardous effect of fungicides.

MATERIALS AND METHODS

The experiment was conducted in the Department of Plant Pathology and in the field laboratory of Bangladesh Agricultural University during November'14 to April'15 following a Randomized Complete Block Design (RCBD) with three replications. Soybean variety Sohag (V$_1$), BARI Soybean-6 (V$_2$), BINA Soybean-1 (V$_3$), BINA Soybean-2 (V$_4$), BINA Soybean-3 (V$_5$) was used in the experiment. The seeds of Sohag and BARI Soybean-6 were collected from Regional Oil Research Center, Bangladesh Agricultural Research Institute (BARI) Joydebpur, Gazipur and seeds of BINA Soybean-1, BINA Soybean-2 and BINA Soybean-3 were colleced from Bangladesh Institute of Nuclear Agriculture (BINA), Mymensingh. Six treatments viz. T$_0$ = Control, T$_1$ = IPM Lab Bio-Pesticide @ 2%, 1 spray, T$_2$ = IPM Lab Bio-Pesticide @ 2%, 2 spray, T$_3$ = BAU Bio Fungicide @ 2%, 1 spray, T$_4$ = BAU Bio Fungicide @ 2%, 2 spray and T$_5$ = Bavistin @ 0.2%, 2 spray (used as a positive control) were assessed in this experiment. BAU-Biofungicide and *Trichoderma harzianum* were collected from the Eco-friendly Disease Management Laboratory and IPM Lab, Department of Plant Pathology, Bangladesh Agricultural University, respectively.

The PDA media was prepared according to Islam (2009) and poured in 500ml glass bottles and sterilized in an autoclave at 121^{0}C, 15PSI for 15 min. The media were acidified with 30 drops of 50% lactic acid per 250ml medium to avoid the contamination of bacteria. It was then cultured on the same medium for multiplication through incubation at room temperature. Ten PDA plates of 7 days old culture were taken. Two g of *Trichoderma harzianum* was added to 100 mL water to make the concentration 2% (10^6 cFu/ml). About 2.5 ml of Tween-20 was added for uniform mixing of *Trichoderma* spores in the suspension. BAU-Biofungicide

(20g) was taken in a 1000 ml beaker and water was added up to the mark. The material was filtered through cheese cloth. The filtrated liquid was used as 2% BAU-Biofungicide solution. Then two g of Bavistin was taken in a beaker and 1000 ml of water was added up to the mark. The material was then stirred properly with the help of a spoon. As a result, 0.2% Bavistin solution was prepared.

The field was prepared properly with the application of fertilizer and manure at recommended rate. The size of the unit plot was 2.0 m × 1.5 m. Plot to plot distance was 40cm. Row to Row distance was 40 cm and plant to plant distance was 5 cm. Seeds of 5 different soybean varieties were sown in lines at about 5 cm depth on the 19th November, 2014. Tricho-suspension, BAU-Biofungicide and Bavistin were sprayed at 15 days interval during 3 February, 18 February, 2015, respectively. Different intercultural operations like-shading, irrigation, gap filling, weeding and insecticides spray were done timely.

Five plants were randomly selected from each plot for recording data Plant height (cm), No. of infected leaves/plant, Total no. of infected plants /plot, Total no. of pods/plant, No. of infected pods/plant, Seed weight/plant (g), Yield/plot (kg), Total yield (ton ha[-1]). The visual symptoms of anthracnose were critically observed and infected plants were identified comparing the symptoms with those of Commonwealth Mycological Institute (CMI) description. The incidence of anthracnose was recorded thrice at 80, 95 and 110 days after sowing (DAS). The incidence of anthracnose was calculated by following the formula (Ansari, 1995):

$$\text{Disease Incidence (\%)} = \frac{\text{No. of infected plants /plot}}{\text{Total number of plants /plot}} \times 100$$

Collected data were analyzed by Mstat-c and minitab 15 statistical softwares.

RESULTS AND DISCUSSION

Symptoms and causal organism
Symptoms were typically appeared during the early reproductive stage on pods as irregular shaped, brown areas and resembled pod blight. In late reproductive stage, infected tissues were covered with black fruiting bodies (acervuli) that produced minute black spines (setae) that could be seen with the unaided eyes. The setae are diagonistic for preliminary identification of the pathogen. The most common pathogen associated with anthracnose was *Colletotrichum dematium* var *truncatum*. *Colletotrichum dematium* var *truncatum* was characterized by crowded, black acervuli borne on well-developed stroma. The acervuli were oval to elongate, hemi-spherical to truncate conical in shaped with numerous black, needle like, intermixed long and short setae. The infected pods were used according to Botta *et al.* (1994) for the identification of *Colletotrichum dematium* var *truncatum*.

Effect of different varieties on the growth, disease incidence of anthracnose and yield of soybean

Plant height
In this experiment, among the five soybean varieties the tallest variety was found BINA Soybean-3 (67.94cm) followed by BINA Soybean-1(62.35cm). BINA Soybean-2 was found the most dwarf variety among the five varieties (35.25cm) (Table 1).

Number of pods/plant
There was a significant variation among the varieties in respect of number of pods/plant shown in Table 1. The highest number of pod was recorded in BINA Soybean-1 (30.34) and the lowest no. of pods/plant was observed in BINA Soybean-2 (25.48) (Table 1).

Table 1. Effect of varieties on the growth and development of soybean against anthracnose

Vaiety	Plant height (cm)	No. of pod/ Plant	Infection (%)	Pod infection (%)			Seed wt./plant 2.44 (g)	Yield/plot (kg)	Total yield (ton ha^{-1})
				80DAS	95DAS	110DAS			
Sohag	56.00	29.49	39.28	9.51	12.7	16.23	4.35	0.33	2.18
BARI Soybean-6	55.89	25.24	33.62	10.24	13.48	19.47	5.92	0.44	2.96
BINA Soybean-1	62.35	30.34	28.16	7.76	10.15	12.16	8.64	0.65	4.32
BINA Soybean-2	35.25	25.48	31.01	13.09	17.04	23.18	6.93	0.52	3.45
BINA Soybean-3	67.94	27.43	33.18	11.35	13.75	18.08	7.74	0.58	3.87
LSD$_{(0.05)}$	0.902	0.111	2.44	0.774	0.959	0.384	0.018	0.002	0.032

Table 2. Effect of treatments on the growth and development of soybean against anthracnose

Treatment	Plan height (cm)	No. of pod/ Plant	Infection (%)	Pod infection (%)			Seed wt./plant (g)	Yield/plot (kg)	Total yield (ton ha^{-1})
				80DAS	95DAS	110DAS			
T$_0$ (Contol)	45.40	12.04	54.62	21.84	29.25	50.61	3.29	0.25	1.65
T$_1$ (IPM Lab Biopesticide @ 2%, 1spray)	50.83	18.47	43.66	14.78	19.15	21.24	4.05	0.30	2.04
T$_2$(IPM Lab Biopesticide @ 2%,2spray)	62.03	34.71	28.31	5.94	7.18	7.75	9.02	0.68	4.51
T$_3$(BAU-Biofungicide @ 2%, 1 spray)	52.07	25.69	38.34	11.42	14.33	15.78	5.16	0.38	2.58
T$_4$(BAU-Biofungicide @ 2%, 2 spray)	67.22	45.17	20.42	3.83	5.05	5.64	11.41	0.85	5.71
T$_5$ (Bavistin @2%, 2 sapray)	55.37	29.46	12.97	4.55	5.71	5.92	7.34	0.55	3.64
LSD(0.05)	0.988	0.121	1.224	0.847	1.051	0.420	0.019	0.002	0.035

Incidence

The effects of different soybean varieties on the disease incidence were found statistically significant (Table 1). The highest plant infection (39.28%) was found in the variety of Sohag and the lowest incidence was found in BINA Soybean-1 (28.16%).

Pod infection

The effects of different soybean varieties on percent infected pods/plant were observed significantly at 80 days after sowing (DAS), 95 DAS and 110 DAS (Table 1). At 80 DAS, the highest infection of pod was recorded in BINA Soybean-2 (13.09%) and the lowest pod infection was observed in BINA Soybean-1 (7.76%). Percent pod infection was increased at 95 DAS compared to 80 DAS ranged from 10.15% (BINA Soybean-1) to 17.04% (BINA Soybean-2) and the highest percent of pod infection was observed at 110 DAS (Table 1). At 110 DAS, the highest pod infection was recorded maximum in BINA Soybean-2 (23.18%) and minimum infection was in BINA Soybean-1 (12.16%) (Table 1).

Seed weight/plant, Yield/plot and Yield/ha

Seed weight per plant, yield/plot and yield/hectare were significantly influenced by different varieties. The maximum seed weight/plant (8.64 g), yield/plot (0.52 kg) and total yield (4.32 ton ha^{-1}) were found in the variety BINA Soybean-1. On the other hand, the minimum seed weight/plant (4.35 g), yield/plot (0.33 kg) and total yield (2.18 ton ha^{-1}) were obtained from the variety Sohag (Table 1).

These findings are in accordance with BINA Annual report (2011-2012) where they reported that BINA Soybean-1 gave the higher production than BARI Soybean-6 in case of yield/ha in every trial at different places (Mymensingh, Rangpur and Magura) in Bangladesh and BINA website where they stated that BINA Soybean-3 is taller than BINA Soybean-1 and BINA Soybean-2 is the most dwarf variety of soybean. Again it stated that BINA Soybean-1 has the higher yield potential than both BINA Soybean-2 and BINA Soybean-3.

Effect of treatments on the growth, and development of different agronomic characteristics of soybean against anthracnose were discussed in Table-2.

Plant height

A significant variation of plant height was found in soybean varieties in respect to different treatments (Table 2). The highest plant height 67.22cm was recorded in the plots where BAU-Biofungicide was sprayed two times followed by the plots sprayed two times with IPM Lab Biopesticide (62.03cm).

Number of pods/plant

The highest number of pod (45.17) was found in the plots where BAU-Biofungicide was sprayed two times followed by the plots sprayed two times with IPM Lab Biopesticide (34.71). The moderate no. of pods/plant was recorded in the plots treated by Bavistin (29.46) followed by the plots sprayed one time with BAU-Biofungicide (25.69) and the plots sprayed one time with IPM Lab Biopesticide (18.47). However, the lowest no. of pod was recorded in control plots (12.04) (Table 2).

Plant infection

Significantly the highest plant infection was found in T$_0$ (control) (54.52%) and the lowest plant infection was found in treatment T$_5$ (Bavistin @ 0.2%) (12.97%) which was followed by two spraying of BAU Biofungicide @ 2% (20.42%) (Table 2).

Percent Pod infection

The maximum percent pod infection found in control plot and the values were 50.61% at 110 DAS, 29.25% at 95 DAS and 21.84% at 80 DAS. The minimum infection was recorded in BAU-Biofungicide two sprayed plots (at 80 DAS-3.83%, at 95 DAS-5.05% and 110 DAS-5.64%) (Table 2).

Seed weight/plant, yield/plot and yield/ha

The maximum seed weight/plant (11.41 g), yield/plot (0.85 kg) and yield/ha (5.71 ton ha^{-1}) were obtained in the plot where BAU-Biofungicide @ 2% was sprayed two times followed by IPM Lab Biopesticide @ 2%, 2 spray (9.02 g, 0.68 kg and 4.51 ton ha^{-1}, repectively). The lowest values of seed weight/plant (3.29g), yield/plot (0.25kg) and yield/ha (1.65 ton ha^{-1}) was found in control plots (Table 2).

These findings are in an accordance with the findings of Hasan (2012) who stated that BAU-Biofungicide gave higher plant height, maximum weight of pods/plant (20.05 g) and weight of mature pods/plant (18.55 g) and minimum % of leaf area diseased than Bavistin in groundnut and also with Hannan *et al.* (2011) who observed the effectiveness of BAU Bio-fungicide for the foot rot disease of chickpea under field condition and found an excellent increased yield in his experimental field.

Percent reduction of anthracnose infection of soybean against different treatments

The highest % reduction of infected plants/soybean plot was observed in treatment T$_5$ (Bavistin @ 2%, two spray) 76.25% followed by treatment T$_4$ (BAU-Biofungicide @ 2%, two spray) 62.61%, treatment T$_2$ (IPM Lab Biopesticide @ 2%, two spray) 48.17%. However, the lowest % reduction of infected plants/plot was found in T$_1$ (IPM Lab Biopesticide @ 2%, one spray) (20.07%) (Figure 1). These findings supported by Mostofa (2009) who found remarkable reduction of Cercospora leaf spot, Rust and Anthracnose disease severity in soybean were observed in seed treatment with BAU-Biofungicide and foliar spray of Bavistin @ 0.1% followed by seed treatment with BAU-Biofungicide and foliar spray of BAU-Biofungicide @ 2%.

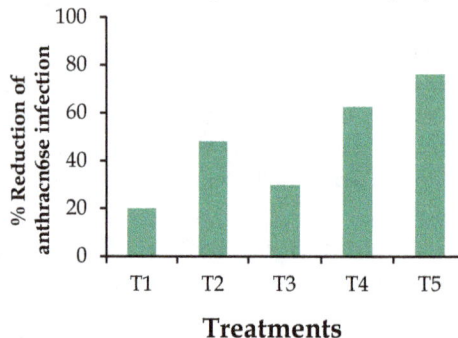

Figure 1. Percent of reduction of infected plants/plot over control

Percent increased yield (ton ha^{-1}) over control

The percent increased yield (ton ha^{-1}) over control is presented in Figure 2. The highest % increased yield was (246.06%) over control was observed in treatment T$_4$ (BAU-Biofungicide @ 2%, two spray) followed by treatment T$_2$ (IPM Lab Biopesticide @ 2%, two spray) (173.33). However, the lowest % increased yield (23.03%) over control was found in treatment T$_1$ (IPM Lab Biopesticide @ 2%, one spray) (Figure 2). These findings are in an agreement with Hossain (2003) who reported that the yield of vegetables and pulses can be increased by 25 to 80% over the untreated control by using BAU-Biofungicide.

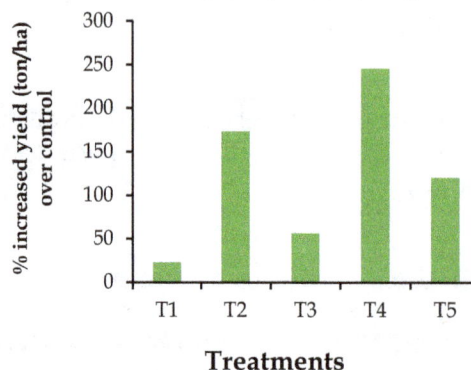

Figure 2. Percent increased yield (ton ha^{-1}) over control

The findings of the present study indicate the efficacy of BAU-Biofungicide over IPM Lab Biopesticide and Bavistin in promoting both vegetative and reproductive growth of soybean and in reducing incidence of anthracnose of soybean. BAU-Biofungicide may have some antagonistic activities that inhibit or suppress the growth of *Colletotrichum dematium* var *truncatum*.Jaime Alioscha Cuervo-Parra *et al.* (2011) stated that, *T. harzianum* VSL291 produced lytic enzymes: *β*-1, 3-glucanases, chitinases, proteases, xylanases and lipases, when grown in minimal medium, with fungal cell walls as the sole carbon source. The highest proteolytic activities detected in *T. harzianum* VSL291 broth with *M. roreri, Penicillium expansum* and *Byssochlamys spectabilis* cell walls appear to be associated with increased activities of 1, 3 glucanases, chitinases, lipases, proteases and xylanases and bio-control index derived from the experiments of confrontation. These results suggest that proteolytic enzymes according to their degree of induction could participate in the antagonistic effect of *T. harzianum* VSL291 against the fungi tested. Harman *et al.*(2004) reported that The presence of T22 (a strain of *Trichoderma herzianum)* increased protein levels and activities of β-1,3 glucanase, exochitinase, and endochitinase in both roots and shoots, even though T22 colonized roots well but colonized shoots hardly at all. With some enzymes, the combination of T22 with *P. ultimum* gave the highest enzymatic activity. On the other hand, plants grown from T22-treated seed had reduced symptoms of anthracnose following inoculation of leaves with *Colletotrichum graminicola*, which indicates that root colonization by T22 induces systemic resistance in maize. These previous results suggested that the application of *Trichoderma herzianum* might promote the activity of some defence related protein such as β-1,3 glucanase, exochitinase, endochitinase, peroxidases etc. in plant body. This biological agent also promotes the function of some cell wall degrading enzymes such as chitinases, proteases, xylanases which could degrade the cell wall of *Colletotrichum dematium* var *truncatum* infected soybean plants.

In addition, Shoresh *et al.* (2008) stated that *Trichoderma* spp. are effective bio-control agents for numerous foliar and root phytopathogens, and some are also known for their abilities to enhance systemic resistance to plant diseases as well as overall plant growth. In their study they also found that some proteins were up-regulated in shoot and root after application of *T. harzianum* strain T22 on roots which are involved in carbohydrate metabolism and photosynthesis activity. Increasing in these protein classifications suggests enhanced respiratory and photosynthetic rates. These changes may be required for the enhanced growth response induced by colonization of *Trichoderma* following seed or soil treatments. These findings also indicate that foliar treatment might also enhances the activity of proteins related to carbohydrate metabolism and photosynthesis activity and systemic resistance of plants.

CONCLUSION

The result of the present study indicates that the BAU-Biofungicide (*Trichoderma harzianum* suspension) showed better result compared to the IPM Lab Biopesticide in reducing incidence of anthracnose of soybean and promoting both vegetative and reproductive growth of soybean. This result might indicate that foliar application of *Trichoderma harzianum* in soybean plants might triggers some defence related and cell wall degrading proteins which can suppress the growth and degrade the cell wall of *Colletotrichum dematium* var. *truncatum* and some enzymes related to carbohydrate metabolism and photosynthesis activities and these physiological activities very much related to plant growth and development. BINA soybean-1 variety performed lowest disease incidence and highest yield than the other four soybean varieties used in this study, it might be due to its physiological and morphological features.

The present experiment was conducted for one season (rabi season) in a limited scale. So, this study needs to be carried out under different agro-ecological zones in the country before drawing a sound conclusion. However, farmers can be advised to use two times spray of BAU-Biofungicide with 15 days interval and variety BINA Soybean-1 to fulfill the requirement of biological control of anthracnose of soybean with maximum benefit (yield).

REFERENCES

1. Annual report, 2012: Determination of optimum spacing and seed rate for growth and yield of soybean lines, Bangladesh Institute of Neuclear Agriculture. pp. 286-287.

2. Ansari MM, 1995 Control of sheath blight of rice by plant extracts. Indian Phytopathology, 3: 268-270.

3. Botta G, Annone J and Ivancovich A, 1994. Predominance and importance of anthracnose of soybean in the Northern Buenos province Argentina. Abst. Papers, Soybean Wrold Conference V. pp. 11.

4. Cuervo-Parra J A, Ramirez-Suero M,Sanchez-Lopez V and Ramirez-Lepe M, 2011. Antagonistic effect of *Trichoderma harzianum* VSL29 on phytopathogenic fungi isolated from cocoa (Theobroma cacao L.) fruits.African Journal of Biotechnology, 10: 10657-10663.

5. Hannan M. A., Hasan M. M., Hossain I., Rahman S. M. E., Park M. S., Oh D. H. 2011. Integrated management of foot-rot disease of chickpea under field condition. JournalAgricultural Science Chungbuk National University, 27: 215-220.

6. Harman G E, Petzoldt R, Comis A, Chen J, 2004. Interactions between *trichoderma harzianum* strain t22 and maize inbred line Mo17 and effects of these interactions on diseases caused by *Pythium ultimum* and *Colletotrichum graminicola*. Phytopathology, 94: 147-53.

7. Hasan MM, 2012. Biological control of leaf of groundnut, MS thesis, Department of Plant Pathology. Bangladesh Agricultural University, Mymensingh, Bangladesh. pp. 7.

8. Hossain I, 2003. Use of BAU-Biofungicide to control seed borne diseases of vegetables and pulses. Agricultural Technologies, Bangladesh Agricultural University, Mymensingh. pp. 18.

9. http://www.bina.gov.bd/index.php?option=com_content&view=article&id=78&Itemid=92.

10. Hymowitz T and Shurtleff WR, 2005. Debunking soybean myths and legends in the historical and popular literature. Crop Science, 45: 473-476.

11. Islam T, 2009. Population dynamics of *Phomopsis vexans, Sclerotium rolfsii, Fusarium oxysporum f.sp lycoperci and Trichoderma* in the soil of eggplant field, MS thesis, Department of Plant Pathology, Bangladesh Agricultural University, pp. 48-57.

12. McGee DC, 1992. Soybean diseases. A reference source for seed technologist. APS press. The American Phytopathological Society St. Paul, Minnesota, USA. Napoles M. C., Guevara E., Montero F., Rossi A., Ferreira A. 2009. Role of Bradyrhizobium japonicum induced by genistein on soybean stressed by water deficit. Spanish Journal of Agricultural Research, 7: 665-671.

13. Osho SM, 2003. The Processing and Acceptability of a Fortified Cassava-based Product (gari) with Soybean. Nutrition and Food Science, Vol. 33, No.6.

14. Oyekan PO, 1985. Report of the nationally coordinated research projects on soybeans. Proceedings of the Fifth National Meeting of Nigerian Soybean Scientists, Publication, 5: 7-9.

15. Rahman Z, 2007. Biological control of sheath blight of rice using antagonistic *Trichoderma*, MS Thesis. Department of Plant Pathology, Bangladesh Agricultural University, Mymensingh. pp. 1-58.

16. SAIC 2007. SAARC Agricultural Statistics of 2006-07. SAARC Agric. Inform. Centre, Farmgate, Dhaka-1215. pp. 23

17. Satter MA, 2001. Biofertilizers in Bangladesh: Problem and prospect. In: Proc. 3rd Nat. Wrokshop on Pulses, 11-12 June, 2001. BARC, Farmgate, Dhaka-1207. pp. 95-102.

18. Shoresh M, Harman G E, 2008. The relationship between increased growth and resistance induced in plants by root colonizing microbes. Plant Signal Behaviour,3: 737-9.

19. Sinclair JB, 1994. Reducing losses from plant disease. World soybean research conference V. abstracts. 10p.

4

CORRUPTION AND THE AGRICULTURAL PRODUCTION EFFICIENCY OF THE EUROPEAN COUNTRIES DURING THE RECENT ECONOMIC CRISIS

Mohammad Monirul Hasan[1,*] and József Tóth[2]

[1]Department of Economic and Technological Change, Center for Development Research, University of Bonn, Walter-Flex-Str. 3, D-53113, Germany; [2]Corvinus University of Budapest, Department of Agriculture Economics and Rural Development, H-1093, Budapest, Fovam ter 8, Hungary

*Corresponding author: Mohammad Monirul Hasan, E-mail: mhasan@uni-bonn.de

---

ARTICLE INFO

ABSTRACT

Key words

Technical efficiency
Corruption control
Govt. effectiveness
Economic crisis
Regional disparity
European agriculture

This paper examines the association between controls of corruption and the agricultural production efficiency of 23 European Union Member States during the recent economic crisis. Production efficiency, measured in terms of technical efficiency, is the effectiveness of a given set of inputs that is used to produce an output. Owing to climate and geographical location agriculture in European countries is diverse. The economic downturn led by the financial crisis which started in mid-2007, is still prevailing across European countries. Control of corruption along with the existing economic crisis of the member states are affecting agriculture production efficiency. This study used the national level production data for the period of 2003-2009. It shows that the technical efficiency of most Member States have declined over the years and that it was significantly lower in austere economic crisis time 2007-09 than 2003-06 for all countries. It is also found that the declining trend of technical efficiency is significantly lower for central and eastern European countries than for the western European countries. Study finds that the control of corruption in the presence of high government effectiveness, decreases the technical efficiency of agricultural production in the Member States.

INTRODUCTION

Production efficiency, measured in terms of technical efficiency, is the effectiveness of a given set of inputs that is used to produce an output. A firm is supposed to be technically efficient if the firm produces the maximum output from the minimum quantity of inputs, such as land, labor, capital and technology. Agriculture in European countries is diverse due to climate and geographic locations. Furthermore, some European countries are restricting agricultural production due to market adjustment while some others are trying to produce as much as they can. The economic crisis that started in 2007, and is still ongoing, is spread across Europe and is going to impact the agricultural sector a lot. European Commission estimated that GDP will contract by about 4% in 2009 and by 0.1% in 2010, both in the EU-27 and in the Eurozone (EC, 2009). The longer-term consequences on potential output are however less clear. So there is an unclear impact on the overall economy including agricultural production in both the short and the long term. Our aim is to measure how technical efficiency of agriculture is affected by the control of corruption in the presence of government effectiveness in this period of economic crisis especially in 2003-09 and which part of Europe is being challenged by this declining technical efficiency.

The financial crisis has had a large negative impact on potential output in the short term and there is the prospect of a longer period of slow growth which could lead to economic depression (EC, 2009). But, according to literature, it is too soon to draw a complete and strong conclusion about the fate of the economy as a whole. The slow process of industrial restructuring due to credit constraints, the impaired system of capital allocation and structural rigidities can hurt the level and growth rate of Total Factor Productivity (TFP) in the medium to long term by locking up resources in relatively unproductive activities. Agriculture is the most important industry in Europe and the output of agricultural sector is being controlled in so many ways to meet the challenges of lower prices and higher payments to the farmers. But the production efficiency is in question year by year. What is going on in this particular issue? Does control of corruption affect technical efficiency due to economic recession in European Union (EU) Member States? In this paper we observe technical efficiency in recent years. We also use the control of corruption and government effectiveness variables to test whether the technical efficiency changes over. Furthermore we analyze the statistical significance of the changes in the technical efficiencies of the EU Member States during the period 2007-2009 using panel data econometric regression model.

Literature review

Some authors calculated the technical efficiency in European countries for specific products. Zhu et al. (2008a) calculated technical efficiency during the period 1995-2004 for dairy farms in EU Member States. Carroll et al. (2009) reported that in the period 1996-2006 the technical efficiency changed for several production types in Ireland. Some authors (e.g. Morrison Paul et al., 2000; Brümmer et al., 2006; and Lambarra et al., 2009) compared the impact of policy reform on technical efficiency.

The effect of farm size on technical efficiency is being investigated using various indicators of size as there is no consensus on the best measure for size in agriculture. Total output produced (e.g. Latruffe et al., 2004); utilized agricultural area (e.g. Nasr et al., 1998; Munroe, 2001; Helfand and Levine, 2004; Hadley, 2006; Rios and Shively, 2006; Latruffe et al., 2008a; Carroll et al., 2009); number of cows or pigs (Weersink et al., 1990; Sharma et al., 1999; Brümmer and Loy, 2000; Hadley, 2006; Tonsor and Featherstone, 2009); farm value added (Hallam and Machado, 1996); labour used or assets (e.g. Bojnec and Latruffe, 2007) are all being used as inputs for the technical efficiency score calculation.

To capture the effect of agricultural policy reform on technical efficiency, some authors (e.g. Morrison Paul et al., 2000; Brümmer et al., 2006; Carroll et al., 2009; Lambarra et al., 2009) used year-specific dummy variables to capture policy changes. Several authors, such as Weersink et al. (1990), Hallam and Machado (1996), Sharma et al. (1999), O'Neill and Matthews (2001), Rezitis et al. (2003), Helfand and Levine (2004), Latruffe et al. (2008b), Zhu et al. (2008b), Tonsor and Featherstone (2009) and Bakucs et al. (2010) included location dummies such as state, regional or country level in their regression of farm technical efficiency scores to see the regional disparities. Zhu et al. (2008b) found that being located in Less Favourable Areas (LFA) decreased the technical efficiency of Greek olive farms during the period 1995-2004. The fact that a location

600 metres above sea level reduced Slovenian farms' technical efficiency in 1995-1996 was shown by Brümmer (2001). Technical efficiency is found consistently negative in most studies such as Giannakas et al. (2001), Rezitis et al. (2003), Hadley (2006), Zhu et al. (2008a), Latruffe et al. (2008b) and Bakucs et al. (2010).

Corruption works to 'grease the economy' in highly regulated and bureaucratic economies. Corruption could increase economics development because illegal practices works as "speed money" could surpass bureaucratic delays which in turn increases government effectiveness (Hasan, A. et al., 2014 and Wei, 1998). Corruption starts working well for the low level of economic development and becomes harmful to growth for the high level of economic settings (Méndez and Sepúlveda, 2006); Neeman et al. (2008); Méon and Weill (2010). In highly regulated countries corruption can increase the enterprise efficiency and productivity. Dreher and Gassebner (2011) found that the existence of a larger number of procedures and higher minimum capital requirements are detrimental to entrepreneurship. So corruption reduces the negative impact of regulations on entrepreneurship in highly regulated economies. Similar results are also found by Leff (1964); Leys (1965); Huntington (1968); Beck and Mahler (1986); Lien (1986) and Vial and Hanoteau (2010), establishing that corruption increases efficiency. So it is more likely that being a country with less and lesser corruption with effective governance may reduce the aggregate technical efficiency of agricultural sector. Government effectiveness is positively associated with technical efficiency scores which is found from analysis of Evans and others done in 2000 (Grigoli, Francesco and Ley, Eduardo, 2012).

Economic recession reduces production-specific organization capital by unsettling normal production, distribution, marketing and inventory strategies which lead to the reduction of production efficiency (Ohanian, 2002). 'Organizational capital' refers to the knowledge firm use to organize production (Prescott and Visscher, 1980). Changes in organizational capital are important for production technology which may fall during Economic depression resulting the reduction of production efficiency (Ohanian, 2002). The reduction of efficiency occurs by chief managers to shift time away from production and into search activities. Reduced amount of managerial labor input to organizing and planning production lowers efficiency. The economic crisis during 2007-09 also has created the same vibration to the agricultural technical efficiency in the European countries.

Economic literatures is focusing on technical efficiency by investigates its nature and the determinants of technical efficiency. Many papers highlighted on the specific country or a product to see the technical efficiency but less focus on the cross country comparison especially in the time of austere economic crisis. Most papers talk about corruption and government effectiveness and also efficiency. But role of control of corruption to agricultural technical efficiency in the presence of high government effectiveness is scant in the literature. Hence it becomes more relevant especially in the context of austere economics crisis of Europe. This paper contributes to the study of how control of corruption is associated with the declining trend of agricultural technical efficiency during the economic crisis time especially in the cross country comparison.

Data

For our analysis of agriculture productivity we used World Bank data. The aggregate agricultural output and inputs of each EU Member State during the period 2003-2009 were used. There are some limitation of the data usage in the study. The analysis is confined to 2009 due to unavailability of data for the period 2010-2013. Twenty-three EU Member States were considered because data for some countries of the EU-27 are not available in the World Bank data set. We also incorporated the control of corruption index and government effectiveness index from the World Bank dataset. This gave us a short panel of 7 years and 23 EU Member States. Our dataset was classified into three regional groups, namely western European countries (Austria, Denmark, Finland, France, Germany, Ireland, Luxembourg, Netherlands and Sweden), central and eastern European countries (Bulgaria, Czech Republic, Estonia, Hungary, Latvia, Lithuania, Poland, Romania, Slovakia and Slovenia) and southern European countries (Italy, Malta, Portugal and Spain). In our analysis of technical efficiency we collected the following variables:

- Inputs: Total agricultural land (square kilometer) ; Total agricultural labor (number of agricultural workers); Average fertilizer use (kilogram per hector);
- Output: Aggregate output of agriculture (value added, constant 2000; USD)

From World Bank dataset we have taken total agricultural land which is measured in square kilometer of European Member States. Agricultural land consist of arable land area under permanent crops and pastures, temporary meadows, land under kitchen garden and land temporary fellow. Abandoned land is excluded. Another factor of production is labor which is measured in agricultural workers has taken into account in this analysis. We could get other factors of production such as capital, entrepreneurship, financial loans and activities etc. but due to data unavailability it was not possible. We used the two governance indicators such as control of corruption and government effectiveness to see the changes in the technical efficiency. Government effectiveness is defined by the perception of the quality of public services, civil service and the degree of its independence from political pressures. It also captures the quality of policy formulation, implementation and the credibility of the government's commitment to such policies. Control of corruption characterizes the perceptions of the extent to which public power is exercised for private gain (including petty and grand forms of corruption) by state bureaucrats (government officials, elites) and private interests. Both of these indicators such as government effectiveness and control of corruption is taken from World Governance Indicators (WGI) 2013 of the World Bank. The World Bank made this index.

METHODOLOGY

The efficiency was measured using the Data Envelopment Analysis (DEA) method which deals with the frontier analysis of the given data. DEA is a non-parametric technique which makes no assumptions about the form of the production technology. It is a non-stochastic approach and the efficiency of each firm is measured by the distance of its input-output vectors to the frontier. It fits a piece-wise linear frontier using a linear programming technique. In this method we used the input oriented method and constant return to scale. The analysis was carried out in two stages. Firstly we calculated the technical efficiency of the 23 EU Member States for the period 2003-2009, and then we analyzed the association of corruption and governance with technical efficiency for this specified period. . In this second part we used econometric regression for seeing the significant association between corruption and technical efficiency along with regional dummies and time dummies.

Data Envelopment Analysis (DEA)

The mathematical explanation of technical efficiency was first framed by Farrell (1957), who described the efficiency in an input-orientation framework, which is to say in terms of potential input reduction while holding the output level unchanged.

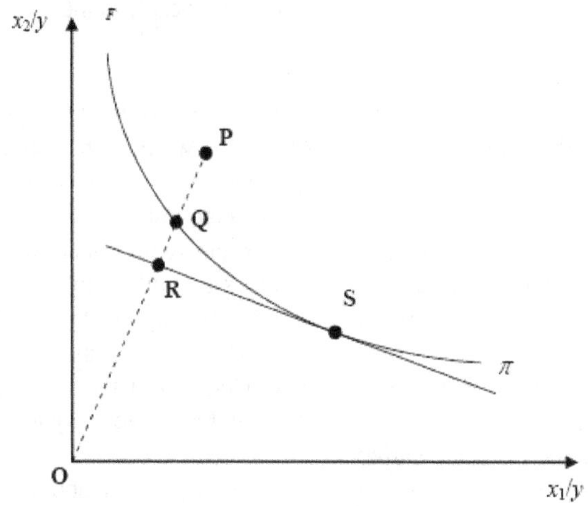

Figure 1: Input-oriented representation of technical efficiency; Source: Farrell (1957).

Figure 1 portrays the case of a firm producing one output *y* from two inputs, *x1* and *x2*. The production frontier πF illustrates the isoquant defining the minimum possible combinations of two inputs that firms can take for producing one unit of output. The frontier confines the observations, in the sense that the observed firms should lie on or beyond it. The point Q is on the frontier but point P is not. F is the technical efficiency frontier which says that any firm lying on the frontier has no possibility of reducing one input without increasing another input, meaning that such a firm is therefore technically efficient. Firm p is not technically efficient although firm P and Q are using the same proportion of inputs shown by the constant ratio ray OP. P could reduce both inputs by PQ and still can produce the same level of output. Proportionally the potential input reduction is:

$$\frac{PQ}{OP} = \frac{OP - OQ}{OP} = 1 - \frac{OQ}{OP} \qquad (1)$$

The technical efficiency of firm P is defined by:

$$TE = \frac{OQ}{OP} \qquad (2)$$

The measure of TE is bounded by minimum value 0 and maximum value of 1 which represent 100% or fully technically efficient where the potential input requirement is zero. Charnes *et al.* (1978) introduced the most popular method of DEA analysis with multiple input and outputs using linear programming technology. DEA is a simple and sophisticated tool to get consistent estimates of changes in technical efficiency even in the presence of noise. Evidences (Kloss and Petrick, 2014) showed that the assumption of constant technical returns to scale is confirmed. Under the Constant Return to Scale (CRS) assumption, the technical efficiency score, θ for the *i-th* firm in the input-orientation framework brings the solution for the following linear programming model (Coelli *et al.*, 2005):

$$min_{\theta, \lambda} \, \theta,$$
$$St \quad -q_i + Q\lambda \geq 0,$$
$$\theta x_i - X\lambda \geq 0, \qquad (3)$$
$$\lambda \geq 0,$$

Assuming K inputs, M outputs and I number of firms, we can say that x_i is the K×1 vector of inputs of i-th firm, and qi is the M × 1 vector of outputs of i-th firm. X is a K×I input matrix, Q is an M×I output matrix. θ is a scalar (=TE), λ is a I×1 vector of constants.

Panel data regression: Random Effect Tobit model

After calculating the technical efficiency score of 23 EU Member States in different years, we sought to analyse the pattern of these efficiency scores. We chose the random effect tobit regression model because out data is truncated between 0 and 1 with panel setting. To justify the use of the fixed effect or the random effect model, we used the Hausman (1978) test. This shows that the probability of chi-square value is greater than 0.05 which prescribes to use the random effect model. The Random effect assumes that the entity's error term is not correlated with the predictors which allows for time-invariant variables to play a role as explanatory variables. In equation, it can be described as:

$$y_{it} = \beta x_{it} + u_i + e_{it} \qquad (4)$$

The regression model

The regression model can be simply described as:

$$TE_{it} = \beta GE_{it} + \gamma CC_{it} + \delta T_{it} + \rho R_{it} + u_i + e_{it} \qquad (5)$$

Where TE represents the technical efficiency of country *i* in time *t*, *β* represents the coefficient of governance effectiveness, *γ* represents the coefficient of control of corruption. *T* denotes the time dummies of each country of a specific year and *R* represents the regional dummies of the specific region.

DISCUSSION

The aggregate output of agriculture shows an increasing pattern from 2005 to 2009 (Figure 2). But this increased production value might be due to increased input use. Here, it is observed that some EU Member States experienced declining technical efficiency scores over time, especially in the period 2007-2009. Many factors could contribute to this decline such as government policies including the economic crisis. In the regression result we tried to capture this hypothesis.

EU Member States such as France, Italy, Netherlands and Malta could consistently maintain their technical efficiency scores at the highest level, i.e. 1 (Table 1). This means that in our analysis these countries are fully efficient in production. On the other hand, other EU Member States show a volatile trend of technical efficiency over time which means that they might using high rates of inputs to produce the same level of output. We need to test this hypothesis. The changes in the technical efficiency in 2003-2009 are mostly found in the member states like Denmark, Hungary, Ireland, Luxemburg, Poland, and Romania. Other member states also follow the same trend. In 2003, Denmark had Technical efficiency score 1 but it fails to keep it in the later stages (Table 1).

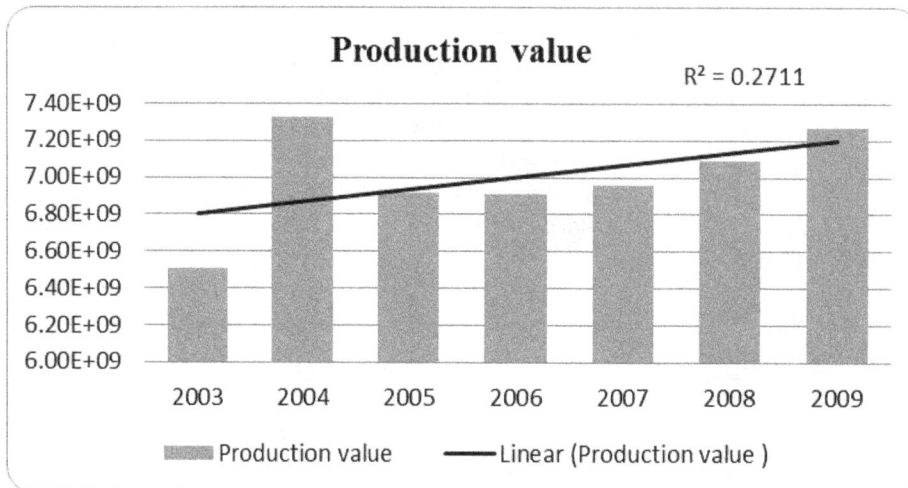

Figure 2. Aggregate output of agriculture (value added, constant 2000; USD); Source: Authors' calculation

EU Member States such as Denmark, Finland, France, Germany, Italy, Malta, Netherlands, Romania, Slovenia, Spain and Sweden could consistently maintain their technical efficiency scores at the highest level, which is almost more than 0.75 (Table 1). This means that in our analysis these countries are efficient in production. On the other hand, other EU Member States such as Estonia, Latvia and Lithuania have technical efficiency score almost below 0.2 which is also lower than other Central and East European member states such as Bulgaria, Czech Republic and Hungary. There are 4 member states like Luxembourg, Poland, Portugal and Slovakia have technical efficiency in between 0.3 to 0.5.

The technical efficiency of most countries in our study is volatile. There is no specific pattern by bare observation. The economic crisis began in the USA and spread throughout Europe very quickly in this period 2007-09. The financial crises led to liquidity constraints and also to economic recession. The European Commission sought to increase the output and make necessary steps to keep robust agriculture production. The financial crisis can be observed also in our analysis by assuming that technical efficiency will fall significantly in 2007-2009 in all countries. This analysis can be seen in the econometrics panel regression part of our study.

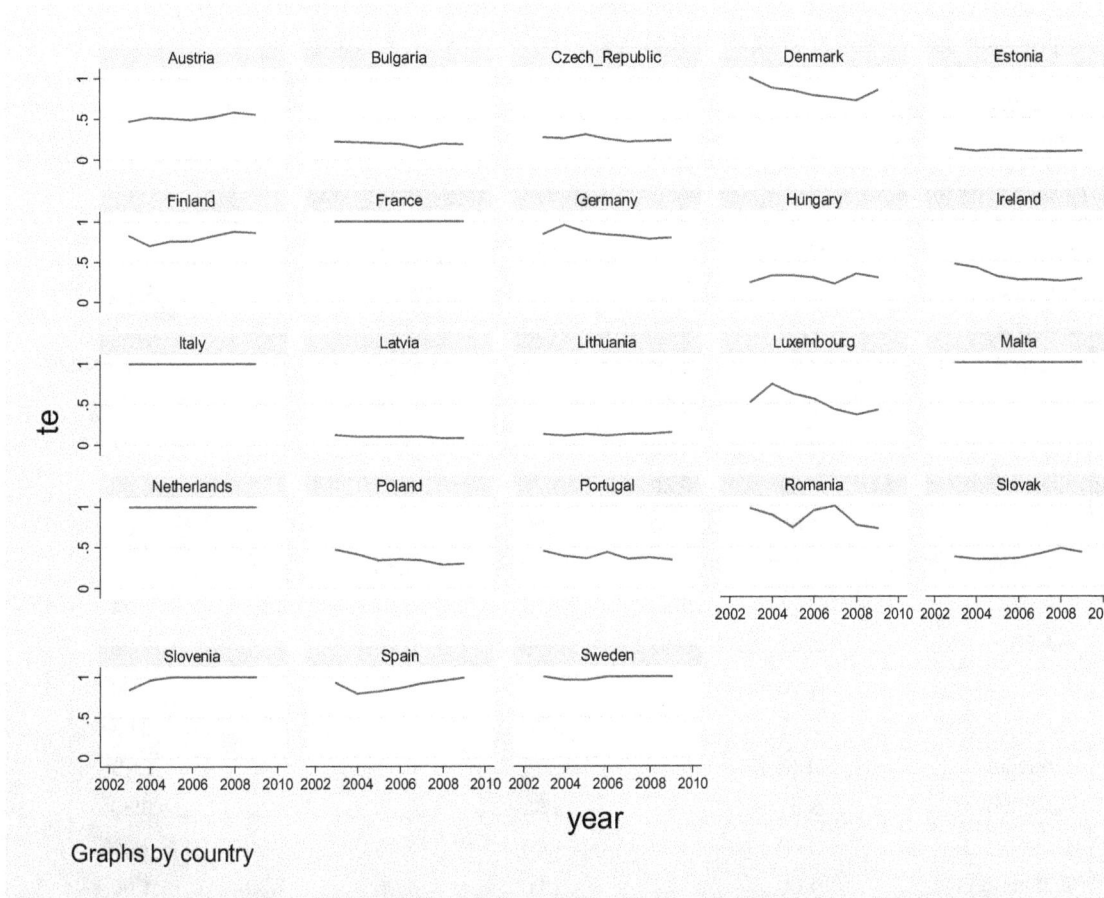

Figure 3. The technical efficiency score of 23 EU Member States, 2003-2009; Source: Authors' calculation

Econometric analysis

In our random effect tobit regression model (Table 2) we assumed that the economic crisis of the period 2007-2009 had a negative effect on the technical efficiency score of the EU Member States. When we analysed the model the results were significant. In our strongly balanced panel data of technical efficiency, we regressed the variable technical efficiency over all year dummies, government effectiveness, control of corruption and also on the regional dummies of the three regions, western European countries, central and eastern European countries, and southern European countries. After the random effect regression analysis, it is found that the period 2007-2009 gave significant result in the reduction of technical efficiency of all 23 EU Member States in the study (Table 2). In the period 2007-09 the technical efficiency of the countries declined significantly with 5% level of significance.

In the regional dummies it was found that the technical efficiency of the central and eastern European country declined significantly (1% level) in all years whereas the decline in the western European countries was not statistically significant (Table 2). That means that the central and eastern European countries is being less efficient.

In 2007, the technical efficiency score was significantly reduced by 0.037, followed by reductions of around 0.042 in 2008 and 0.039 in 2009. In a central and eastern European country the technical efficiency declined by 0.527 units. Although Western Europe also has an increased value it is not statistically significant in our analysis (Table 2).

Although *government effectiveness* is not significant, the *control of corruption* is highly significant at the 1% level (Table 2). This implies that if control of corruption increases by 1 per cent then the technical efficiency is likely to decline by 0.12. May be there are some other factors that work for this relationship but our present data doesn't allow us to investigate the causality how control of corruption affect technical efficiency. But it gives us the sign of negative association which in a sense supports the 'grease the wheel' hypothesis which is, if a country increases the control of corruption with high government effectiveness, the technical efficiency of agricultural production is likely to decrease.

Table 1. The technical efficiency score of 23 EU Member States, 2003-2009.

Country	2003	2004	2005	2006	2007	2008	2009
Austria	0.47	0.52	0.51	0.49	0.53	0.58	0.55
Bulgaria	0.22	0.21	0.20	0.19	0.15	0.19	0.18
Czech Republic	0.27	0.26	0.30	0.25	0.21	0.22	0.23
Denmark	1.00	0.87	0.84	0.78	0.75	0.71	0.84
Estonia	0.12	0.09	0.10	0.09	0.08	0.08	0.09
Finland	0.82	0.70	0.76	0.76	0.82	0.88	0.86
France	1.00	1.00	1.00	1.00	1.00	1.00	1.00
Germany	0.84	0.95	0.86	0.83	0.81	0.77	0.79
Hungary	0.24	0.32	0.32	0.29	0.22	0.34	0.29
Ireland	0.46	0.41	0.30	0.26	0.26	0.25	0.27
Italy	1.00	1.00	1.00	1.00	1.00	1.00	1.00
Latvia	0.12	0.10	0.10	0.10	0.10	0.09	0.09
Lithuania	0.13	0.11	0.13	0.11	0.13	0.13	0.15
Luxembourg	0.52	0.74	0.62	0.56	0.43	0.36	0.42
Malta	1.00	1.00	1.00	1.00	1.00	1.00	1.00
Netherlands	1.00	1.00	1.00	1.00	1.00	1.00	1.00
Poland	0.47	0.42	0.34	0.35	0.34	0.29	0.30
Portugal	0.45	0.39	0.36	0.44	0.35	0.37	0.34
Romania	0.97	0.89	0.73	0.95	1.00	0.76	0.72
Slovakia	0.37	0.34	0.34	0.35	0.41	0.47	0.43
Slovenia	0.84	0.96	1.00	1.00	1.00	1.00	1.00
Spain	0.93	0.80	0.82	0.86	0.92	0.95	0.99
Sweden	1.00	0.96	0.96	1.00	1.00	1.00	1.00
Source: own calculation							

CONCLUSION

The technical efficiency of many European countries or member states has declined over the period 2003-2009 and it is significantly lower in 2007-2009. Most countries of Europe show this tends with some exceptions. Some countries are having trend of stagnant for years and some are even better. However, to find the relationship of control of corruption with this technical efficiency scores we find that control of corruption is negatively associated with the technical efficiency score. Government effectiveness is also negatively associated but it is not significant in our analysis. We added other control variables such as the year dummies and the regional dummies to see the robustness of our results and we have found that control of corruption still significantly impacting the technical efficiency. We have also noticed that in the regression results technical efficiency is significantly lower in the years of 2007-09which signifies the austere economic crisis in Europe. So we can say that economic crisis is also impacting technical efficiency negatively although the impact is less than 10% reduction of technical efficiency. The central and eastern European countries are becoming less efficient than the western European countries according to the regression model result.

Technical efficiency of central and eastern European countries is reduced by 0.52 point which is significant at 1% level. So central and eastern European countries are facing hardship of technical efficiency loss especially in the time of austere economic crisis for the agricultural production.

The result in this paper shows basically the association among these variables. Other factors related to productivity also affect the efficiency which cannot be observed from this paper due to data limitation and also methodological stringency. Other methodologies could establish the causality among these variables and establish the relationships. However, this paper acknowledges these limitation invites more research on impact assessment. The produced results in this paper is robust and justified with panel data regression model which clearly reveals the association of the variables with the technical efficiency of the agriculture production in Europe.

Table 2. Random effect tobit model regression analysis of panel data

Dependent variable: Technical Efficiency score	Co-efficient	Standard error
Control of Corruption	-0.123***	0.045
Government Effectiveness	-0.018	0.037
Year 2004	-0.008	0.016
Year 2005	-0.033	0.016
Year 2006	-0.026	0.017
Year 2007	-0.037**	0.017
Year 2008	-0.042**	0.017
Year 2009	-0.039**	0.016
central and eastern Europe	-0.527***	0.162
western Europe	0.091	0.170
Observation	161	
Number of countries	23	
Wald chi2(10)	30.81	
Prob> chi2	0.0006***	
Integration points	12	
rho	0.960398	
/sigma_u	0.27***	0.040
/sigma_e	0.05***	0.003
note: *** p<0.01, ** p<0.05, * p<0.1		
Source: own calculation		

ACKNOWLEDGMENTS

The authors acknowledge the comments of Professor Imre Fertő and Dr. Tamas Mizik from Corvinus University of Budapest when the paper was presented in the conference of Hungarian Association of Agriculture Economics organized by Corvinus University of Budapest, May 3, 2013. The authors are indebted to the comments received from the NJF Seminar 467, Economic framework conditions, productivity and competitiveness of Nordic and Baltic agriculture and food industries, 12-13 February 2014, Tartu, Estonia. The authors acknowledge the financial support of the Hungarian Scientific Research Fund (OTKA, K 84327) 'Integration of small farms into the modern food chain'.

CONFLICTS OF INTEREST

The authors declare that there is no conflict of interest regarding the publication of this paper.

REFERENCES

1. EC, 2009. Impact of the current economic and financial crisis on potential output. Occasional Papers No. 49. Brussel: European Commission.
2. Zhu X, Demeter R and Oude Lansink A, 2008a. Competitiveness of Dairy Farms in Three Countries: The Role of CAP Subsidies, Paper presented at the 12th EAAE Congress, Gent, Belgium.
3. Carroll J, Greene S, O'Donoghue C, Newman C and F Thorne, 2009. Productivity and the Determinants of Efficiency in Irish Agriculture (1996-2006). Paper presented at the 83rd AES Conference, Dublin, Ireland.
4. Morrison Paul C, Johnston W and G Frengley, 2000a. Efficiency of New Zealand sheep and beef farming: The impact of regulatory reform. The Review of Economics and Statistics, 82: 325-337.
5. Brümmer B, Glauben T and W Lu, 2006. Policy reform and productivity change in Chinese agriculture: A distance function approach. Journal of Development Economics, 81: 61-79.
6. Lambarra F, Stefanou S, Sarra T and J Gil, 2009. The impact of the 1999 CAP reforms on the efficiency of the COP sector in Spain. Agricultural Economics, 40: 355-364.
7. Latruffe L, Balcombe K, Davidova S and K Zawalinska, 2004. Determinants of technical efficiency of crop and livestock farms in Poland. Applied Economics, 36: 1255-1263.
8. Nasr R, Barry P and P Ellinger, 1998. Financial structure and efficiency of grain farms. Agricultural Finance Review, 58: 33-48.
9. Munroe D, 2001. Economic efficiency in Polish peasant farming: An international perspective. Regional Studies, 35: 461-471.
10. Helfand S and E Levine, 2004. Farm size and the determinants of productive efficiency in the Brazilian Center-West. Agricultural Economics, 31: 241-249.
11. Hadley D, 2006. Efficiency and Productivity at the Farm Level in England and Wales 1982 to 2002, Final Report to Defra. London: Defra.
12. Rios A and G Shively, 2006. Farm size and nonparametric efficiency measurements for coffee farms in Vietnam. Forests, Trees, and Livelihoods, 16: 397-412.
13. Latruffe L, Balcombe K and S Davidova, 2008a. Productivity change in Polish agriculture: An application of a bootstrap procedure to Malmquist indices. Post-Communist Economies, 20: 449-460.
14. Weersink A, Turney C and A Godah, 1990. Decomposition measures for technical efficiency for Ontario dairy farms. Canadian Journal of Agricultural Economics, 38: 439-456.
15. Sharma K, Leung P and H Zaleski, 1999. Technical, allocative and economic efficiencies in swine production in Hawaii: A comparison of parametric and nonparametric approaches. Agricultural Economics, 20: 23-35.
16. Brümmer B and Loy J P, 2000. The technical efficiency impact of farm credit programmes: A case study in Northern Germany. Journal of Agricultural Economics, 51: 405-418.
17. Tonsor G and A Featherstone, 2009. Production efficiency of specialized swine producers. Review of Agricultural Economics, 31: 493-510.
18. Hallam D and D Machado, 1996. Efficiency analysis with panel data: A study of Portuguese dairy farms. European Review of Agricultural Economics, 23: 79-93.
19. Bojnec S and L Latruffe, 2007. Farm Size and Efficiency: The Case of Slovenia. Paper presented at the 100th seminar of the EAAE, 'Development of Agriculture and Rural Areas in Central and Eastern Europe. Novi Sad, Serbia.
20. O'Neill S and A Matthews, 2001. Technical efficiency in Irish agriculture. The Economic and Social Review, 32: 263-284.
21. Rezitis A, Tsiboukas K and S Tsoukalas, 2003. Investigation of factors influencing the technical efficiency of agricultural producers participating in farm credit programs: The case of Greece. Journal of Agricultural and Applied Economics, 35: 529-541.

22. Latruffe L, Davidova S and K Balcombe, 2008b. Application of a double bootstrap to the investigation of determinants of technical efficiency of farms in Central Europe. Journal of Productivity Analysis, 29: 183-191.

23. Zhu X, Karagiannis G and A Oude Lansink, 2008b. Analysing the Impact of Direct Subsidies on the Performance of the Greek Olive Farms with a Non-Monotonic Efficiency Effects Model. Paper presented at the 12th EAAE Congress, Gent, Belgium.

24. Bakucs L, Latruffe L, Fertö I and J Fogarasi, 2010. Impact of EU accession on farms' technical efficiency in Hungary. Post-Communist Economies, 22: 165-175.

25. Brümmer B, 2001. Estimating confidence intervals for technical efficiency: The case of private farms in Slovenia. European Review of Agricultural Economics, 28: 285-306.

26. Giannakas K, Schoney R and V Tzouvelekas, 2001. Technical efficiency, technological change and output growth of wheat farms in Saskatchewan. Canadian Journal of Agricultural Economics, 49: 135-152.

27. Latruffe L, Guyomard H, and C Mouël, 2009. The role of public subsidies on farms' managerial efficiency: An application of a five-stage approach to France. INRA- LERECO, Working Paper SMART – LERECO No. 09-05.

28. Dreher A and M Gassebner, 2011. Greasing the wheels?-The impact of regulations and corruption on firm entry. Public Choice, 155: 413-432.

29. Leff NH, 1964. Economic development through bureaucratic corruption. American Behavioral Scientist, 8: 8–14.

30. Leys C, 1965. What is the problem about corruption? Journal of Modern African Studies, 3: 215–230.

31. Huntington SP, 1968. Political order in changing societies. New Haven CT: Yale University Press.

32. Beck PJ and MW Mahler, 1986. A comparison of bribery and bidding in thin markets. Economics Letters, 20: 1–5.

33. Lien DHD, 1986. A note on competitive bribery games. Economics Letters, 22: 337–341.

34. Vial V and J Hanoteau, 2010. Corruption, manufacturing plant growth and the Asian paradox: Indonesian evidence. World Development, 38: 693–705.

35. Farrell MJ, 1957. The Measurement of Productive Efficiency. Journal of the Royal Statistical Society, Series A (General), 120: 253-290.

36. Charnes A, Cooper W W and E Rhodes, 1978. Measuring the efficiency of decision making units. European Journal of Operational Research, 2: 429-444.

37. Coelli TJ, Rao DSP, O"Donnell C J and G E Battese, 2005. An introduction to Efficiency and productivity analysis. Springer, USA.

38. Morrison Paul C, 2000b. Modeling and measuring productivity in the agri-food sector: Trends, causes and effects. Canadian Journal of Agricultural Economics, 48: 217-240.

39. Hausman J A, 1978. Specification tests in econometrics. Econometrica, 46: 1251–1271

40. Ohanian, Lee E, 2002. Why did productivity fall so much during the great depression? Quarterly Review, 26(2). Federal Reserve Bank of Minneapolis, ISSN 0271-5287.

41. Prescott Edward C and Visscher Michael, 1980. Organizational capital. Journal of Political Economy, 88(June): 446-61.

42. Hasan Ayaydın and Pınar Hayaloglu, 2014. The effect of corruption on firm growth: evidence from firms in Turkey. Asian Economic and Financial Review, 4: 607-624.

43. Méndez F and F Sepúlveda, 2006. Corruption, growth and political regimes: Cross country evidence. European Journal of Political Economy, 22: 82–98.

44. Méon P and L Weill, 2010. Is corruption an efficient grease? World Development, 38: 244–259.

45. Neeman Z, D Paserman and A Simhon 2008. Corruption and openness. The B.E. Journal of Economic Analysis & Policy, 8: 1-38.

46. Wei SJ, 1998. Foreign, quasi-foreign, and false-foreign direct investment in China, Ninth East Asian Seminars On Economics, June 25-27, 239 - 265, Osaka, Japan.

47. Grigoli Francesco and Ley Eduardo, 2012. Quality of Government and Living Standards: Adjusting for the Efficiency of Public Spending. IMF working paper 12/182.

48. Kloss M and M Petrick, 2014. The productivity of family and hired labor in EU arable farming. GEWISOLA 2014. Göttingen, 17.-19. September 2014.

5

GENETIC DIVERGENCE OF INDIGENOUS PUMMELO GENOTYPES

Md. Sarowar Alam[1], Md. Sultan Mia[2], Md. Salim[3], Jubair Al Rashid[4] and Md. Saidur Rahman[5]

[1]Scientific Officer, Regional Agricultural Research Station, BARI, Akbarpur, Moulvibazar, Bangladesh; [2]Scientific Officer, Regional Agricultural Research Station, BARI, Hathazari, Chittagong, Bangladesh & PhD fellow, School of Plant Biology , University of Western Australia, Perth, WA, Australia; [3]Scientific Officer, Hill Agricultural Research Station, BARI, Ramgarh, Khagrachari; [4]Executive, R & D, MATEX Bangladesh Ltd. Dhaka, Bangladesh; [5]Senior Scientific Officer, RARS, BARI, Akbarpur, Moulvibazar and PhD fellow, Department of Horticulture, BAU, Mymensingh, Bangladesh

*Corresponding author: Md. Sarowar Alam; E-mail: asarowar04bau@gmail.com

ARTICLE INFO

Key words
Genetic divergence,
Cluster analysis,
D^2 analysis,
Pummelo

ABSTRACT

The genetic divergence was studied in 33 pummelo genotypes using D^2 statistics and principal component analysis at Regional Agricultural Research Station, BARI, Akbarpur, Moulvibazar during 2012 to 2014. The genotypes were grouped into 5 clusters and the maximum number of genotypes was included in cluster IV and V and the minimum number in cluster I. The inter cluster distance in all of the cases were higher than the intra cluster distance indicating wider genetic diversity among the accessions of different groups. The highest inter-cluster distance was observed between cluster I and II followed by cluster II and V and the lowest between III and IV. The highest intra- cluster distance was observed for the cluster II and the lowest for the cluster III. For cluster II, the highest mean values for plant height (6.13m), individual fruit weight (1141.67g), fruit length (13.03 cm) and breadth (13.15 cm), number of segments per fruit (14.41), number and weight of seeds per fruit (123.67 and 50.41g), yield per plant (50.94 kg) were observed. The first axis largely accounted for the variation among the pummelo accessions (26.16%) followed by second axis (18.75%). The first 8 axes accounted 90.56 % of the total variation. The characters individual fruit weight (g) and weight of seeds per fruit (g) showing positive value in both the vectors contributed maximum towards divergence. Considering magnitude of genetic distance, contribution of different traits toward the total divergence, magnitude of cluster means for different traits and performance the genotypes of cluster I, II, and IV may be considered as parents for future hybridization program.

INTRODUCTION

Pummelo(*Citrus maxima*) is known as one of the important commercial fruit tree under the genus *Citrus* (Verdi, 1988). It is a native plant species to tropical and subtropical regions in Asia and has been cultivated in China for over 2000 years (Corazza-Nunes *et al.*, 2002; Yong *et al.*, 2006). *C. maxima* was originated from South East Asia,which in the western regions is familiar as shaddock (Uzun and Yesiloglu, 2012). Pummelo generally produces fruits twice a year, grows on varioussoil types at the altitude of 100-400 m above sea level (Dinesh and Reddy, 2012). *C.maxima* is one of three true *Citrus* species together with *C. medica and C. reticulata* (Barrett and Rhodes,1976;Hynniewta *et al.*, 2011). Its status as true or basic species with in *Citrus* is confirmed by other researchers (Barkley *et al.*, 2006; Uzun *et al.*, 2009;Froelicher *et al.*,2011; Garcia-Lore*t al.*, 2013).Therefore, pummelo has been regarded as a parent of many citrus fruits, such as lemons, oranges and grapefruits. It is characterized for distinguished features of huge leaves borne on broadly winged petioles, very large and fragrance flowers and big fruits with a single embryo, while most of other *Citrus* species are polyembryonic (Uzun and Yesiloglu, 2012). Pummelo is one of the popular and the biggest citrus fruit of Bangladesh. In Bangladesh, it is cultivated in an area of around 7460 ha with total production of 59198 metric tons and average yield per plant is around 38.0kg (BBS, 2011). For obtaining varieties with desired traits, hybridization is a very effective tool if diversed parents with promising features are available. Pummelo is a cross pollinated crop and there is a wide variability within the species which create good opportunity to plant breeder to utilize for its improvement (Janick and Moore, 1996).

Genetic diversity is an important factor for crop improvement with desirable traits. Multivariate analysis such as D^2 cluster and factor analysis have been proved to be useful for selecting genotypes for hybridization. Mahalanobis (1949) D^2 analysis has been successfully used in measuring the diversity in several crops. An understanding of nature and magnitude of variability among the existing pummelo germplasm is a prerequisite for its improvement. Precise information on the nature and degree of genetic divergence helps the plant breeder in choosing the diverse parents for purposeful hybridization (Arunachalam, 1981; Samsuddin, 1985). Since published work on pummelo is scanty, the present study has been undertaken with 33 pummelo accession to understand the nature and magnitude of genetic divergence and the characters contributing genetic diversity by D^2 analysis.

MATERIALS AND METHODS

The experiment was conducted at the pummelo orchard and laboratory of the the Regional Agricultural Research Station, Bangladesh Agricultural Research Institute, Akbarpur, Moulvibazar during February 2012 to October 2014.Thirty three pummelo genotypes were included in the present study and data were recorded from three different plants of an accession where each plant was considered as a replication. Individual plants were fertilized with cowdung (20 kg), urea (500 g), TSP (500 g), MP (500 g), gypsum (200 g) , Zinc (4 g), boron (4 g) in two equal installments one at the onset and other at the end of rainy season. (FRG,2012). Irrigation, weeding and other crop management practices were followed as recommended by Ullah *et al.* (2006) to have a good healthy plant. Data on plant height (m), canopy spreading (m), number of flowers per cluster, number of fruits per plant, individual fruit weight (g), fruit length (cm) and breadth (cm), number of segments per fruit, number and weight (g) of seeds per fruit, edible percentage, % Brix (TSS) and yield per plant(kg)were recorded.

Genetic diversity was studied following Malanobis's (1949) generalized distance (D^2) extended by Rao (1952). Clustering of genotypes was done according to Tocher's Method (Rao, 1952) and principal component analysis was done according to Rao (1964). All the statistical analysis were carried out using GENSTAT-5 computer software. Average intracluster distance was calculated by the following formula as suggested by Singh and Chaudhury (1985).

$$\text{Average intracluster } D^2 = \frac{\sum D^2 i}{n}$$

Where, $\sum D^2 i$ = Sum of distances between all possible combination (n) of the varieties/lines included in a cluster and n = All possible combinations

RESULTS AND DISCUSSION

In this study the 33 accessions of pummelo were grouped into five clusters based on D^2 values (Table 1). The distribution pattern indicate that the maximum number (9 each) of genotypes were included in cluster I V and V followed by cluster II and III. The minimum numbers of genotypes were included cluster I. Rahman and Al Munsur(2009) grouped 40 genotypes of lime into 6 clusters.

Table 1. Distribution of 33 pummelo accessions in five clusters with location

Cluster	Numbers	Accessions	Percentage
I	3	CM-Akb-141, CM-Akb-171, CM-Akb-178	9.09
II	6	CM-Akb-144, CM-Akb-147, CM-Akb-148, CM-Akb-158, CM-Akb-159, CM-Akb-166	18.18
III	6	CM-Akb-145, CM-Akb-150, CM-Akb-156, CM-Akb-162, CM-Akb-164, CM-Akb-165	18.18
I V	9	CM-Akb-134, CM-Akb-139, CM-Akb-146, CM-Akb-151, CM-Akb-153, CM-Akb-160, CM-Akb-161, CM-Akb-163, CM-Akb-184	27.27
V	9	CM-Akb-136, CM-Akb-137, CM-Akb-138, CM-Akb-170, CM-Akb-172, CM-Akb-173, CM-Akb-176, CM-Akb-177, CM-Akb-182	27.27

A two dimensional scatter plotting diagram (Z_1-Z_2) constructed using component score 1 on X axis and component score 2 on Y axis exhibited that the genotypes were fallen into five clusters (Figure 1).

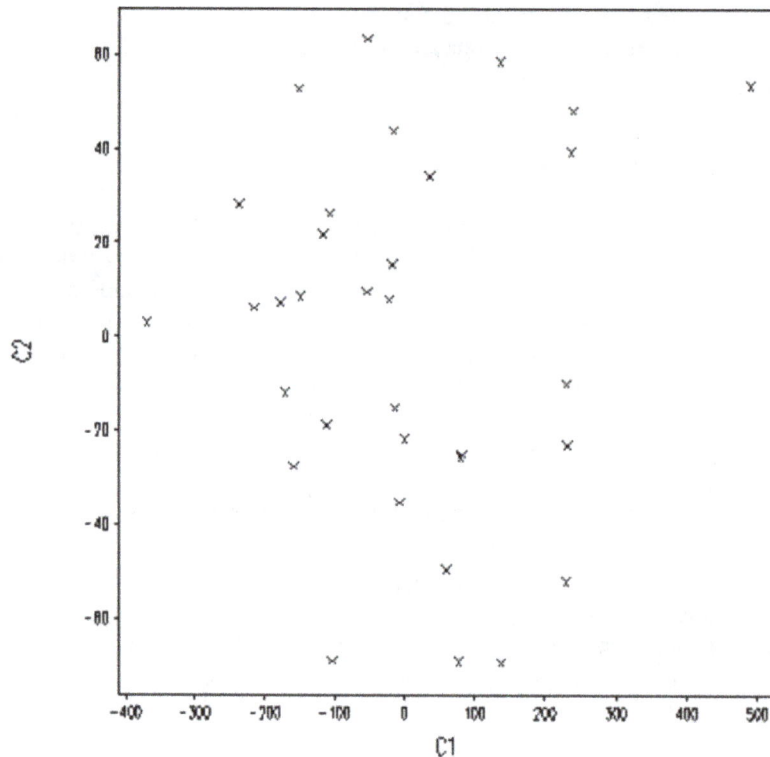

Figure 1. Scattered distribution of 33 pummelo genotypes based on the principal component score

Intra and inter cluster distances are presented in Table 2. The inter-cluster distances were higher than the average intra-cluster distances, which indicated wide genetic diversity among the pummelo accessions of different groups than those of same cluster. Rahman and Munsur (2009) found same result in case of lime. The highest inter cluster distance was observed between cluster I and II (16.638) and followed by cluster II and V (13.374) and the lowest between III and IV (4.482) (Table 2). The highest intra- cluster distance was observed for the cluster II (1.766) and the lowest for the cluster III (1.121). The highest values for intercluster distance indicated that the accessions belonging to cluster I was far away from those of cluster II. The minimum intercluster divergence was observed between IIIand IV indicating that the genotype of these clusters was genetically closer. Hybridization among the genotypes drawn from widely divergent clusters with high yield potential would likely to manifest maximum heterotic combinations as well as new recombination with desired traits. Similar findings were observed in lemon, sweet gourd and jackfruit (Ismail 2008; Rahman *et al.* 2006 and Saifullah*et al.* 1999).

Table 2. Average intra (bold) and inter-cluster distances for 33 pummelo accessions

Clusters	I	II	III	I V	V
I	**1.3**				
II	16.638	**1.766**			
III	12.11	5.51	**1.121**		
I V	8.093	8.894	4.482	**1.359**	
V	4.624	13.374	8.952	4.909	**1.513**

Table 3. Cluster mean values for yield and yield contributing characters of 33 pummelo accessions

Characters	Cluster means				
	I	II	III	IV	V
Plant height (m)	5.30	6.13	5.71	5.27	5.15
Canopy spreading at north-south direction (m)	5.97	6.15	5.95	6.24	5.23
Canopy spreading at east-west direction (m)	5.33	5.30	5.45	6.0	6.38
Number of flowers per cluster	5.88	4.75	5.16	5.55	5.64
Number of fruits per plant	41.0	43.17	39.83	43.77	33.34
Individual fruit weight (g)	593.33	1141.67	965.0	850.92	730.56
Fruit length (cm)	11.37	13.03	12.05	11.75	12.07
Fruit breadth (cm)	11.0	13.15	12.65	12.12	11.31
Number of segments per fruit	13.33	14.41	14.33	14.34	13.17
Number of seeds per fruit	102.33	123.67	77.67	111.53	97.11
Weight of seeds per fruit (g)	35.0	50.41	29.5	44.61	37.22
Edible percentage	53.54	49.37	52.73	49.53	48.12
% Brix (TSS)	8.33	9.17	9.50	9.37	8.97
Yield per plant (kg.)	24.68	50.94	38.49	37.14	24.18

Cluster mean values of 14 different characters are shown in Table 3. Difference in cluster means existed for almost all the characters studied. The highest mean value for number of flowers per cluster (5.88),edible part of fruit (53.54%) were observed in cluster I. For cluster II, the highest mean values for plant height

(6.13m), individual fruit weight (1141.67g), fruit length (13.03 cm) and breadth (13.15 cm), number of segments per fruit (14.41), number and weight of seeds per fruit (123.67 and 50.41g), yield per plant (50.94 kg) were recorded. It was revealed that parental lines fallen in this cluster having the genetic potentiality to contribute better for yield maximization of improved pummelo varieties. In case of cluster III, only the character % Brix (TSS) showed the highest mean value (9.50).Cluster IV showed highest mean values for canopy spreading (6.24 m) for north-south direction, and number of fruits per plant (43.77).Only canopy spreading (6.38 m) for east-west direction was highest mean value in Cluster V.

The results of principal component analysis (Table 4) revealed that the first axis largely accounted for the variation among the pummelo accessions (26.16%) followed by second axis (18.75%). The first 8 axes accounted 90.56 % of the total variation among 14 characters of describing 33 pummelo genotypes.The rest six characters contributed remaining 9.44% of total variation. In a study of diversity of acid lime, Ranpise and Desai (2003) found that fruits per tree, yield per plant, juice volume and juice percentage were major contributing traits towards divergence. The character with maximum contribution to the divergence should be given more emphasis for selection in breeding (Jagadev et al., 1991).

Table 4. Latent root (Eigen value) and percent of variation in respect of fourteen characters in 33 pummelo accessions

Plant characters	Eigen value	Percent of Variance	Cumulative Percentage
Plant height (m)	3.6622	26.16	26.16
Canopy spreading at north-south direction (m)	2.6244	18.75	44.91
Canopy spreading at east-west direction (m)	2.1726	15.52	60.43
Number of flowers per cluster	1.2808	9.15	69.58
Number of fruits per plant	1.08	7.71	77.29
Individual fruit weight (g)	0.7334	5.24	82.53
Fruit length (cm)	0.6289	4.49	87.02
Fruit breadth (cm)	0.4958	3.54	90.56
Number of segments per fruit	0.4531	3.24	93.8
Number of seeds per fruit	0.3416	2.44	96.24
Weight of seeds per fruit (g)	0.3197	2.28	98.52
Edible percentage	0.1331	0.95	99.47
% Brix (TSS)	0.0660	0.47	99.94
Yield per plant (kg.)	0.0083	0.06	100

Contributions of different characters responsible for genetic divergence are presented in Table 5. The canonical variate analysis (CVA) revealed that invector I (Z_1), the important characters responsible for genetic divergence in the major axis of differentiation number of segments per fruit, number of flowers per cluster, number of fruits per plant, edible percentage, fruit breadth (cm),individual fruit weight (g),weight of seeds per fruit(g). In vector II (Z_2), fruit length (cm), canopy spreading at east-west direction (m), weight of seeds per fruit (g),plant height (m), yield per plant (kg.),individual fruit weight (g), number of seeds per fruit had positive impact towards divergence. The characters individual fruit weight (g) and weight of seeds per fruit (g) showing positive value in both the vectors contributed maximum towards divergence. So, the divergence in the present materials due to these two traits will offer a good scope for improvement of pummelo varieties through selection of parents.

Table 5. Latent vectors for fourteen characters of 33 pummelo genotypes

Characters	Vector-I (Z_1)	Vector-II (Z_2)
Plant height (m)	0.0110	0.0214
Canopy spreading at north-south direction (m)	-0.0098	-0.0355
Canopy spreading at east-west direction (m)	-0.0460	0.0712
Number of flowers per cluster	0.3353	-0.2807
Number of fruits per plant	0.2379	-0.0444
Individual fruit weight (g)	0.0427	0.0031
Fruit length (cm)	-0.0090	0.3826
Fruit breadth (cm)	0.0536	-0.6583
Number of segments per fruit	0.3362	-0.3755
Number of seeds per fruit	-0.0148	0.0003
Weight of seeds per fruit (g)	0.0208	0.0670
Edible percentage	0.0613	-0.0709
% Brix (TSS)	-0.4291	-0.0778
Yield per plant (kg.)	-0.2619	0.0041

CONCLUSION

Crosses involving parents belonging to most diverse genotypes are expected to exhibit maximum heterosis and create wide variability in genetic architecture. Considering magnitude of genetic distance, contribution of different traits toward the total divergence, magnitude of cluster means for different traits and performance the genotypes of cluster I, II, and IV may be considered as parents for future hybridization program.

REFERENCES

1. Arunachalam VA, 1981. Genetic distances in plant breeding. Indian Journal of Genetics. 4:226-236.
2. BarkleyNA,ML Roose, RR Krueger and CT Federici, 2006.Assessing genetic diversity and population structure in a citrus germplasm collection utilizing simple sequence repeat markers (SSRs). Theoretical and Applied Genetics, 112: 1519–1531.
3. Barrett HC and AM Rhodes, 1976. A numerical taxonomic study of affinity relationships in cultivated Citrus and its close relatives. Systematic Botany, 1: 105–136.
4. BBS, 2011. Yearbook of Agricultural Statistics of Bangladesh, 2011. Bangladesh Bureau of Statistics, Statistics and Informatics Division, Ministry of Planning, Government of the People's Republic of Bangladesh. Pp: 130.
5. Corazza-Nunes MJ, MA Machado, WMC Nunes, M Cristofani and MLPN Targon, 2002. Assessment of genetic variability in grapefruits (Citrus paradisiMacf.) and pummelos (C. maxima Burm.Merr.) using RAPD and SSR markers. Euphytica, 126: 169–176.
6. Dinesh MR and BMC Reddy, 2012.Physiological Basis of Growth and Fruit Yield Characteristics of Tropical and Sub-tropical Fruits to Temperature. In: Tropical Fruit Tree Species and Climate Change, Sthapit BR, VR Rao and SR Sthapit (Eds.), Bioversity International, New Delhi, India, pp: 45-70.
7. FRG, 2012. Fertilizer Recommendation Guide, Bangladesh Agricultural Research Council (BARC), Farmgate, Dhaka1215.p: 78-79.
8. Froelicher Y, W Mouhaya, JB Bassene, G Costantino and M. Kamiri, 2011. New universal mitochondrial PCR markers reveal new information on maternal citrus phylogeny. Tree Genetics and Genomes, 7: 49-61.

9. Garcia-Lor A, F Curk, H Snoussi-Trifa, R Morillon and G Ancillo, 2013. A nuclear phylogenetic analysis: SNPs, indels and SSRs deliver new insights into the relationships in the 'true citrus fruit trees' group (Citrinae, Rutaceae) and the origin of cultivated species. Annals Botany, 111: 1-19.

10. Hynniewta M, SM Malik and SR Rao, 2011.Karyological studies in ten species of *Citrus* (Linnaeus, 1753) (Rutaceae) of North-East India. Comp. Cytogen, 5: 277-287.

11. Ismail KM, 2008. Genetic diversity and molecular characterization of lemon.Ph D Dissertation. Dept. of Horticulture, Bangladesh Agricultural University, Mymensingh. 233 P.

12. Jagadev PN, KM Samal and L Lenka, 1991. Genetic divergence in rape mustard. Indian Journal of Genetics and Plant Breeding, 51: 465-466.

13. Janick J and JN Moore, 1996. Fruit Breeding (Volume I): Tree and Tropical Fruits. [Eds.]. John Wiley & Sons, Inc. New York.

14. Mahalanobis PC, 1949. Historical note on the D^2 Statistics. Sankhya, 19: 237-239.

15. Rahaman EHMS, MG Rabbani and EJ Garvey, 2006. Genetic diversity of sweet gourd through multivariate analysis. Bangladesh Journal of Agricultural Science, 33: 197-204.

16. Rahman MM and MAZ Al Munsur, 2009 .Genetic divergence analysis of lime, Journal of Bangladesh Agricultural University, 7: 33–37.

17. Ranpise SA and UT Desai, 2003. Genotypic and phenotypic variability in acid lime (*C. aurantifolia* Swingle). Journal of Maharashtra Agriculture University, 28: 21-23.

18. Rao CR, 1964. The use and interception of principal analysis in applied research. Sankhya, 22: 317-318.

19. Rao CR, 1952. Advanced Statistical Method in Biometric Research. Ednl. John Wiley and Sons, New York.

20. Saifullah M, AK Azad, MI Nazrul, MR Islam and MA Hossain, 1999. Genetic diversity in jackfruit (*Artocarpus heterophyllus* Lam) grown in Bangladesh. Bangladesh Journal of Plant Breeding and Genetics, 12: 01-06.

21. Samsuddin AKM, 1985. Genetic diversity in relation to heterosis and combining analysis in spring wheat. Theoretical and Applied Genetics 70: 306-308.

22. Singh RK and BD Chaudhary, 1985. Biometrical methods in quantitative genetic analysis. Kalyani Publ., New Delhi.

23. Ullah MA, MA Hoque and MAI Khan, 2006. Modern cultivation technique of pummelo (*In Bengali*). Regional Agricultural Research Station, BARI, Akbarpur. Moulvibazar. Pp: 1-13.

24. Uzun A and T Yesiloglu, 2012.Genetic Diversity in Citrus. In: Genetic Diversity in Plants, Caliskan, M. (Ed.), InTech, pp: 213-231.

25. Uzun A, T Yesiloglu, Y Aka-Kacar, O Tuzcu, and O Gulsen, 2009. Genetic diversity and relationships within Citrus and related genera based on sequence related amplified polymorphism markers (SRAPs). Scientia Horticulturae, 121: 306–312.

26. Verdi A, 1988. Application of recent taxonomical approaches and new techniques to citrus breeding: Balaban Publishers.

27. Yong L, L De-Chun, W Bo and S Zhong-Hai, 2006. Genetic diversity of pummelo (*Citrus grandis*Osbeck) and its relatives based on simple sequence repeat markers. Chinese Journal of Agricultural Biotechnology, 3: 119-126.

6

EFFECTS OF ORGANIC AND INORGANIC FERTILIZERS ON THE GROWTH, YIELD AND NITROGEN UPTAKE BY BRRI dhan28

Md. Rafiqul Islam[1*], Aurunima Kanchi Suprova Shawon[1], Most. Lutfun Nesa Begum[2] and Azmul Huda[1,3]

[1]Department of Soil Science, Bangladesh Agricultural University, Mymensingh-2202, Bangladesh;
[2]Practical Skill Development Training Department, Proshika, Mirpur-2, Dhaka-1216, Bangladesh;
[3]Department of Soil Science, Sylhet Agricultural University, Sylhet-3100, Bangladesh

*Corresponding author: Md. Rafiqul Islam, E-mail: rafiqss69@bau.edu.bd

---

ARTICLE INFO

Key words

Organic,
Inorganic fertilizers,
BRRI dhan28,
Yield,
Nitrogen uptake

ABSTRACT

A study was conducted at the Soil Science Field Laboratory of Bangladesh Agricultural University, Mymensingh during Boro season of 2014 to evaluate the effect of integrated use of manures and fertilizers for maximizing the growth and yield of BRRI dhan28. The experiment was laid out in a randomized complete block design with six treatments and four replications. The treatments include T_0 [Control], T_1 [Soil Test Basis-Chemical Fertilizer], T_2 [(Cowdung) + STB-CF] on IPNS basis, T_3 [(Poultry Manure) + STB-CF] on IPNS basis, T_4 [(Compost) + STB-CF] on IPNS basis, and T_5 [Farmer's practice]. The maximum grain yield of 4340 kg ha^{-1} (95.59% increase over control) and straw yield of 4024 kg ha^{-1} (56.42% increase over control) were recorded in T_3 [(PM) + STB-CF]. The lowest grain and straw yields were found for T_0 (Control) treatment. The N, P, K and S contents and uptake by BRRI dhan28 were profoundly influenced due to combined application of manures and fertilizers. The performance of the treatment T_3 was better than T_1, T_2 and T_4 in producing the yield of grain and straw of BRRI dhan28 although they received the same amount of nutrients. The results indicate that application of fertilizers in combination with poultry manure could be considered more effective in rice production. So, the treatment T_3 can be used for the successful cultivation of BRRI dhan28.

INTRODUCTION

Rice is considered as staple crop in Bangladesh. It is not only the main source of carbohydrate but it also provides 69.61% of calories and 56.15% of the proteins in the average daily diet of the people (FAO, 2009). In the last three to four decades, great efforts in rice research and farming innovations were made to boost up rice production. Furthermore, rice alone contributes about 9.5% of the total agricultural GDP in the country. Among all crops, rice is the driving force of Bangladesh agriculture. In fact, food production in Bangladesh is dominated by a single crop (rice) and a single season boro, which accounts for over 60% of total rice production. The average yield of rice is 4.22 t/ha in Bangladesh whereas the yield in some other rice producing countries of the world such as Australia, Korea Republic, Japan and China are 9.54, 7.38, 5.33 and 6.69 ton per hectare, respectively (FAO, 2011).

High yield goal with higher cropping intensity demands more inputs, particularly fertilizer nutrients but raising productivity through chemical fertilizer nutrients is likely to have harmful effect on the environment – soil, water and climate. The cropping intensity of Bangladesh is 179% (BBS, 2011). The increasing cropping intensity without adequate and balanced use of chemical fertilizers and with little or no use of organic manures have caused severe fertility deterioration of our soils resulting in stagnating or even declining of crop productivity. The organic matter content of most of our soils is below 1.5% and in many cases it is less than 1% and decreasing day by day (BARC, 2012). Depletion of soil organic matter in many of Bangladesh soils has reached such an alarming stage that it is caused serious concern among the crop production specialists. It is now feared that unless drastic measures are taken to improve and maintain organic matter reserves, many soils would soon become unproductive. Moreover, tropical monsoon climate with high temperature and abundant rainfall provide favorable condition for enhanced microbial decomposition of soil organic matter. So, it is very difficult rather impossible to conserve and maintain high level of organic matter in the soils. Hence, management of soil organic matter has now become major issue in dealing with the problem of soil fertility and productivity in Bangladesh.

A suitable combination of organic and inorganic sources of nutrients is necessary for sustainable agriculture that will provide food with good quality. Nambier (1991) stated that combined use of organic manures and inorganic NPK fertilizers would be quite promising not only in providing greater stability in production, but also in higher soil fertility status. Cowdung, poultry manure and compost supply the macro and micronutrients especially N, P and S for crop production. The long term research of BRRI reveals that the addition of cowdung @ 5 t ha^{-1} yr^{-1} improves the rice productivity as well as prevents the soil resources from degradation (Bhuiyan, 1994). An improvement and maintenance of a good supply of organic matter is essential for sustenance of soil fertility and crop productivity. The present research work was, therefore, undertaken to evaluate the effects of integrated use of manures and fertilizers on the yield and nitrogen uptake by BRRI dhan28 and to determine the suitability of different sources of organic materials for using as manures for rice cultivation.

MATERIALS AND METHODS

The experiment was conducted at the Soil Science Field Laboratory of Bangladesh Agricultural University, Mymensingh during the Boro season of 2013-2014. The study was performed to evaluate the effect of combined use of manures and fertilizers for maximizing the growth and yield of BRRI dhan28. The soil belongs to Sonatala series under the AEZ of the Old Brahmaputra Floodplain. The experimental soil was silt loam in texture having pH 6.3, organic matter 2.05%, total nitrogen 0.14%, available phosphorus 6.2 mg/kg, exchangeable K 0.068 me/100g soil and available sulphur 12.4 mg/kg. BRRI dhan28, a high yielding boro rice variety of rice was used in this experiment as the test crop. The experiment was laid out in a randomized complete block design (RCBD), where the experimental area was divided into 4 blocks. There were 6 different treatment combinations: T_0 - Control, T_1 - STB-CF (HYG), T_2 - STB-CF (HYG) + CD (5 t/ha) on IPNS basis, T_3 - STB-CF (HYG) + PM (3 t/ha) on IPNS basis, T_4 - STB-CF (HYG) + COM (5 t/ha) on IPNS basis, T_5 - FP. Here, STB = Soil Test Basis; CF = Chemical Fertilizer; HYG = High Yield Goal; OM = Organic matter; PM = Poultry Manure; COM = Compost; FP = Farmers' practice. Well decomposed cowdung, compost and poultry manure were applied to the plots as per the treatments by mixing with the soil well before 7 days of sowing.

The nutrient contents of the manures are depicted in Table 1. The amounts of nitrogen, phosphorus, potash and sulphur fertilizers required per plot were calculated as per the treatments. Cowdung, poultry manure, and compost were applied before one week of transplanting. The full dose of TSP, M_oP and Gypsum was applied one day before transplanting. Urea was applied in three equal splits. The first split of urea was applied after 15 days of transplanting. The second split was applied as top dressing after 30 days of transplanting i.e at maximum tillering stage. The third split was applied at panicle initiation stage. Thirty day old seedlings were carefully uprooted from a seedling nursery and transplanted on the experimental plots maintaining plant spacing of 20 cm x 20 cm. Three seedlings were transplanted in each hill. Intercultural operations such as irrigation and weeding were done as and when necessary. The crop was harvested at full maturity. Grain yield was recorded on 14% moisture basis and straw yield on sundry basis. Five hills were randomly selected from each plot at maturity to record the growth and yield contributing characters. Grain and straw samples were analyzed for N content following the method outlined by Bremner and Mulvaney (1982). The N uptake by grain and straw was calculated from its content and yield data using the equation, Nitrogen uptake = [Nitrogen content (%) x Yield (kg ha^{-1})]/100. All the data were statistically analyzed by F-test and the mean differences were ranked by DMRT at 5% level (Gomez and Gomez, 1984).

Table 1. Nutrient contents in cowdung, compost and poultry manure

Manure	Nutrient contents			
	% N	%P	%K	%S
Cowdung	0.57	0.47	0.69	0.23
Compost	0.89	0.30	0.45	0.46
Poultry manure	1.18	1.13	0.81	0.35

RESULTS

Growth and yield contributing characters

Growth and yield contributing characters such as plant height, effective tillers hill^{-1}, panicle length, grains panicle^{-1} and 1000-grain weight were influenced significantly due to the application of poultry manure, compost and NPKS fertilizers. The tallest plant of 89.70 cm was found in T_3 [(PM) + STB-CF] while the shortest plant of 77.30 cm was observed in control treatment. Cowdung when applied @ 5 t ha^{-1} with [STB-CF] (T_2) produced taller plants compared to the application of compost @ 5 t ha^{-1} with same doses of chemical fertilizers (T_2). The highest number of effective tillers hill^{-1} of 13.85 was found in T_3 [(PM) + STB-CF] and the lowest value of 10.55 was observed in T_0 (control). The treatments T_1 [STB-CF], T_2 [(CD) + STB-CF], T_2 [(CD) + STB-CF], T_4 [(COM) + STB-CF] and T_5 [Farmer's practice] demonstrated statistically similar effective tillers hill^{-1}. The highest panicle length (23.30 cm) was found in T_3 [(PM) + STB-CF] while the lowest panicle length (19.25 cm) was observed in T_0. The number of grains panicle^{-1} varied from 115 to 83.25 with the highest value in T_3 [(PM) + STB-CF]. The treatments T_1, T_2 and T_4 were statistically similar with respect to grains panicle^{-1}. The lowest number of grains panicle^{-1} (83.25) was found in control. The 1000-grain weight ranged from 19.95 g in T_0 (control) to 20.90 g in T_1 [STB-CF].

Grain yield

The grain yield of BRRI dhan28 varied significantly due to application of cowdung, poultry manure, compost and NPKS fertilizers (Table 3). The grain yield ranged from 2220 to 4340 kg ha^{-1}. The highest grain yield (4340 kg ha^{-1}) was observed in T_3 [(PM) + STB-CF] and the lowest value (2220 kg ha^{-1}) was recorded in T_0 (control). Again the treatments T_1, T_2, T_4 and T_5 produced identical grain yields. Based on grain yield, the treatments may be ranked in order of $T_3 > T_2 > T_4 > T_5 > T_1 > T_0$. In association with same recommended fertilizer doses poultry manure treated plots gave better grain yield than cowdung and compost treated plots indicating the superior effect of poultry manure. The increase in grain yield over control ranged from 18.92% to 95.59% where the highest increase was obtained in T_3 [(PM) + STB-CF] and the lowest one was obtained in control as shown in Table 3.

Table 2. Effects of manures and fertilizers on the yield attributes of BRRI dhan28

Treatment combinations	Plant height (cm)	Effective tillers/hill (No.)	Panicle Length (cm)	Grains panicle^{-1} (No.)	1000-grain weight (g)
T_0 [Control]	77.30 c	10.55 b	19.25 c	83.25 b	19.95
T_1 [STB-CF]	79.70 bc	11.50 b	21.33 b	98.00 ab	20.90
T_2 [(CD) + STB-CF]	81.20 b	11.55 b	21.56 b	97.50 ab	19.90
T_3[(PM) + STB-CF]	89.70 a	13.85 a	23.30 a	115.00 a	20.55
T_4[(COM)+STB-CF]	79.60 bc	10.95 b	22.40 ab	102.25 ab	20.15
T_5:[FP]	80.25 bc	11.20 b	21.84 ab	91.00 b	20.85
CV (%)	2.35	11.64	4.42	12.22	5.03
SE (±)	0.957	0.675	0.477	5.976	0.513
P value	0.000	0.047	0.001	0.036	0.000

The figure(s) having common letter(s) in a column do not differ significantly at 5% level of significance.
CV = Coefficient of variation, SE = Standard error of means; STB = Soil Test Basis, CF = Chemical Fertilizer, OM = Organic Manure, CD = Cowdung, PM = poultry manure, COM = Compost, FP = Farmers' practice

Straw yield

Straw yields of BRRI dhan28 also varied significantly by different treatments under study. The yields of straw ranged from 2572 to 4024 kg ha^{-1} (Table 3). The highest straw yield of 4024 kg ha^{-1} was obtained in T_3 [(PM) + STB-CF] and the lowest value of 2572 kg ha^{-1} was noted in T_0 (control). The treatment may be ranked in the order of $T_3 > T_2 > T_5 > T_4 > T_1 > T_0$ in terms of straw yield. Poultry manure exerted comparatively better performance in producing straw yields as compared to cowdung and compost. Regarding the percent increase of straw yield, maximum increase (56.42%) was noted in T_3 [(PM) + STB-CF] and the minimum value (1.95%) was found in T_1 [STB-CF] as shown in Table 3.

Table 3. Effects of fertilizers and manures on the yield of BRRI dhan28

Treatment	Grain		Straw	
	Yield (kg ha^{-1})	% increase over control	Yield (kg ha^{-1})	% increase over control
T_0: (Control)	2220 c	-	2572 c	-
T_1: STB-CF (HYG)	2643 bc	18.92	2622 c	1.95
T_2: STB-CF (HYG) + CD (5t/ha)	3218 b	44.59	3225 b	25.29
T_3: STB-CF (HYG) + PM (3t/ha)	4340 a	95.59	4024 a	56.42
T_4:STB-CF (HYG) + Compost (5 t/ha)	3099 b	39.19	3150 b	22.57
T_5: (Farmers' Practice)	2958 b	33.33	3224 b	25.29
CV (%)	13.85		10.91	-
SE (±)	213.32		171.05	-
P value	0.0001		0.0003	-

The figure(s) having common letter(s) in a column do not differ significantly at 5% level of significance.
CV = Coefficient of variation, SE = Standard error of means; STB = Soil Test Basis, CF = Chemical Fertilizer, OM = Organic Manure, CD = Cowdung, PM = poultry manure, COM = Compost, FP = Farmers' practice

Nitrogen uptake by grain and straw of BRRI dhan28

Results in Table 4 indicate that the N uptake by grain of BRRI dhan28 varied significantly due to application of manures and fertilizers. The N uptake by grain ranged from 42.69 to 74.18 kg ha^{-1}. The highest N uptake (74.18 kg ha^{-1}) was recorded in T_4 [(COM) + STB-CF] which was identical with T_3 [PM + STB-CF] and T_5: FP and the lowest N uptake (42.69 kg ha^{-1}) was found in T_1 [STB-CF]. The straw N uptake of BRRI dhan28 was not significant due to application of manures and fertilizers (Table 4). The straw N uptake ranged from 28.81 to 40.72 kg ha^{-1}. The highest N uptake (40.72 kg ha^{-1}) was recorded in T_0 [Control] and the lowest N uptake (28.81 kg ha^{-1}) was found in T_5 [FP]. The Table 4 reports that the total N uptake by grain of BRRI dhan28 varied significantly due to application of manures and fertilizers. The total N uptake ranged from 73.14 to 109.59 kg ha^{-1}. The highest total N uptake (109.59 kg ha^{-1}) was recorded in T_3 [PM + STB-CF] which was statistically similar with T_4 [COM + STB-CF] and the lowest N uptake (73.14 kg ha^{-1}) was found in T_1 [STB-CF].

Table 4. Effects of fertilizer and manure on nitrogen uptake by BRRI dhan28

Treatment	N uptake (kg ha^{-1})		
	Grain	Straw	Total
T_0 [Control]	42.92 c	40.72	76.18 c
T_1 [STB-CF]	42.69 c	30.45	73.14 c
T_2 [(CD) + STB-CF]	54.45 bc	31.14	85.58 bc
T_3 [(PM) + STB-CF]	71.22 ab	38.34	109.59 a
T_4 [(COM) + STB-CF]	74.18 a	26.15	100.33 ab
T_5: FP	58.42 abc	28.81	87.23 bc
CV (%)	18.86	27.35	15.01%
SE (±)	5.40	4.46	6.6557
Level of significance	**	NS	**
P value	0.003	0.213	0.011

The figure(s) having common letter(s) in a column do not differ significantly at 5% level of significance.
CV = Coefficient of variation, SE = Standard error of means; STB = Soil Test Basis, CF = Chemical Fertilizer, OM = Organic Manure, CD = Cowdung, PM = poultry manure, COM = Compost, FP = Farmers' practice

DISCUSSION

The highest grain and straw yield of BRRI dhan28 was observed in the treatment T_3 where poultry manure was applied in combination with chemical fertilizers on IPNS basis. Poultry manure performed better than other manures in producing grain yield of BRRI dhan28 as it contains more uric acid which enhances the decomposition process. So plants can uptake more nutrients from soil. These results are well corroborated with Rahman *et al.* (2009) and Islam *et al.* (2014). The N uptake by grain and straw was also influenced significantly due to application of manures and fertilizers. The total N uptake was enhanced with the application of poultry manure in combination with chemical fertilizers. Chandel *et al.* (2003), Parvez *et al.* (2008) and Hoque *et al.* (2014) reported that application of nitrogen from manures and fertilizers increased the N uptake both by grain and straw of rice. Islam *et al.* (2014) also recommended poultry manure 3 t ha^{-1} + STB-CF for the yield maximization of BRRI dhan49 in Aman season.

CONCLUSIONS

It can be concluded that integrated use of manures and fertilizers improves rice yield to a significant extent than the single application of chemical fertilizers. Based on the overall results of the study, the application of poultry manure @ 3 ton/ha with chemical fertilizers on IPNS basis can be recommended for obtaining the maximum yield of BRRI dhan28 in Boro rice production with some further trials.

REFERENCES

1. BARC (Bangladesh Agricultural Research Council), 2012. Fertilizer Recommendation Guide 2012, Soils Pub. No. 45 Bangladesh Agricultural Research Council, Farmgate, Dhaka.
2. BBS (Bangladesh Bureau of Statistics), 2011. Statistical Year Book of Bangladesh. Statistical Division, Ministry of Planning, Government of People's Republic of Bangladesh, Dhaka pp 45-99.
3. Bhuiyan NI, 1994. Crop production trends and need of Sustainability in Agriculture, Paper presented at the workshop on Integrated Nutrient Management of Sustainable Agriculture held at SRDI, Dhaka, Bangladesh, during June 26-28, 1994.
4. Bremner JM, and Mulvaney CS, 1982. Nitrogen- total, In Methods of Soil Analysis Part 2, Page AL, Miller RH, Keeney DR (Editors): American Society of Agronomy, Inc, Publisher, Madison, Wisconsin USA. pp. 595-624.
5. Chandel RS, Singh K, Singh AK, Sudhakar PC, 2003. Effect of sulphur nutrition in rice (*Oryza sativa L.*) and mustard (*Brassica juncea L.*Czern and Coss) grown in sequence. Indian Physiology, 8: 155-159.
6. FAO, 2009. Production Year Book. Food and Agriculture Organization of the United Nations, Rome, Italy.
7. FAO, 2011. Production Year Book. Food and Agriculture Organization of the United Nations, Rome, Italy.
8. Gomez KA and Gomez AA, 1984. Statistical Procedures for Agricultural Research. John Wilely and Sons. New York.
9. Hoque A, Islam MR, Siddique AB, Afroz H, Yeasmen N, 2014. Integrated use of manures and fertilizers for maximizing the growth and yield of Boro rice (cv. BRRI dhan 28) *Journal of Soil and Nature*, 7: 7-11.
10. Islam MR, Karim MS, Siddique AB, Rubel MH, Rahman MT 2014. Yield maximization of Aman rice (cv. BRRI dhan49) through integrated use of manures and fertilizers. Bangladesh journal of progressive science and technology, 12: 55-58.
11. Nambiar KKM 1998: Long term fertility effects on wheat productivity. Proceedings of International Conference. Mexico. DF. CIMMYT. pp. 516-560.
12. Parvez MS, Islam MR, Begum MS, Rahman MS, Miah MJ 2008. Integrated use of manure and fertilizers for maximizing the yield of BRRI dhan30. *Journal of Bangladesh Society of Agricultural Science and Technology* 5: 257-260.
13. Rahman MS, Islam MR, Rahman MM, Hossain MI 2009. Effect of cowdung, Poultry manure and Urea-N on the yield and nutrient uptake of BBRI dhan29. *Bangladesh Research Publication Journal* 2: 12-16.

7

CLIMATE CHANGE AND CROP PRODUCTION CHALLENGES: AN OVERVIEW

*Sushan Chowhan[1], Shapla Rani Ghosh[2], Tushar Chowhan[3], Md. Mahmudul Hasan[4] and Md. Shyduzzaman Roni[5]

[1]Bangladesh Institute of Nuclear Agriculture (BINA), Sub-station, Khagrachari, Bangladesh; [2]Department of ICT, Mawlana Bhashani Science and Technology University, Tangail, Bangladesh; [3]Department of Geography and Environment, University of Dhaka, Bangladesh; [4]Apex Organic Soya Industries Ltd., Sapmara, Gaibandha, Bangladesh; [5]Department of Horticulture, Bangabandhu Sheikh Mujibur Rahman Agricultural University, Gazipur, Bangladesh

*Corresponding author: Sushan Chowhan; E-mail: sushan04@yahoo.com

ARTICLE INFO	ABSTRACT

Key words

Climate change,
Simulation models,
Stress,
Coping strategies

Climate change has heterogeneous effect on crop production. Potential yield of some crops were found to be decreasing in different simulation models. High temperature, drought, salinity, excessive rain fall are the major stresses faced by crops in a changing climatic condition. Coastal areas of Bangladesh are highly vulnerable to climate change. It was found that a total of 1,405.57 MT yield are lost in different crops. Data shows the production trends of many crops remaining in a steady state or their increase is very slow compared to elapse of time. Some possible adaptation measures such as sorjan system, floating bed agriculture, growing crops in raised beds, harvesting rain water, cultivation of salt and flood tolerant crop varieties etc. were suggested to reduce possible climate change risk and to cope up with the current situation.

INTRODUCTION

Climate change and crop production are interrelated processes, both of which take place on a global scale. It describes how weather patterns will be affected around the globe. On the other hand, global warming describes an average temperature increase of the earth over time. These changes could be manifested in changes in climate averages as well as changes in extremes of temperatures and precipitation. It is likely that the changes will vary depending on region. Global warming is projected to have significant impacts on conditions affecting agriculture, including temperature, carbon dioxide, glacial run-off, precipitation and the interaction of these elements. These conditions determine the carrying capacity of the biosphere to produce enough food for the human population and domesticated animals. The overall effect of climate change on agriculture will depend on the balance of these effects.

At the same time, agriculture has been shown to produce significant effects on climate change, primarily through the production and release of greenhouse gases such as carbon dioxide, methane, and nitrous oxide, but also by altering the earth's land cover, which can change its ability to absorb or reflect heat and light, thus contributing to radiative forcing. Land use change such as deforestation and desertification, together with use of fossil fuels, are the major anthropogenic sources of carbon dioxide; agriculture itself is the major contributor to increasing methane and nitrous oxide concentrations in earth's atmosphere. Despite technological advances, such as improved varieties, genetically modified organisms, and irrigation systems, weather is still a key factor in agricultural productivity, as well as soil properties and natural communities. The effect of climate on agriculture is related to variabilities in local climates rather than in global climate patterns. The earth's average surface temperature has increased by 1.5°F (0.83°C) since 1880 (Anonymous, 2012). There is unequivocal evidence that the global climate is warming because of an increased concentration of greenhouse gases (GHG) in the earth atmosphere. According to the IPCC 4th assessment report (2007), continued GHG emission at or above current rates would cause further warming and induce changes in global climate system during the 21st century that would very likely be larger than those observed during the 20th century. A 0.1 to 0.5 m rise in sea-level by the middle of this century (as predicted by most of the estimate) will pose a great threat to the livelihoods and agriculture in low-lying coastal areas of the world including about 1/5th of the total land area of Bangladesh. With a population of about 148 million, it is one of the poorest and most vulnerable countries in the world to disaster and climate change impacts. Different types of natural hazards including floods(e.g. river flood, urban flood and flash flood), cyclone and storm surges, tidal surges/intrusion of saline water, salinity, water-logging/submergence, drought, river bank erosion, tornadoes etc affect the country almost every year (Rahman, 2013). These catastrophic events significantly hinder the crop production systems. In Bangladesh, over 30 % of the net cultivable area is in the coastal region. Out of 2.85 million hectares of the coastal and off-shore areas about 0.828 million hectares of the arable lands, which constitutes about 52.5 percent of the net cultivable area in 64 upazilas of 13 districts (Miah, 2010). But these vast cultivable areas is under great threat of vulnerabilities of the climate change and crop production is rapidly declining due to climate risk factors. Saline water intrusion, sea level rise, water logging, cyclone and storm surges are climatic hazards affecting the low lying coastal areas. Impacts of climate change and sea-level rise should have real consequences on the livelihoods of the coastal people as it would be affected by salinity intrusion, flooding, drainage congestion, cyclones, heavy storms and erosion of the land masses (WB, 2000; Agarwala et al., 2003).

The climate of Bangladesh is influenced by monsoon climate and characterized by high temperature, heavy rainfall, often-excessive humidity and marked seasonal variations. Although more than half of the area is north of the tropics, the effect of the Himalayan mountain chain is such as to make the climate more or less tropical throughout the year. The climate is controlled primarily by summer and winter winds, and partly by pre-monsoon (March to May) and post-monsoon (late October to November) circulation. The southwest monsoon originates over the Indian ocean, and carries warm, moist and unstable air. The easterly trade winds are also warm, but relatively drier. The northeast monsoon comes from the Siberian desert, retaining most of its pristine cold, and blows over the country, usually in gusts, during dry winter months.

Environmental stress is the primary cause of crop losses worldwide, reducing average yields for most major crops by more than 50% (Bray et al., 2000). The tropical crop production environment is a mixture of conditions that varies with season and region. Climatic changes will influence the severity of environmental stress imposed on crops. Moreover, increasing temperatures, reduced irrigation water availability, flooding, and salinity will be major limiting factors in sustaining and increasing vegetable productivity. Extreme climatic conditions will also negatively impact soil fertility and increase soil erosion. Thus, additional fertilizer application or improved nutrient-use efficiency of crops will be needed to maintain productivity or harness the potential for enhanced crop growth due to increased atmospheric CO_2. The response of plants to environmental stresses depends on the plant developmental stage and the length and severity of the stress (Bray, 2002). Plants may respond similarly to avoid one or more stresses through morphological or biochemical mechanisms (Capiati et al., 2006). Environmental interactions may make the stress response of plants more complex or influence the degree of impact of climate change. Temperature limits the range and production of many crops. In the tropics, high temperature conditions are often prevalent during the growing season and, with a changing climate, crops in this area will be subjected to increased temperature stress.

Drought, a slow onset disaster is the single most important factor affecting world food security and the catalyst of the great famines of the past (CGIAR 2003). The world's water supply is fixed, thus increasing population pressure and competition for water resources will make the effect of successive droughts more severe (McWilliam, 1986). Inefficient water usage all over the world and inefficient distribution systems in developing countries further decreases water availability. Water availability is expected to be highly sensitive to climate change and severe water stress conditions will affect crop productivity, Crop production is threatened by increasing soil salinity particularly in irrigated cropland which provide 40% of the world's food (FAO 2004). Excessive soil salinity reduces productivity of many agricultural crops, including most vegetables which are particularly sensitive throughout the ontogeny of the plant. Vegetable production occurs in both dry and wet seasons in the tropics. However, production is often limited during the rainy season due to excessive moisture brought about by heavy rain. For instance, most vegetables are highly sensitive to flooding.

MATERIALS AND METHODS

Scientific approach requires a close understanding of the subject matter. This paper mainly depends on the secondary data. Different published reports of different journals and reports mainly supported in providing data in this paper. This paper is completely a review paper. Therefore no specific method has been followed in preparing this paper. It has been prepared by Internet search, comprehensive studies of various articles published in different journals, books and proceedings available in the libraries of BSMRAU, BARI, BRRI and BARC. Valuable information has been collected through personal contact with respective resource personnel to enrich the paper. It compiled the all related information to prepare this paper.

RESULTS AND DISCUSSION

The impacts of climate change on crop production are expected to be widespread across the globe, although studies suggest that African agriculture is likely to be most affected due to heavy reliance on low-input rainfed agriculture and due to its low adaptive capacity (Mertz et al., 2009). Broadly speaking, climate change is likely to impact crop productivity directly through changes in the growing environment, but also indirectly through shifts in the geography and prevalence of agricultural pests and diseases, associated impacts on soil fertility and biological function, and associated agricultural biodiversity. While many impact predictions tend towards the negative, increased CO_2 will also contribute to enhanced fertilization, although there is significant debate as to the extent to which this may increase plant growth. The Inter-governmental Panel on Climate Change (IPCC, 2007) concluded that 'in mid- to high latitude regions, moderate warming benefits crop and pasture yields, but even slight warming decreases yields in seasonally dry and low-latitude regions (medium confidence)'. In IPCC language, moderate warming is in the range of 1–3°C. Smallholder and subsistence farmers, pastoralists and artisanal fisher-folk will suffer complex, localized impacts of climate change (high confidence). Food and forestry trades are projected to increase in response to climate change with increased dependence on food imports for most developing countries (medium to low confidence). Warming beyond 2–3°C will likely result in yield declines in all areas.

IPCC (2009) indicates that rising temperatures, drought, floods, desertification and weather extremes will severely affect crop production, especially in the developing world. Developing countries will be affected most for three reasons:

I.   Climate change will have its most negative effects in tropical and subtropical regions;
II. Most of the predicted population growth to 2030 will occur in the developing world (United Nations Population Division DoEaSA, 2009); and
III. More than half of the overall work force in the developing world is involved in agriculture (FAO, 2005).

While anthropogenic effects on climate have been apparent for several decades, modeling future climate change is not an exact science due to the complexity and incomplete understanding of atmospheric processes. None the less, there is broad agreement that, in addition to increased temperatures, climate change will bring about regionally dependent increases or decreases in rainfall, an increase in cloud cover and increases in sea level. Extreme climate events will also increase in intensity or frequency, such as higher maximum temperatures, more intense precipitation events, increased risk and duration of drought, and increased peak wind intensities of cyclones. Predictions in sea level rise indicate that this will continue for centuries after temperatures stabilize, causing flooding of coastal lands and salinization of soils and subsurface water in coastal regions.

Models of crop response to climate change mainly consider temperature, soil moisture and increased carbon dioxide. However, many other processes not easily incorporated into models could potentially have significant effects including: pests and diseases, brief exposures of crops to very high temperatures, elevated ozone, loss of irrigation water, and increase in inter-annual climate variability associated with monsoons and phenomena like El Niño. The model outputs, while encompassing a wide range of potential outcomes, tend to have the following in common:

- The yield potential of staple foods will decline in most production environments and commodity prices will rise
- While projections for a few countries with northerly latitudes indicate net positive impacts of climate change, projections for most developing countries are negative.
- Only 'best-case' scenarios predict no net effect of climate change on global cereal yields by 2030 but predictions beyond that time frame are much more pessimistic.

A. Simulation Models

Models are a mathematical representation of a real world system. The use of models is very common in other disciplines, including the airplane industry, automobile industry, civil, industrial and chemical engineering etc. The use of models in agriculture and environmental sciences is not very common.

Crop simulation models integrate current scientific knowledge from many different disciplines, including crop physiology, plant breeding, agronomy, agro meteorology, soil physics, soil chemistry, pathology and entomology. Crop simulation models in general calculate or predict crop yield as a function of weather conditions, soil conditions and crop management scenarios

SRES (Special Report Emissions Scenarios)

It is a report by IPCC that was published in 2000. The greenhouse gas emissions scenarios described in the report have been used to make projections of possible future climate change.

GCM (Global Circulation Models)

GCMs and global climate models are widely applied for weather forecasting, understanding the climate, and projecting climate change.

RCM (Regional Climate Models)

This modeling technique consists of using initial conditions, time-dependent lateral meteorological conditions and surface boundary conditions to drive high-resolution RCMs. The driving data is derived from GCMs (or analyses of observations) and can include GHG and aerosol forcing. One of the primary advantages of these techniques is that they are computationally inexpensive, and thus can easily be applied to output from different GCM experiments. Another advantage is that they can be used to provide local information, which can be most needed in many climate change impact applications.

PRECIS (Providing Regional Climate for Impacts Studies)

It is a regional climate modeling system designed to run on a Linux based PC. PRECIS can be applied to any area of the globe to generate detailed climate change projections.

VAR (Vector Auto Regression)

It is a statistical model used to capture the linear interdependencies among multiple time series. VAR models generalize the univariate auto regression (AR) models. All the variables in a VAR are treated symmetrically; each variable has an equation explaining its evolution based on its own lags and the lags of all the other variables in the model. It does not require expert knowledge.

DSSAT (Decision Support System for Agro technology Transfer)

The DSSAT-CSM simulates growth, development and yield of a crop growing on a uniform area of land under prescribed or simulated management as well as the changes in soil water, carbon, and nitrogen that take place under the cropping system over time. The DSSAT-CSM is. The DSSAT helps decision makers by reducing the time and human resources required for analyzing complex alternative decisions (Tsuji et al., 1998).

APSIM (Agricultural Production Systems Simulator)

It is used to simulate biophysical processes in farming systems, particularly to the economic and ecological outcomes of management practices in the face of climate risk.

Data Sets Required for Model Calibration and Validation

Daily weather data: Rainfall, max and min temperatures, solar radiation or sunshine hours.

Soil profile characterization data (one time activity—data to be collected from soil survey reports): Soil texture by horizon, bulk density, fraction stones, organic carbon, soil pH (water), horizon thickness and depth, root growth distribution, surface characteristics such as soil color, slope, permeability, drainage class, soil series name.

Management data: Crop, cultivar, planting date, seedling rate, plant spacing, row spacing, planting depth, irrigation (dates, amounts, type and method of irrigation), fertilizer applied (dates, amounts, type of material, method of application), chemical application (date, amount, type, method of application), tillage/ intercultural operations (dates, depth, equipment used), organic fertilizer (date, amount, type and method of application), thinning and weeding (date and method).

Initial soils data (to be collected at sowing of a crop): Initial soil water measurements up to maximum soil depth (1.5 m) and soil sampling at 30 cm depth intervals, Initial soil fertility up to 1.5 m at 30 cm depth intervals. Soil samples should be analyzed for NH_4, NO_3, P, K, pH and organic C and N, Soil surface residue, amount and composition (N and C content).

Soil water measurements: At every 10 to 15 day interval up to maximum soil depth (1.5 m) at 30 cm depth interval using gravimetric method if others not available

Vegetative and reproductive development (crop-specific visual observations only): Observations taken at every 2 or 3 days. Emergence date, vegetation stages (V_1, V_2, V_3 ---------- V_n) and reproductive stages (R_1, R_3, R_5, R_6 -------- R_8).

Crop growth analysis (crop-specific, could change with crop): Plants sampled every two weeks and sample harvested area is determined (1 m2), no. of plant sampled, total above ground biomass, weights of leaf, petioles, pods and seeds; leaf area (can be done on a sub-sample), no. of pods, no. of seeds and nitrogen concentration (optional) etc. data are collected.

Yield and yield components at harvest (crop specific, could change with crop): Harvest date, harvest density (plants/m2), harvest area, total above ground biomass, pod yield (seed + shell), seed yield, 1000 seed weight, 1000 pod weight recorded.

B. Climatic change scenarios

The climate in Bangladesh is changing and it is becoming more unpredictable every year. The impacts of higher temperatures, more variable precipitation, more extreme weather events, and sea level rise are already felt in Bangladesh and will continue to intensify. Climate change poses now-a-days severe threat mostly in crop sector and food security among all other affected sectors. Crop yields are predicted to fall by up to 30%, creating a very high risk of hunger and only sustainable climate-resilient crop production is the key to enabling farmers to adapt and increase food security (Climate Change Cell, 2007). The coastal area of Bangladesh is naturally susceptible to disaster.

Table 1. Climate change scenario for Bangladesh

Model	Year	Temperature change (°C) Mean (standard deviation)			Precipitation change (%) Mean (standard deviation)			Sea Level Rise (cm)
		Annual	DJF	JJA	Annual	DJF	JJA	
GCM	2030	1.0	1.1	0.8	5	-2	6	
PRECIS	2030 (Max)	0.3	-0.02	1.3*	4	-8.7	3.8	14
	2030 (Min)	1.18	0.65	1.78*				
GCM	2050	1.4	1.6	1.1	6	-5	8	
PRECIS	2050 (Max)	0.2	0.07	0.89*	2.3	-4.7	3.0	32
	2050 (Min)	1.24	0.59	1.65*				

Note: * JJAS (June, July, August, September); DJF= December January February, JJA= June July August.
(Source: Miah, 2010)

Drought

Unpredictable drought is the single most important factor affecting world food security and the catalyst of the great famines of the past (CGIAR 2003). The world's water supply is fixed, thus increasing population pressure and competition for water resources will make the effect of successive droughts more severe (McWilliam 1986). Inefficient water usage all over the world and inefficient distribution systems in developing countries further decreases water availability. Water availability is expected to be highly sensitive to climate change and severe water stress conditions will affect crop productivity, particularly that of vegetables. In combination with elevated temperatures, decreased precipitation could cause reduction of irrigation water availability and increase in evapotranspiration, leading to severe crop water-stress conditions (IPCC 2001). Vegetables, being succulent products by definition, generally consist of greater than 90% water (AVRDC 1990). Thus, water greatly influences the yield and quality of crops; drought conditions drastically reduce crop productivity. Drought stress causes an increase of solute concentration in the environment (soil), leading to an osmotic flow of water out of plant cells. This leads to an increase of the solute concentration in plant cells, thereby lowering the water potential and disrupting membranes and cell processes such as photosynthesis. The timing, intensity, and duration of drought spells determine the magnitude of the effect of drought.

Table 2. Drought prone areas (in mha) of Bangladesh

Drought Class	Rabi	Pre-Kharif	Kharif
Very Severe	0.446	0.403	0.344
Severe	1.71	1.15	0.74
Moderate	2.95	4.76	3.17
Slight	4.21	4.09	2.90
No Drought	3.17	2.09	0.68
Non-T. aman			4.71

Source: Drought Manual, BARC, 2003

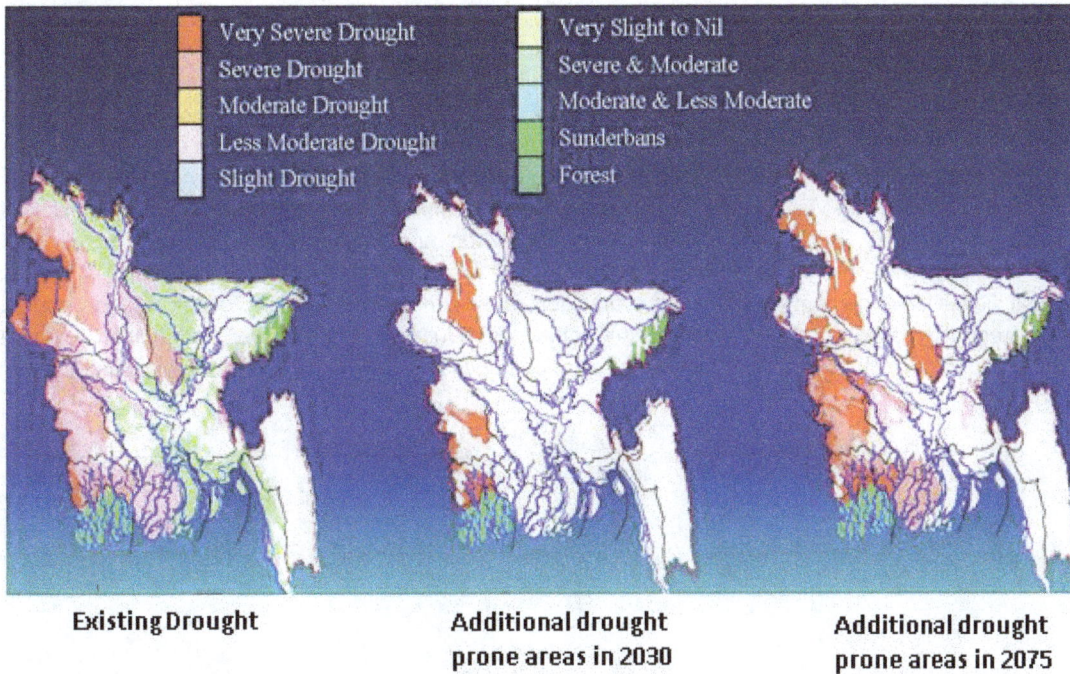

Figure 1. Maps showing existing drought and drought in the year 2030 and 2075.
(Source: Sultana *et. al.*, 2008)

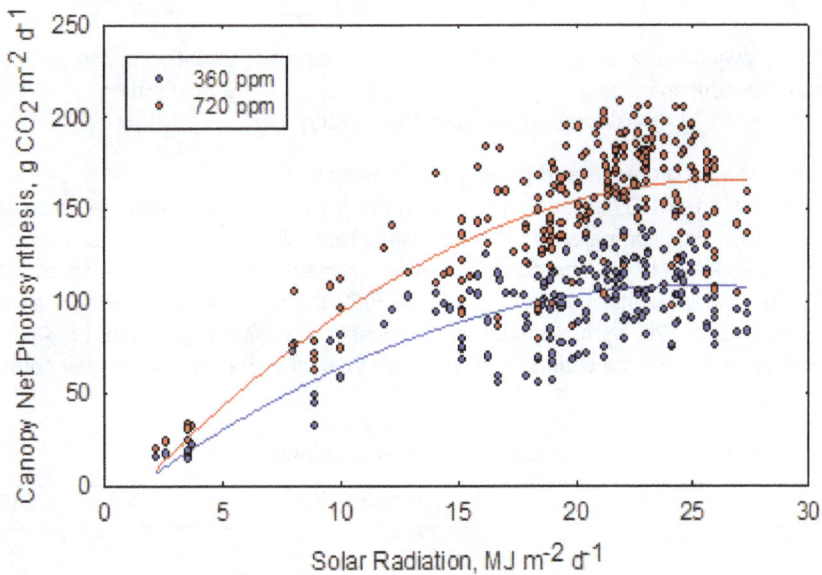

Figure 2. Cotton photosynthesis-solar radiation.
(Source: Baker and Allen, 1993)

High temperatures

Temperature limits the range and production of many crops. In the tropics, high temperature conditions are often prevalent during the growing season and, with a changing climate, crops in this area will be subjected to increased temperature stress. Analysis of climate trends in tomato growing locations suggests that temperatures are rising and the severity and frequency of above optimal temperature episodes will increase in the coming decades (Bell et al., 2000). Vegetative and reproductive processes are strongly modified by temperature alone or in conjunction with other environmental factors (Abdalla and Verkerk, 1968).

High temperature stress disrupts the biochemical reactions fundamental for normal cell function in plants. It primarily affects the photosynthetic functions of higher plants (Weis and Berry, 1988). High temperatures can cause significant losses in productivity due to reduced fruit set, and smaller and lower quality fruits (Stevens and Rudich, 1978). Pre-anthesis temperature stress is associated with developmental changes in the anthers, particularly irregularities in the epidermis and endothesium, lack of opening of the stromium, and poor pollen formation (Sato et al., 2002). In pepper, high temperature exposure at the pre-anthesis stage did not affect pistil or stamen viability, but high post-pollination temperatures inhibited fruit set, suggesting that fertilization is sensitive to high temperature stress (Erickson and Markhart, 2002). Hazra et al., (2007) summarized the symptoms causing fruit set failure at high temperatures in tomato; this includes bud drop, abnormal flower development, poor pollen production, dehiscence, and viability, ovule abortion and poor viability, reduced carbohydrate availability, and other reproductive abnormalities. In addition, significant inhibition of photosynthesis occurs at temperatures above optimum, resulting in considerable loss of potential productivity.

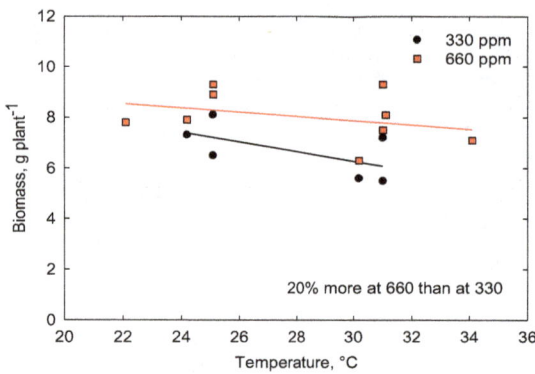

Figure 3. Rice growth in different temperature and CO_2 concentration.

Figure 4. Variation in rice yield in different CO_2 concentration.

(Source: Baker and Allen, 1993)

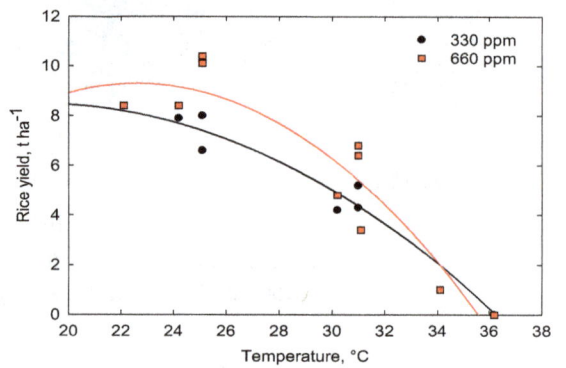

Effect of climate change on biomass and grain yield of wheat

Without the direct effect of CO_2 the model predicted that all three climate change scenarios significantly increases biomass production compared to the baseline (Table 3), with A1B increasing the most (28%) followed by B1 (16%) and A2 (12%). The combined effect (climate and CO_2) increased biomass production much more, with A1B increasing the most (74%) and A2 (55%) and B1 (41%) being similar. Increased CO_2 concentrations of 220 and 120 ppm resulted in increased biomass production of 43-45% and 25%, respectively. The effect of the climate change scenarios on grain yield shows the same trend as biomass with a slightly higher increase rate.

Table 3. Effects of climate change on biomass and grain production of wheat

Scenario	CO2 ppm	Biomass (Kg/ha)	Grain Yield (Kh/ha)
Baseline	330	5039f	2467f
A1B	330	6463d	3167d
A1B	550	8753a	4349a
A2	330	5651e	2834e
A2	550	7813b	3978b
B1	330	5856e	2880e
B1	450	7104c	3520c

Note: Baseline= 1961-1990, A1B = 3.2^0C, A2= 3.6^0C, B1= 2.7^0C more than the baseline temperature

(Source: Wang et. al., 2011)

Table 4. Rice and wheat production under different climate change scenarios

Simulation	HYV Aus ('000 tones)	HYV Aus Percent Change	HYV Aman ('000 tones)	HYV Aman Percent Change	HYV Boro ('000 tones)	HYV Boro Percent Change	Wheat ('000 tones)	Wheat Percent Change
Baseline (1994-95)	702	0	4,484	0	6,200	0	890	0
CCCM	512	-27	4,170	-7	6,014	-3	712	-20
GFDL	512	-27	3,901	-13	5,766	-7	347	-61
330 ppmv CO_2+20°C	569	-19	3,901	-13	5,952	-4	561	-37
330 ppmv CO_2+40°C	435	-38	3,363	-25	5,766	-7	285	-68
580 ppmv CO2+00°C	920	31	5,605	-25	7,626	23	1,228	38
580 ppmv CO_2+20°C	793	13	4,977	11	7,440	20	881	-1
580 ppmv CO_2+40°C	660	-6	4,529	1	7,192	16	534	-40
660 ppmv CO_2+00°C	983	40	5,964	33	8,060	30	1,317	48
660 ppmv CO_2+20°C	856	22	5,336	19	7,874	27	970	9
660 ppmv CO_2+40°C	730	4	4,888	9	7,626	23	614	-31

CCCM= Canadian Climate Centre Model.   (Source: Karim et al., 1998)
GFDL= Geophysical Fluid Dynamics laboratory.

Table 5. Temperature effects on crop yield

Crop	T opt, °C	T max, °C	Yield at T opt, (t/ha)	Yield at 28 °C, (t/ha)	Yield at 32°C (t/ha)	% decrease (28 to 32 °C)
Rice	25	36	7.55	6.31	2.93	54
Soybean	28	39	3.41	3.41	3.06	10
Dry bean	22	32	2.87	1.39	0.00	100
Peanut	25	40	3.38	3.22	2.58	20
Grain sorghum	26	35	12.24	11.75	6.95	41

Source: Reddy et. al., 2005

Table 6. Climate change responses of sorghum (Location: Parbhani, Maharashtra)

Variety: CSH 15		Rainfall: 790 mm			
Mean seasonal temperature: 29°C		AWC of soil: 120 mm			
Scenarios	Crop duration (days)	Grain Yield t ha^{-1}	% Change	CV (%)	HI (%)
Control	94	4.12	0	11	35
+ Temp.	82	2.90	-27	18	33
+ Temp.+RF	82	3.05	-26	16	33

Data period: 1969-2007   Source: Singh et. al., 2009

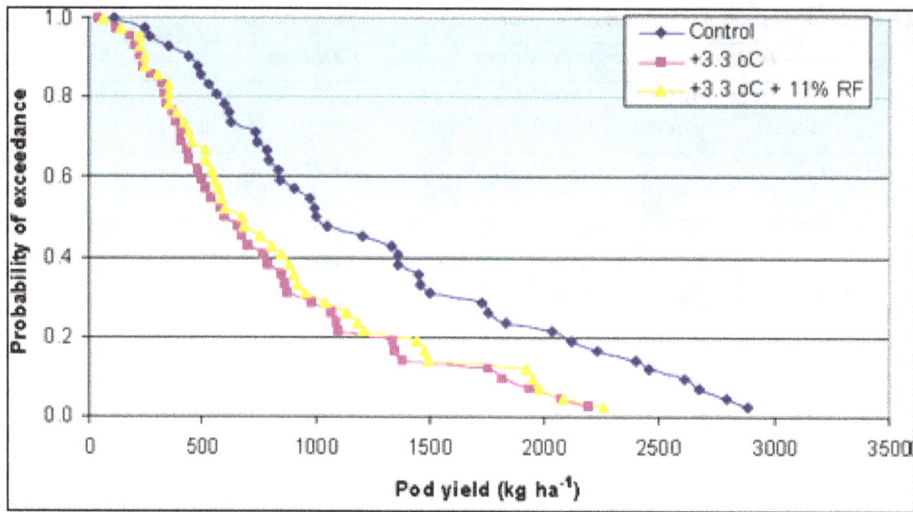

Figure 5. Probability distribution of yield under climate change (Groundnut - Anantapur).
(Source: Singh et. al., 2009)

Table 7. Simulated effect of climate changeon ICRISAT crops

Crop	% change in grain yield		
	+Temp.	+ CO_2	Net change
Sorghum	- (27 to 55%)	+ (0 to 10%)	- (22 to 50%)
P. millet	- (38 to 56%)	+ (0 to 10%)	- (33 to 51%)
Groundnut	- (38 to 44%)	+ (10 to 20%)	- (23 to 29%)
Pigeon pea	- (23 to 26%)	+ (10 to 20%)	- (8 to 11%)
Chick pea	- (22 to 24%)	+ (10 to 20%)	- (7 to 9%)

Source: Tubiello et al., 2007

When adaptation is not considered, most of the major potato producing countries would suffer great losses in potential potato yield. Bolivia is the only country where potential yield would increase without adaptation, and with adaptation it is predicted to increase a staggering 77%. In most other major potato producing countries, adaptation mitigates a large part of the climate change induced yield loss. In Iran, for example, yield loss decreases from 48% to 13%. China, Peru, Russia, and the USA are other notable examples of countries where adaptation could mitigate much of the negative effects of global warming. When considering adaptation, Bangladesh, Brazil, Colombia, and Ukraine have the largest decrease in potential yield (more than 20% in 2040-59). The percentage of area with yield increase (Table 8) reflects the possibility to mitigate the effect of climate change by shifting the location of production with existing potato growing regions. It is particularly high (>30%) in Argentina, Canada, China, Japan, UK, Russia, and Spain.

Effect of climate change on rice production in Bangladesh

Tables 9 and 10 show predicted yields of BR3 and BR14 boro rice varieties, respectively at 12 locations of Bangladesh in the years 2008, 2030, 2050 and 2070. These predictions have been made using a fixed concentration of atmospheric CO_2 of 379 ppm (the value reported for the year 2005 in the fourth assessment report of IPCC) and for planting date of 15 January. The tables show significant reduction in rice yield in the future due to predicted changes in climatic condition. Compared to 2008, predicted average reductions of BR3 variety for the 12 selected locations are about 11% for the year 2030, 21% for 2050 and 54% for 2070. The corresponding reductions for BR14 variety are about 14%, 25% and 58% for the years 2030, 2050 and 2070, respectively. Some regional variation could also be observed in the predictions, with somewhat higher reductions predicted for central, southern and south-western regions.

Table 8. Potato area and changes in potential potato yield induced by climate change in the 2040-59 and the percentage of the potato area in a country where potential potato yield will increase

Country	Potato area (1000 ha)	Change in potential yield (%)		Areas with yield increase (% of cells)	
		Without Adaptation	With Adaptation	Without Adaptation	With Adaptation
China	3430	-22.2	-2.5	8.5	30.7
Russia	3289	-24.0	-8.8	12.4	48.4
Ukraine	1534	-30.3	-24.8	0.0	2.7
Poland	1290	-19.0	-16.1	0.0	2.4
India	1253	-23.1	-22.1	0.4	2.0
Belarus	692	-18.8	-16.6	0.0	0.0
United States	548	-32.8	-5.9	1.4	20.1
Germany	300	-19.6	-15.5	0.0	0.0
Peru	263	-5.7	5.8	8.3	13.9
Romania	262	-26.0	-9.9	0.0	19.2
Turkey	207	-36.7	-17.1	0.0	10.4
Netherlands	181	-20.0	-10.9	0.0	0.0
Brazil	177	-23.2	-22.7	0.0	0.0
United Kingdom	169	-6.2	8.1	50.0	57.1
France	168	-18.7	-6.9	4.5	29.9
Colombia	167	-32.5	-30.6	4.5	4.5
Kazakhstan	165	-38.4	-12.4	2.3	9.4
Iran	161	-48.3	-13.3	0.0	21.4
Canada	155	-15.7	4.6	17.9	55.5
Spain	142	-31.4	-6.6	0.0	37.5
Bangladesh	140	-25.8	-24.0	0.0	0.0
Bolivia	131	8.4	76.8	22.6	29.0
Lithuania	126	-13.7	-9.2	0.0	0.0
Argentina	115	-12.9	0.5	11.4	35.2
Nepal	115	-18.3	-13.8	0.0	16.7
Japan	102	-17.4	-0.9	8.8	41.2

Source: Hijmans, 2003

Table 9. Predicted yield of BR 3 variety of boro rice (kg ha^{-1}) at 12 selected locations for the years 2008, 2030, 2050 and 2070

Station Name	Cultivar	2008	2030	2050	2070	% change in yield for 2030	% change in yield for 2050	% change in yield for 2070
Rajshahi	BR3	3063	4083	3265	1785	33.3	6.59	-41.7
Bogra	BR3	5741	5119	4070	2036	-10.8	-29.1	-64.5
Dinajpur	BR3	6848	4824	4364	2692	-29.6	-36.3	-60.7
Mymensingh	BR3	5995	5275	4455	2739	-12.0	-25.7	-54.3
Tangail	BR3	5487	5160	3874	1938	-5.95	-29.4	-64.7
Jessore	BR3	5571	4432	4583	1997	-20.4	-17.7	-64.2
Satkhira	BR3	4700	4364	3603	2066	-7.14	-23.3	-56.0
Barisal	BR3	6043	4006	3971	2091	-33.7	-34.3	-65.4
adaripur	BR3	4582	4017	3647	2186	-12.3	-20.4	-52.3
Chandpur	BR3	5975	5455	4039	2772	-8.70	-32.4	-53.6
Comilla	BR3	6115	5987	4456	3075	-2.09	-27.1	-49.7
Sylhet	BR3	5960	5117	5750	3595	-14.1	-3.52	-39.7

Source: Basak et.al., 2010

Table 10. Predicted yield of BR 14 variety of boro rice (kg ha^{-1}) at 12 selected locations for the years 2008, 2030, 2050 and 2070

Station Name	Cultivar	2008	2030	2050	2070	% change in yield for 2030	% change in yield for 2050	% change in yield for 2070
Rajshahi	BR14	2334	2771	2392	1148	18.7	2.48	-50.8
Bogra	BR14	4306	3668	2637	1398	-14.8	-38.8	-67.5
Dinajpur	BR14	5047	3374	3023	1656	-33.1	-40.1	-67.2
Mymensingh	BR14	4353	3790	3186	1873	-12.9	-26.8	-57.0
Tangail	BR14	4104	3883	2565	1297	-5.38	-37.5	-68.4
Jessore	BR14	4032	3160	3153	1305	-21.6	-21.8	-67.6
Satkhira	BR14	3153	3171	2434	1377	0.57	-22.8	-56.3
Barisal	BR14	4397	2889	2705	1457	-34.3	-38.5	-66.9
Madaripur	BR14	3229	2606	2578	1491	-19.3	-20.2	-53.8
Chandpur	BR14	4389	3981	2801	1842	-9.29	-36.2	-58.0
Comilla	BR14	4678	4368	3063	1978	-6.62	-34.5	-57.7
Sylhet	BR14	4596	3764	4240	2378	-18.1	-7.74	-48.3

Source: Basak et.al., 2010

Flooding

Crop production is often limited during the rainy season due to excessive moisture brought about by heavy rain. Most crops are highly sensitive to flooding and genetic variation with respect to this character is limited. In general, damage to crops by flooding is due to the reduction of oxygen in the root zone which inhibits aerobic processes. Flooded crop plants accumulate endogenous ethylene that causes damage to the plants (Drew 1979). Low oxygen levels stimulate an increased production of an ethylene precursor, 1-aminocyclopropane-1-carboxylic acid (ACC), in the roots. The rapid development of epinastic growth of leaves is a characteristic response to water-logged conditions and the role of ethylene accumulation has been implicated (Kawase 1981). The severity of flooding symptoms increases with rising temperatures; rapid wilting and death of tomato plants is usually observed following a short period of flooding at high temperatures (Kuo et al., 1982).

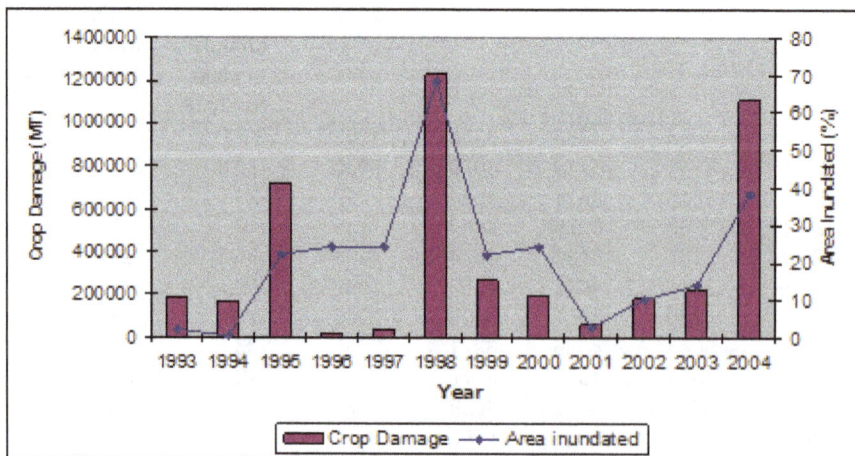

Figure 6. Crop damage (MT) due to historical flood in Bangladesh
(Source: Madhu, 2009)

Salinity

Crop production is threatened by increasing soil salinity particularly in irrigated croplands which provide 40% of the world's food (FAO 2002). Excessive soil salinity reduces productivity of many agricultural crops, including most vegetables which are particularly sensitive throughout the ontogeny of the plant. Onions are sensitive to saline soils, while cucumbers, eggplants, peppers, and tomatoes are moderately sensitive. In hot and dry environments, high evapotranspiration results in substantial water loss, thus leaving salt around the plant roots which interferes with the plant's ability to uptake water. Physiologically, salinity imposes an initial water deficit that results from the relatively high solute concentrations in the soil, causes ion-specific stresses resulting from altered K^+/Na^+ ratios, and leads to a build up in Na^+ and Cl^- concentrations that are detrimental to plants (Yamaguchi and Blumwald 2005). Plant sensitivity to salt stress is reflected in loss of turgor, growth reduction, wilting, leaf curling and epinasty, leaf abscission, decreased photosynthesis, respiratory changes, loss of cellular integrity, tissue necrosis, and potentially death of the plant (Jones, 1986; Cheeseman, 1988). Salinity also affects agriculture in coastal regions which are impacted by low-quality and high-saline irrigation water due to contamination of the groundwater and intrusion of saline water due to natural or man-made events. Salinity fluctuates with season, being generally high in the dry season and low during rainy season when freshwater flushing is prevalent. Furthermore, coastal areas are threatened by specific, saline natural disasters which can make agricultural lands unproductive, such as tsunamis which may inundate low-lying areas with seawater. Although the seawater rapidly recedes, the groundwater contamination and subsequent osmotic stress causes crop losses and affects soil fertility. In the inland areas, traditional water wells are commonly used for irrigation water in many countries. The bedrock deposit contains salts and the water from these wells are becoming more saline, thus affecting irrigated vegetable production in these areas.

Table 11. Influence of salinity on total dry matter production (g/10 plants) of rice cultivars and their classification to salinity tolerance

Variety	Salinity level (ds m-1)			
	0	4	6	12
IR 20	0.068	0.054	0.045	0.027
Pokkali	0.133	0.083	0.054	0.035
MR 33	0.069	0.058	0.039	0.021
MR68	0.060	0.044	0.028	0.018
MR 84	0.094	0.048	0.033	0.024
MR 52	0.075	0.057	0.040	0.028
MR 211	0.081	0.065	0.049	0.035
MR 219	0.066	0.052	0.038	0.024
MR 220	0.101	0.052	0.026	0.014
MR 232	0.077	0.054	0.040	0.031
BR 29	0.053	0.040	0.018	0.004
BR 40	0.075	0.054	0.039	0.023

(Source: Hakim et al., 2010)

Table 12. Tuber yield and harvest index for seven diploid potato clones and the tetraploid cultivar 'Norland' exposed to control conditions or 150 mM salt stress for 7 days at tuber initiation

Clones	Tuber yield (g)	
	Control	Stress
Norland	266.2	98.7
10908-05	241.4	71.5
10909-18	265.8	167.5
F20-ID	74.7	45.9
9506-04	100.0	137.2
11374-01	225.0	59.2
10602-02	70.5	51.3
9788-03	12.5	29.7
Mean	157.0	82.6

(Source: Shaterian et.al., 2008)

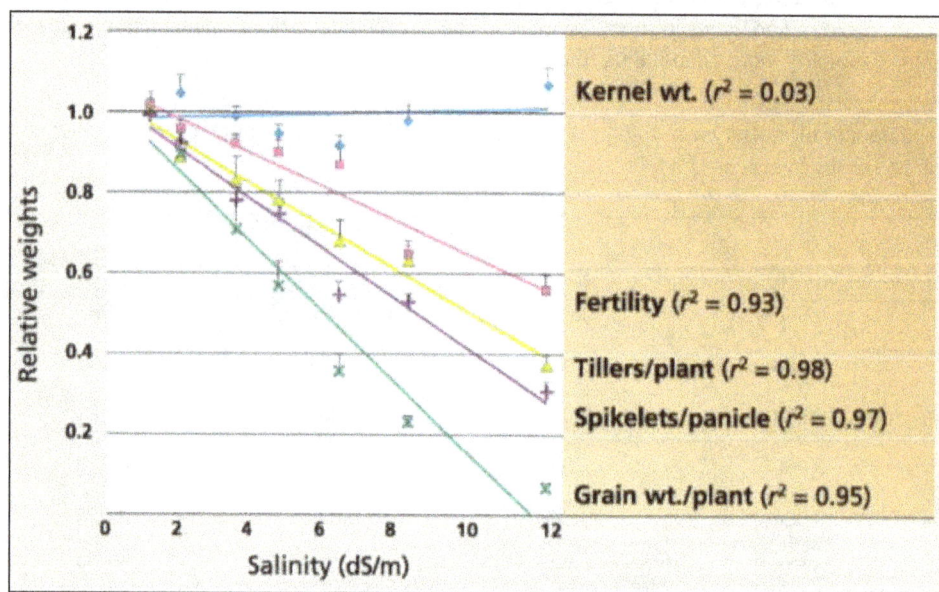

Figure 7. Relationship between salinity and various yield components of rice (*Oryza sativa* L. cv M202).
(Source: Zeng and Shanon, 2002)

From estimates it is seen that the main reasons of yield reduction (20-40 % yield loss) in T.Aman crop are erratic rainfall, increased intensity and frequency of drought, increased salinity, tidal surges, floods, cyclone, use of local varieties, increased incidences of pests and diseases etc in the context of climate change. Total yield loss of T.Aman crop has been estimated to about 6.93 lakh ton per year in 450,320 hectares based on last 5-10 years climate change scenarios. Similarly, average yield level of HYV Boro is being affected (30-40 % yield loss) by high temperature (causing sterility) and increased salinity and that of T.Aus/Aus crop is being affected (20-40 % yield loss) by tidal surges. Vegetables, pulses, oil seed crops and fruit crops are being affected (20-40 % yield loss) by drought, increased salinity, soil wetness, excessive rainfall and water-logging and tidal surges in most coastal districts. From the study, total crop loss for major crops (viz. cereals, potato, pulses, oil seeds, vegetables, spices and fruit crops) due to different climate risks has been estimated to about 14.05 lakh tons per year based on last 5-10 years of climate change scenarios in ten districts. But the people are to live with these climatic vulnerabilities and risks in the coastal region.

Climate change scenarios on the coastal regions of Bangladesh

Table 13. Crop loss/yield reduction due to climate risk factors

Crop	% yield reduction	District	Total areas (in ha) affected	Total yield loss '000' MT	Major Climatic Risks
T. Aman	40-60	Cox's Bazar	35,500	71.00	Drought, flash floods, salinity, erosion, tidal surges, pests & diseases
	20-40	Patuakhali	76,200	114.30	Drought, water-logging, tidal floods, pests and diseases
	20-40	Baraguna	27,500	41.25	Drought, flood, pests & diseases
	20-40	Pirojpur	25,600	38.40	Flood, drought, tidal surge, pests
	20-40	Barisal	57,700	86.62	Flood, water-logging, drought, pests & diseases
	20-40	Noakhali	59,950	89.93	Drought, flood, water, logging, pests & diseases
	20-40	Satkhira	35,700	53.55	Drought, water-logging, erosion, pests
	20-40	Khulna	28,850	43.27	Cyclone, water-logging, salinity, pests
	20-40	Bagerhat	61,270	91.90	Drought, flood, river erosion, pests
	20-40	Bhola	42,000	63.00	Drought, tidal flood, cyclone, pests
T. Aus	20-40	All districts	75,000	112.50	Submergence, drought, salinity, Fe toxicity, river & ground water salinity, cyclone, pests & diseases
HYV Boro	20-40	All districts	150,000	300.00	Drought, salinity, Fe toxicity, river & ground water salinity, cyclone
Potato, Sweet Potato	40-60	All districts	28,055	140.25	Short winter, clayey soils, salinity, fogginess. Average yield = 10-12 t/ha
Pulses (Khesari, M.bean)	20-40	All districts	201,850	60.55	Untimely rainfall, soil wetness, drought, salinity, Pests & diseases
Oilseed Crops (Mustard,sesame, G.nut)	20-40	All districts	63,750	19.12	Late/short winter, salinity, clayey soils, Pests & diseases
Spice Crops (Chilli, Onion, Garlic)	20-40	All districts	53,630	26.81	Early rainfall, soil wetness, salinity, pests and diseases.
Fruit crops (banana,papaya, water melon, amra, guava, amra etc)	20-40	All districts	10,625	53.12	Erratic rain, drought, high temperature, salinity, tidal flood, water-logging, pests & diseases and cyclone
		Total		1,405.57	

All districts= Khulna, Bagerhat, Satkhira, Barisal, Bhola, Barguna, Pirojpur, Patuakhali, Cox's Bazar and Noakhali.
(Source: Miah, 2010)

Table 14. Adaptation practices for sustainable crop production in the context of climate change

District	Recommended adaptation practices
1. Cox's bazaar	Sorjan system of cultivating year round vegetables, spices and fruits on raised beds and creeper vegetables on bed edges, Fish culture in ditches during wet months, Introduction of salt tolerant crop varieties, Encourage fruits and vegetables gardening Introduction of high value vegetable crop varieties (hybrid cucumber, ladies finger, chillis etc) as relay cropping in vegetable growing areas. Embankment repair and introduction of late T.Aman variety
2. Noakhali	Excavation of canals ponds for saline free water and introduction of salt tolerant varieties, Promote introduction of salt-tolerant pulses and oil seed crops (viz. cowpea, soybean, mungbean, ground nut etc.) Introduce floating bed agriculture in water-logged areas. Drainage improvement, introduce short duration and salt-tolerant crop varieties. Introduce standing water (submergence var.) boro cultivation Making high embankment and introduce salt tolerant varieties
3. Barisal	Sorjan system of cultivating year round vegetables, spices and fruits on raised beds and creeper vegetables on bed edges and cultivation of fish in ditches during wet months. Introduce submergence tolerant rice varieties (BRRIdhan-51,52, BINAdhan-11,12) Popularize floating bed agriculture. Zero tillage (potato, maize) and floating bed agriculture. Introduce submergence tolerant rice varieties
4. Barguna	Introduction of salt tolerant pulse crops (mungbean, cowpea, soybean, ground nut, sweet potato, chilli) Sorjan system of cultivating year round vegetables, spices & fruits on raised beds and creeper vegetables on bed edges and cultivation of fish in ditches during wet months. Introduction of drought and salt-tolerant crop varieties creating facilities of irrigation in fallow lands. Introduction of zero tillage (potato) Digging ponds and canals for rain water harvest.
5. Satkhira	Introduction of salt-tolerant crop varieties in salt affected areas - rice crops (BRRIdhan- 44,47 BINAdhan-8,10) -Utilization of fallow bunds under gher areas for year round vegetable cultivation, Popularization of zero tillage (potato, maize) Sorjan system of cultivating year round vegetables, spices and fruits on raised beds and creeper vegetables on bed edges and cultivation of fish in ditches during wet months, Floating bed agriculture in water-logged areas.

Source: Miah, 2010

Trends of Crop production per hectare (Satkhira district) due to climate change are shown graphically below: (Figures 8-11)

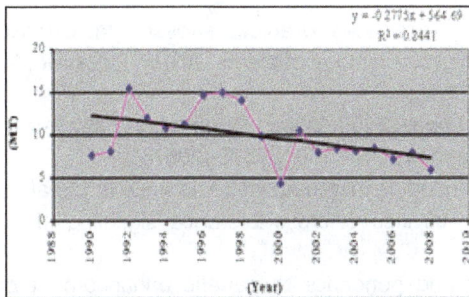

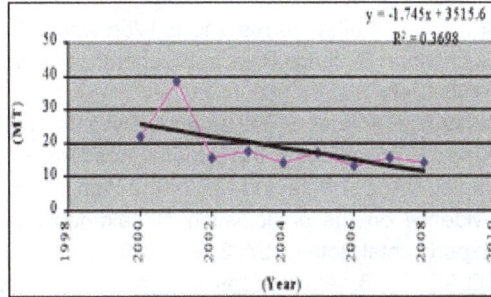

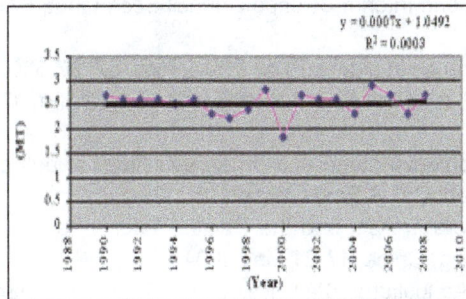

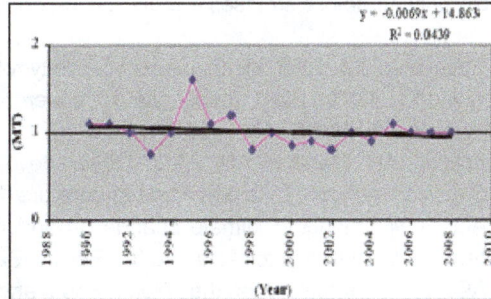

Figure 8. Trend of T Aman (HYV) production.   **Figure 9.** Trend of sesame (HYV) production.

Figure 10. Trend of Onion production.   **Figure 11.** Trend of Winter vegetables production.

(Source: Miah, 2010)

CONCLUSION

Climate change can boost up certain crop's production and also can decrease particulars yield. Therefore it is advised to cultivate stress tolerant crops and change production technology in vulnerable areas such as coastal region and flood prone areas. Bangladesh is not responsible for climate change or global warming, but is severely affected. To cope with the present situation of climate change in crop sector of Bangladesh, research should be conducted for introducing of salinity and flood tolerant variety of crops. Otherwise, the food security will be hampered in the near future.

REFERENCES

1. Abdalla AA and K Verderk, 1968. Growth, flowering and fruit set of tomato at high temperature. The Netharlands Journal of Agricultural Science, 16: 71-76.
2. Agarwala S, T Ota, AU Ahmed, J Smith and M van Aalst, 2003. Development and Climate Change in Bangladesh: Focus on the Coastal Flooding and the Sundarbans, Organization for Economic Cooperation and Development (OECD), Paris. pp. 34-67
3. Anonymous. 2012. Climate change and agriculture- Wikipedia, the free encyclopedia. (Cited from- http://www.en.wikipedia.org/wiki/Climate_change_and_agriculture)
4. AVRDC, 1990. Vegetable Production Training Manual, Asian Vegetable Research and Training Center. Shanhua, Tainan, 447 pp.
5. Baker JT and LH Jr. Allen, 1993a. Contrasting crop species responses to CO_2 and temperature: Rice, soybean, and citrus. *Vegetatio* 104/105: 239-260. Also: pp. 239-260. In: *CO_2 and Biosphere.* (Advances in Vegetation Science 14). J. Rozema, H. Lambers, S.C. van de Geijn and M.L. Cambridge (eds.). Kluwer Academic Publishers, Dordrecht.
6. Baker JT and LH Jr. Allen, 1993b. Effects of CO_2 and temperature on rice: A summary of five growing seasons. Journal of Agricultural Meteorology, (Japan), 48: 575-582.
7. Basak JK, MA Ali, N Islam and A Rashid, 2010. Assessment of the effect of climate change on boro rice production in Bangladesh using DSSAT model. Journal of Civil Engineering (IEB), 38: 95-108.

8. Bell GD, MS Halpert, RC Schnell, RW Higgins, J Lowrimore, VE Kousky, R Tinker, W Thiaw, M Chelliah and A Artusa, 2000. Climate Assessment for 1999. Supplement, Bulletin of the American Meteorological Society, 81: 22-52

9. Bray EA, J Bailey-Serres and E Weretilnyk, 2000. Responses to abiotic stresses. In: Gruissem W, Buchannan B, Jones R (eds) Biochemistry and molecular biology of plants. ASPP, Rockville, MD. pp. 1158-1249.

10. Burke M, E Miguel, S Satyanath, J Dykema and D Lobell, 2009. Warming increases risk of civil war in Africa. Proceedings of the National Academy of Sciences USA 106, 20670–20674.

11. Capiati DA, SM País and MT Téllez-Iñón, 2006. Wounding increases salt tolerance in tomato plants: evidence on the participation of calmodulin-like activities in cross-tolerance signaling. Journal of Experimental Botany, 57: 2391-2400.

12. CGIAR, 2003. Applications of molecular biology and genomics to genetic enhancement of crop tolerance to abiotic stresses – a discussion document. Interim Science Council Secretariat, FAO. pp. 19-58

13. Cheeseman JM, 1988. Mechanisms of salinity tolerance in plants. Plant Physiology, 87: 57-550.

14. Drew MC, 1979. Plant responses to anaerobic conditions in soil and solution culture. Current Advances in Plant Science, 36: 1-14.

15. Erickson, AN. Markhart AH, 2002. Flower developmental stage and organ sensitivity of bell pepper (*Capsicum annuum* L) to elevated temperature. Plant Cell & Environment, 25: 123-130.

16. FAO, 2004. Impact of climate change on agriculture in Asia and the Pacific. Twenty-seventh FAO Regional Conference for Asia and the Pacific. Beijing, China, 17-21 May 2004.

17. FAO, 2005. Summary of the World Food and Agricultural Statistics. FAO, Rome. pp. 334-337. Available at: http://faostat.fao.org (accessed 4 August 2009).

18. Hakim MA, AS Juraimi, M Begum, MM Hanif, MR Ismail and A Selamat, 2010. Effect of salt stress on germination and early seedling growth of rice (*Oryza sativa* L.). African Journal of Biotechnology, 9: 1911-1918.

19. Hazra P, HA Samsul, D Sikder and KV Peter, 2007. Breeding tomato (*Lycopersicon Esculentum* Mill) resistant to high temperature stress. International Journal of Plant Breeding, 1: 21-26

20. Hijmans RJ, 2003. The Effect of Climate Change on Global Potato Production. American Journal of Potato Research, 80: 271-280.

21. IPCC, 2007. IPCC Fourth Assessment Report: Climate Change 2007. IPCC, Geneva.

22. IPCC, 2009. The Intergovernmental Panel on Climate Change. Available at: http://www.ipcc.ch (accessed 22 September 2009).

23. IPCC, 2001. Climate change 2001. Impacts, adaptation and vulnerability. Intergovermental Panel on Climate Change. New York, USA.

24. IPCC, 2007: Summary for Policymakers. In: *Climate Change 2007:* The Physical Science Basis. Contribution of Working Group I to the Fourth Assessment Report of the Intergovernmental Panel on Climate Change [Solomon, S., D. Qin, M. Manning, Z. Chen, M. Marquis, K.B. Averyt, M.Tignor and H.L. Miller (eds.)]. Cambridge University Press, Cambridge, United Kingdom and New York, NY, USA.

25. Jones JW, G Hoogenboom, CH Porter, KJ Boote, WD Batchelor, LA Hunt, PW Wilkens, U Singh, AJ Gijsman and JT Ritchie, 2003. The DSSAT cropping system model. European ournal of Agronomy, 18: 235-236.

26. Jones RA, 1986. The development of salt-tolerant tomatoes: breeding strategies. Acta Horticulturae, 190: Symposium on Tomato Production on Arid Land. pp. 134-156.

27. Karim Z, SG Hussain, and M Ahmed, 1998. Assessing Impact of Climate Variations on Food grain Production in Bangladesh. Water Air and Soil Pollution, 92: 53-62.

28. Kawase M, 1981. Anatomical and morphological adaptation of plants to waterlogging. Horticultural Science, 16: 30-34.

29. Khan, HM, M Sarwar and J Islam, 2008. Climate Change and Strategic Adaptation Provisions for Coastal Bangladesh, IDB Bhaban, Dhaka, Bangladesh. pp. 28-61

30. Kuo DG, JS Tsay, BW Chen and PY Lin, 1982. Screening for flooding tolerance in the genus *Lycopersicon*. Horticultural Science, 17: 76-78.

31. Madhu MK, 2009. Climate Change and Agriculture in the South-West. (Available on line at http://www.teacher.buet.ac.bd/akmsaifulislam/climaterisk/student/Group-2.ppt).
32. McWilliam JR, 1986. The national and international importance of drought and salinity effects on agricultural production. Australian Journal of Plant Physiology, 13: 1-13.
33. Mertz O, K Halsnaes, JE Olesen and K Rasmussen, 2009. Adaptation to climate change in developing countries. Environmental Management, 43: 743–752.
34. Miah M U, 2010. Assessing Long-term Impacts of Vulnerabilities on Crop Production Due to Climate Change in the Coastal Areas of Bangladesh, Bangladesh Center for Advanced Studies, Dhaka. pp: 12-128.
35. Rahman M, 2013. Climate Change, Disaster and Gender Vulnerability: A Study on Two Divisions of Bangladesh. American Journal of Human Ecology, 2: 72-82.
36. Reddy VR, V. Ambumozhi and KR Reddy, 2005. Achieving Food Security and Mitigating Global Environmental Change. (Available On line at www.ars.usda.gov/SP2UserFiles/Place/.../BANGKOK_2005.ppt)
37. Reynolds MP, 2010. Climate change and Crop production. CAB International, Nosworthy Way, Wallingford, Oxfordshire OX10 8DE, UK. pp. 1-37.
38. Sato S, MM Peet and JF Thomas, 2002. Determining critical pre- and post-anthesis periods and physiological process in *Lycopersicon esculentum* Mill. Exposed to moderately elevated temperatures. Journal of Experimental Botany, 53: 1187-1195.
39. Shaterian J, DR Waterer, H De Jong and KK Tanino, 2008. Methodologies and Traits for Evaluating the Salt Tolerance in Diploid Potato Clones. American Journal of Potato Research, 85: 93-100.
40. Singh P, AVR Kesava Rao and K Srinivas, 2009. Role of Crop Simulation Models in Assessing Impacts of Climate Change and Evaluating Adaptation Options. ICRISAT. pp. 124-131.
41. Stevens MA and J Rudich, 1978. Genetic potential for overcoming physiological limitations on adaptability, yield, and quality in tomato. Horticultural Science, 13: 673-678.
42. Sultana W, MA Aziz and F Ahmed, 2008. Climate Change: Impact on Crop Production and its Coping Strategies. Agronomy division, BARI, Gazipur. pp. 2-45. (Available on line at: http://www.wamis.org/agm/meetings/rsama08/Bari103-Sultana_Coping_Strategies.pdf)
43. Tubiello FN and G Fischer, 2007. Reducing climate change impacts on agriculture: Global and regional effects of mitigation, 1990-2080. Technological Forecasting and Social Change, 74:1030-1056.
44. United Nations Population Division Department of Economic and Social Affairs (DoEaSA) (2009) World Population Prospects: the 2008 Revision. pp. 3-86. Available at: http://esa.un.org/unpp
45. Wang H, Y He and B Qian, 2011. Impact of Climate Change on Wheat Production for Ethanol in Southern Saskatchewan, Canada. World Renewable Energy Congress 2011, Linkoping, Sweden. pp 647-648.
46. Weis E and JA Berry 1988. Plants and high temperature stress. Society of Experimental Biology, pp 329-346.
47. World Bank, 2000. Bangladesh, Climate Change and Sustainable Development. Report No 21104-BD. pp. 5-59
48. Yamaguchi T and E Blumwald, 2005. Developing salt-tolerant crop plants: challenges and opportunities. Trends in Plant Science, 10: 616-619.
49. Zeng L and MC Shannon, 2000. Salinity effects on seedling growth and yield components of rice. Crop Science, 40: 996-1003.
50. Zhang DD, P Brecke, HF Lee, YQ He and J Zhang, 2007. Global climate change, war and population decline in recent human history. Proceedings of the National Academy of Sciences USA 104, 19214-19219.

EFFECTS OF DIFFERENT LEVELS OF UREA AND MAGIC GROWTH SPRAY SOLUTION ON THE YIELD AND YIELD ATTRIBUTES OF BRRI dhan29

Monika Nasrin, Md. Abdus Salam, Md. Akhter Hossain Chowdhury[1], Md. Arif Hossain Khan[2] and Md. Muzammel Hoque[3]

Department of Agronomy and [1]Department of Agricultural Chemistry, Bangladesh Agricultural University, Mymensingh-2202, Bangladesh; [2]Joint Director (Fertilizer), Fertilizer Management Division, BADC, Rajshahi; [3]Principal Scientific Officer, BANSDOC, Agargaon Dhaka, Bangladesh

*Corresponding author: Md. Akhter Hossain Chowdhury; E-mail: akhterbau11@gmail.com

ARTICLE INFO

Key words

Urea,
Magic growth spray
solution,
BRRI dhan29

ABSTRACT

Magic growth solution along with prilled urea in the rice leaf as foliar spray may save urea compared to soil application of urea alone. In this regard, an experiment was conducted at the Agronomy Field Laboratory of Bangladesh Agricultural University during *Boro Season* (December-April, 2015) to evaluate the effects of urea and magic growth spray solution on the yield and yield attributes of BRRI dhan29. The experiment consisted of ten treatments viz., T_1 = Control, T_2 = 99 kg urea ha^{-1}, T_3 = 63 kg urea ha^{-1} + 2.16L ha^{-1} magic growth spray solution, T_4 = 99 kg urea ha^{-1} + 5.66L ha^{-1} magic growth spray solution, T_5 = 117 kg urea ha^{-1} + 1.44L ha^{-1} magic growth spray solution, T_6 = 117 kg urea ha^{-1} + 5.66L ha^{-1} magic growth spray solution, T_7 = 126kg urea ha^{-1} + 5.66L ha^{-1} magic growth spray solution, T_8 = 132 kg urea ha^{-1}, T_9 = 132 kg urea ha^{-1} + 5.66L ha^{-1} magic growth spray solution, T_{10}= 132 kg urea ha^{-1}+ 2.16L ha^{-1} magic growth spray solution and laid out in a randomized complete block design (RCBD) with three replications. The results revealed that urea and magic growth spray solution exerted significant influence on the yield contributing characters and yield of BRRI dhan29 except panicle length, sterile spikelets and 1000-grain weight. The highest grain and straw yields (6.16 and 9.33 t ha^{-1}, respectively) were obtained from T_7 treatment which could be the resultant effect of highest number of effective tillers hill^{-1}, highest number of grains panicle^{-1} and lowest number of sterile spikelets panicle^{-1}. Grain yield was significantly and positively correlated with plant height, effective tillers hiil^{-1}, panicle length and grains per panicle. Economic analysis showed that net return and benefit cost ratio (BCR) was the highest (1.41) in T_7 treatment. Thus the overall results suggest that farmers may be advised to apply 126 kg urea along with 5.66L magic growth solution per hectare to produce economically highest grain yield of BRRI dhan29, under the agro-climatic condition of Bangladesh Agricultural University.

INTRODUCTION

Rice (*Oryza sativa* L.) is one of the most extensively cultivated cereals of the world. It is the principal food crop of Bangladesh and constitutes 95% of food grain production in the country. The soil and climate of Bangladesh are favorable for rice cultivation throughout the year. About 75% of the total cropped area and over 80% of the total irrigated area is planted to rice (BBS, 2012). Nitrogen (N) is one of the major nutrients required for plant growth. For maximizing yield of rice, nitrogenous fertilizer is the kingpin in rice farming. The N content of Bangladesh soil is low due to warm climate accompanied by extensive cultivation. The efficiency of applied N use by the rice plant is also low. Farmers of the country usually do not apply N in their fields properly and timely. It is estimated that only about 25% of the added N is recovered by the crops and the rest 75% is lost due to leaching, surface runoff, NH_3 volatilization, decreased nitrification and other processes (Naznin et al., 2013). Besides, at present the N fertilizer is   costly. Under these circumstances, it is important to find out the effective method of application of urea that would give higher yield of crops and also reduce fertilizer cost. Panir (2014) applied magic growth solution in the rice leaf as foliar spray which saved 44% urea compared to soil application, increased absorption rate, improved soil health and ultimately increased rice yield. Broadcast application of urea on the surface soil causes losses up to 50% but application of magic growth solution in leaf may result in negligible loss. The savings in applied N reached 44% of urea when applied as foliar spray during *boro* and *aman* seasons (Crasswell and De Datta, 1980). The objective of this study was, therefore, to investigate the effects of different levels of urea and magic growth spray solution on the yield and yield attributes of BRRI dhan29.

MATERIALS AND METHODS

The experiment was conducted at the Agronomy Field Laboratory, Bangladesh Agricultural University, Mymensingh during the period from January to June 2015. The experimental area belongs to Sonatala soil series under Old Brahmaputra Floodplain (AEZ-9). The region occupies a large area of Brahmaputra sediments which are laid down before the river shifted into its present Jamuna channel about 200 years ago (UNDP and FAO, 1988). The field was a medium high land of silty loam soils having pH around 6.5. The experiment consisted of ten treatments viz., T_1 = Control, T_2 = 99 kg urea ha^{-1}, T_3 = 63 kg urea ha^{-1} + 2.16L ha^{-1} magic growth spray solution, T_4 = 99 kg urea ha^{-1} + 5.66L ha^{-1} magic growth spray solution, T_5 = 117 kg urea ha^{-1} + 1.44L ha^{-1} magic growth spray solution, T_6 = 117 kg urea ha^{-1} + 5.66L ha^{-1} magic growth spray solution, T_7= 126kg urea ha^{-1} + 5.66L ha^{-1} magic growth spray solution,T_8= 132 kg urea ha^{-1}, T_9= 132 kg urea ha^{-1} + 5.66L ha^{-1} magic growth spray solution, T_{10}= 132 kg urea ha^{-1} + 2.16L ha^{-1} magic growth spray solution. The experiment was laid out in a randomized complete block design with three replications. Each of the replication presented a block in the experiment. Each block comprised of 10 unit plots which were designed to assign the above mentioned treatments. Total numbers of unit plot were 30 (10 × 3). Spaces between blocks and unit plots were 1 m and 0.5 m, respectively. The size of unit plot was 20 m^2 (4.0 m × 5 m). The land was puddled with country plough, stubles were removed and leveled by laddering. The plots were fertilized with 100, 75, 60 and 10 kg ha^{-1} of triple super phosphate, muriate of potash, gypsum and zinc sulphate, respectively at the time of final land preparation (BRRI, 2010). Urea with magic growth spray solution was applied as per experimental treatments. Urea was applied at 10 DAT, 25 DAT, 40 DAT and magic growth spray solution was applied for three times. 1st spray was applied at 17 DAT, 2nd spray was applied at 29 DAT and 3rd spray was applied at 44 DAT. Weeding, gap filling, irrigation and pesticide application were done as and when necessary throughout the growth period of the crop. The crop was harvested at full maturity. The date of harvesting was confirmed when 90% of the grains become golden yellow in color. Harvesting was done on 25 April 2015. The harvested crop of each plot was separately bundled, properly tagged and then brought to the threshing floor. Grains were sun dried to a moisture content of 14% and then weighed. Straw was also sun dried and weighed. Yields of both grain and straw were converted to t ha^{-1}. Data were collected on plant height, total tillers $hill^{-1}$, effective tillers $hill^{-1}$, ineffective tillers $hill^{-1}$, panicle length, grains $panicle^{-1}$, sterile spikelets $panicle^{-1}$, 1000-grain weight, grain yield, straw yield and harvest index. Collected data were analyzed statistically and mean differences were adjudged by Duncan Multiple Range Test (Gomez and Gomez, 1984).

RESULTS AND DISCUSSION

Effects of different levels of urea and magic growth spray solution on various parameters of rice cv. BRRI dhan29 have been presented and discussed under the following heads

Yield attributes

Different levels of urea and magic growth spray solution had significant effects on almost all the yield and yield contributing characters of rice cultivar BRRI dhan29 except ineffective tillers hill^{-1}, panicle length, sterile spikelets panicle^{-1} and 1000-grain weight (Table 1 and Fig. 1). The tallest plant (92.33 cm) was obtained from T_7 treatment while the shortest one (81.40 cm) was from control. The increase in plant height due to the application of increased levels of urea and magic growth might be associated with stimulating effects of N on various physiological processes including cell division and cell elongation of the plant. In general, plant height increased with the increasing levels of N as urea and magic growth spray solution. These results explicitly confirmed the results obtained by Singh and Singh (1986) and Alam (2002) who recorded a positive effect of increasing levels of urea super granule (USG) on plant height.

Table 1. Effects of different levels of urea and magic growth spray solution on the yield attributes and harvest index of BRRI dhan29

Treatment	Plant height (cm)	Total tillers hill^{-1} (no.)	Panicle length (cm)	Grains panicle^{-1} (no.)	Sterile spikelets panicle^{-1} (no.)	1000-grain weight (g)	Harvest index (%)
T_1	81.40d	9.67d	20.81	74.33e	18.67	21.00	44.99a
T_2	85.13cd	11.33cd	22.79	82.33d	22.33	21.10	42.38ab
T_3	86.90bc	12.00c	23.51	83.33cd	22.67	21.27	42.87ab
T_4	87.76bc	13.00bc	23.89	84.33c	33.67	21.08	42.05ab
T_5	88.37b	12.33bc	23.87	89.33bc	24.67	21.57	44.76a
T_6	86.47bc	12.33bc	23.02	91.33b	25.67	20.95	43.93ab
T_7	92.93a	15.00a	24.92	100.00a	27.00	21.21	39.38b
T_8	83.60c	14.33ab	23.52	80.00d	33.00	21.84	39.25b
T_9	88.33b	13.67b	23.00	88.67bc	27.00	21.45	37.05c
T_{10}	84.07cd	12.00c	23.19	83.33cd	35.33	21.27	39.13bc
CV (%)	2.28	5.34	4.65	9.96	6.68	2.88	8.39
S_x	0.15	0.11	0.62	0.93	0.72	0.35	0.01
Sig. level	0.01	0.05	NS	0.01	NS	NS	0.05

In a column figures with dissimilar letter/s differ significantly as per DMRT.
Sig. = Significance, NS = Not significant

[T_1= No urea application (control), T_2= 99 kg urea ha^{-1} ,T_3= 63 kg urea ha^{-1} + 2.16L ha^{-1} magic growth spray solution, T_4= 99 kg urea ha^{-1} + 5.66L ha^{-1} magic growth spray solution, T_5= 117 kg urea ha^{-1} + 1.44L ha^{-1} magic growth spray solution, T_6= 117 kg urea ha^{-1} + 5.66L ha^{-1} magic growth spray solution, T_7= 126 kg urea ha^{-1} + 5.66L ha^{-1} magic growth spray solution, T_8= 132 kg urea ha^{-1},T_9=132 kg urea ha^{-1} + 5.66L ha^{-1} magic growth spray solution ,T_{10}= 132 kg urea ha^{-1} + 2.16L ha^{-1} magic growth spray solution]

The highest number of total tillers hill^{-1} (15.00) was produced in T_7 treatment which was identical with the tillers of T_8 treatment but significantly different from all other treatments. On the other hand the lowest number of total tillers hill^{-1} (9.67) was recorded from T_1 (control) treatment. The progressive improvement in the formation of tillers with urea and magic growth levels might be due to the availability of N which could be responsible for enhanced tillering of the plant. These results are in compliance with those of Kamal *et al.* (1991) who recorded increased number of tillers hill[-1] with increased levels of N as urea super granule (USG). Application of treatment T_7 produced the highest number of effective tillers hill[-1] (12.00) which was statistically similar with the treatments T_5 and T_8 but significantly different from other treatments. The T_1 treatment produced the lowest number (9.00) of effective tillers hill[-1] (Fig. 1). Adequacy of N as magic growth probably favored the cellular activities during panicle formation and development which led to increased number of productive tillers hill[-1]. Singh et al. (1983) also agreed to this view. Numerically the highest number (3.67) of ineffective tillers hill[-1] was produced when crop was fertilized with T_9 treatment and the lowest number (1.00) was found in T_5 treatment (Figure 1). Though the treatment effects were not significant on panicle length, application of treatment T_7 numerically produced the longest panicle (23.79 cm) and T_1 (control) produced the shortest one (21.81 cm).

Both urea and magic growth solution exerted significant influence on the number of grains of BRRI dhan29. Grains panicle[-1] increased with the increase in levels of urea and magic growth solution. The highest number of grains panicle[-1] (100) was recorded in T_7 treatment which was significantly different from other treatments. The second highest number of grains was obtained from T_6 treatment which was identical with the number of grains of T_5 and T_9 treatments and the lowest grains (74.33) was recorded from control treatment (Table 1). Adequate supply of N from magic growth contributed to grain formation which probably increased the number of grains panicle[-1] with increasing N level. Numerically maximum number of sterile spikelets panicle[-1] (35.00) was recorded from treatment T_{10} and the lowest number (18.67) was produced from T_1 treatment. Different treatments of urea and magic growth did not influence the weight of the grain (Table 1). However, the heaviest grain was produced by T_8 treatment while the lightest grain was observed in T_6 treatment.

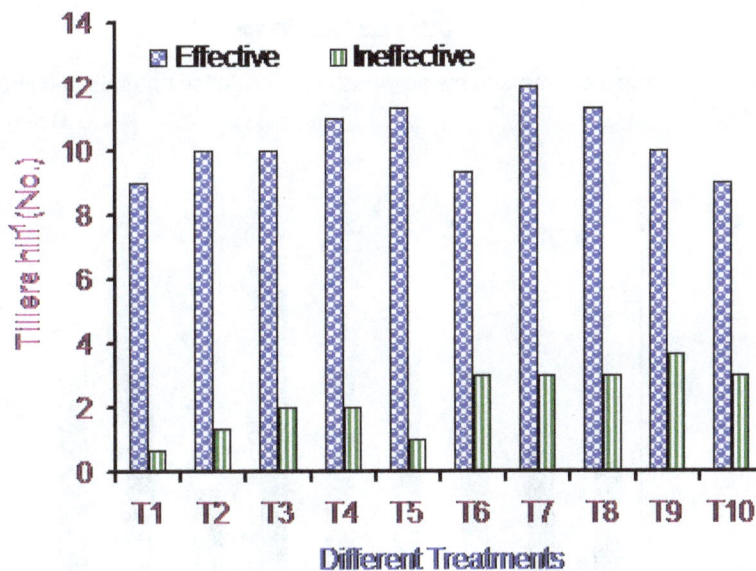

Figure 1. Effects of different levels of urea and magic growth solution on the number of effective and ineffective tillers hill[-1] of BRRI dhan29

Grain and straw yield

Urea and magic growth solution also exerted significance influence on the grain and straw yield of BRRI dhan29 (Fig. 2). The highest grain yield (6.16t ha[-1]) was obtained from T_7 treatment (126kg urea ha[-1]+5.66L ha[-1] magic growth spray solution). Highest yield from that treatment (T_7) might be due to the positive correlations among effective tillers, panicle length, grains per panicle with grain yield (Fig. 4) The lowest grain yield (4.33t ha[-1]) was obtained from T_1 (control) treatment. These results are in agreement with those obtained by Singh and Mahapatra (1989) who recorded the highest grain yield applying 90 kg N ha^{-1} as USG, but not with those reported by Quayum and Prasad (1994). They observed that grain yield increased with N application up to 112.5 kg ha^{-1} as USG. The differences might have been taken place because of dissimilarity in native fertility status of soil. However, when only urea was applied, the grain yield was found to be lower compared to only magic growth solution. Grain yield increase over control was influenced by different levels of urea and magic growth solution (Fig. 3). Yield increase was highest (42.26%) in T_7 treatment and the lowest (11.55%) in T_8 treatment. Grain yield was gradually increased with increasing levels of urea and magic growth solution.

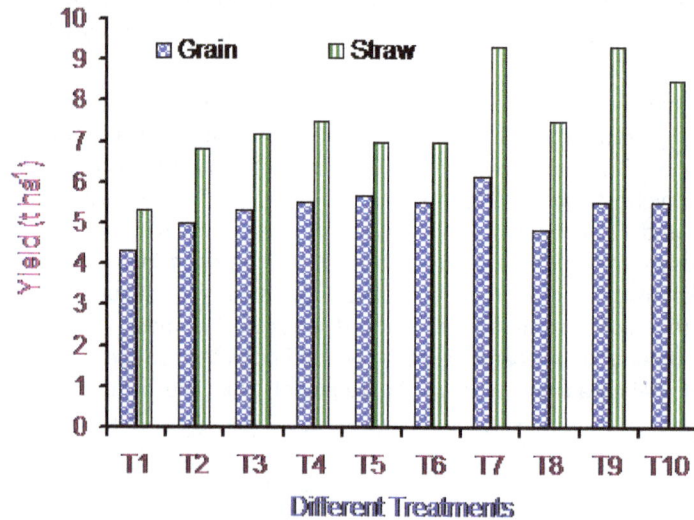

Figure 2. Effects of different levels of urea and magic growth spray solution on grain and straw yields of BRRI dhan29

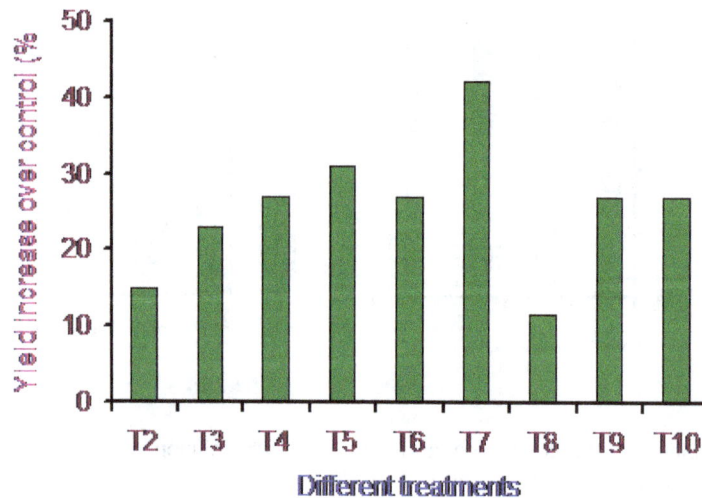

Figure 3. Yield increase of BRRI dhan29 over control as influenced by different levels of urea and magic growth spray solution

Straw yield was also significantly influenced by different treatments (Fig. 2). The highest yield (9.33t ha^{-1}) was produced by T_7 and T_9 treatment and the lowest yield (5.33t ha^{-1}) was obtained from control (Fig. 2). Nitrogen influenced vegetative growth in terms of plant height and number of total tillers hill^{-1} which resulted in differences of straw yield. These findings corroborate with those of Quayum and Prasad (1994). The harvest index (HI), defined as the ratio of grain to total dry matter production in crops, is used as an index of dry matter distribution. The highest harvest index (44.76%) was recorded in T_5 treatment of urea which was statistically similar with other treatments and the lowest (37.05%) HI was recorded in T_9 treatment (Table 1). It is now well established that partitioning of dry matter is related to increased yields in improved cultivars rather than total biomass production (Gifford and Evans, 1981)

Correlation and regression studies

Yield is a complex character which results from the interactions of various yield contributing characters. Hence, an attempt was taken to examine the extent of contribution of various yield attributes to grain yield. The relationship among plant height, effective tillers hill[-1], panicle length, grains panicle[-1] with grain yield of BRRI dhan29 was studied. The correlation and regression lines of these parameters have been shown in Figure 4. Results showed that grain yield had significant positive correlations with plant height (R^2=0.815), effective tillers hill[-1] (R^2=0.8954), panicle length (R^2=0.7265) and grains panicle[-1] (R^2=0.8454). Regression analysis revealed that increase in plant height, effective tillers hill[-1], panicle length and grains panicle[-1] resulted in the corresponding increase in the grain yield of BRRI dhan29. Haque (2002) also found positive correlations among straw yield, number of bearing tillers hill[-1], number of grains panicle[-1], 1000 grain weights of BRRI dhan30 and BRRI dhan31.

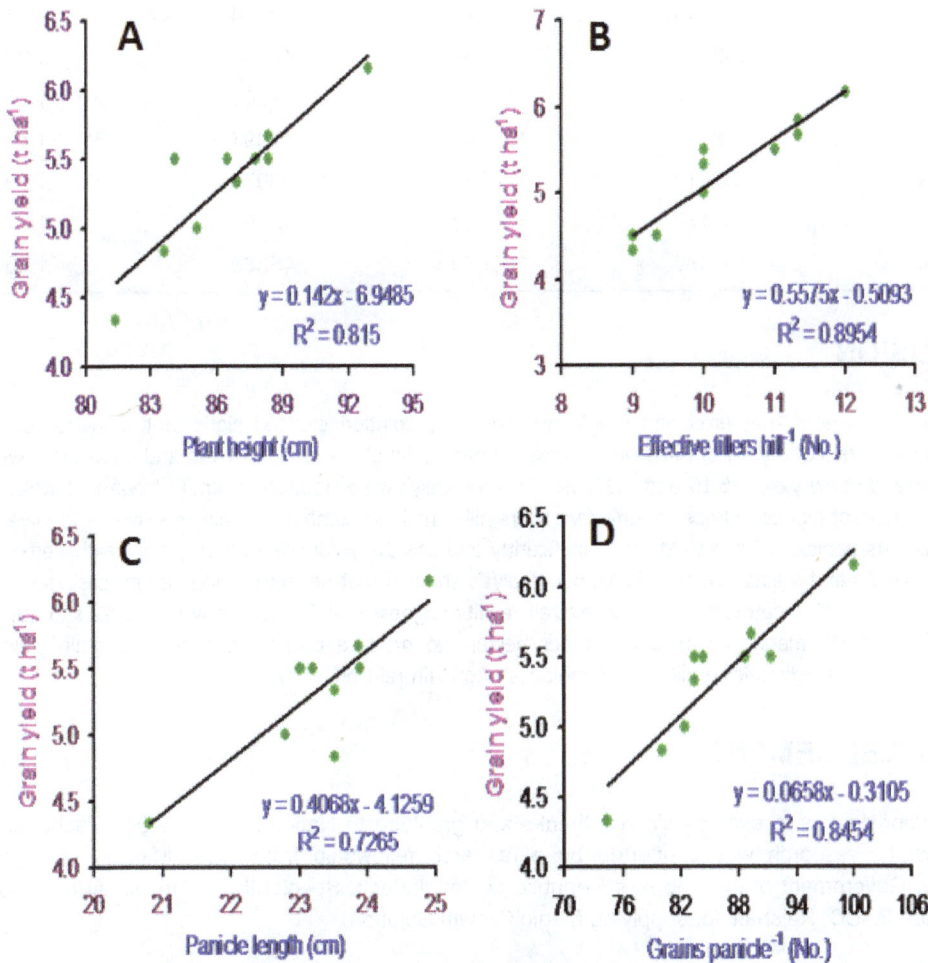

A

y = 0.142x - 6.9485
R^2 = 0.815

B

y = 0.5575x - 0.5093
R^2 = 0.8954

C

y = 0.4068x - 4.1259
R^2 = 0.7265

D

y = 0.0658x - 0.3105
R^2 = 0.8454

Figure 4. Relationships among grain yield vs plant height (A), grain yield vs effective tillers hill[-1] (B), grain yield vs panicle length (C) and grain yield vs grains panicle[-1] (D) of BRRI dhan29 as influenced by different levels of urea and magic growth spray solution.

Net return and benefit cost ratio

From the economic analysis, it is observed that the highest total return (Tk. 195967), the highest net income (Tk. 57891) and the highest BCR (1.41) were recorded from the T_7 treatment which was followed by T_9 treatment (BCR- 1.37). On the other hand, T_1 (control) treatment showed the lowest net return (Tk. 30140) and BCR (1.23) (Table 2).

Table 2. Per hectare cost and returns of BRRI dhan29 as influenced by different levels of urea and magic growth spray solution

Treatment	Total cost (TK.)	Total return (TK.)	Net return (TK.)	BCR
T_1	130400	160540	30140	1.23
T_2	131984	170080	38096	1.28
T_3	133548	171130	37582	1.28
T_4	137644	180998	43354	1.31
T_5	133712	182851	49139	1.37
T_6	137932	180637	42705	1.30
T_7	138076	195967	57891	1.41
T_8	132512	170588	38076	1.28
T_9	131172	180563	49391	1.37
T_{10}	134672	180060	45388	1.33

CONCLUSION

The results revealed that urea and magic growth spray solution exerted significant influence on the yield contributing characters and yield of BRRI dhan29 except panicle length, sterile spikelets and 1000-grain weight. The highest grain and straw yields (6.16 and 9.33t ha[-1], respectively) were obtained from T_7 treatment which could be the resultant effect of highest number of effective tillers hill[-1], highest number of grains panicle[-1] and lowest number of sterile spikelets panicle[-1]. Grain yield was significantly and positively correlated with plant height, effective tillers hiil[-1], panicle length and grains panicle[-1]. Economic analysis showed that net return and benefit cost ratio (BCR) was the highest (1.41) in T_7 treatment. Thus the overall results suggest that farmers may be advised to apply 126 kg urea along with 5.66L magic growth solution per hectare to produce economically highest grain yield of BRRI dhan29, under the agro-climatic condition of Bangladesh Agricultural University.

ACKNOWLEDGEMENT

The author desires to express sincere thanks and gratitude to National Science and Technology (NST) authority as the research was supported by a research fellowship from NST, Ministry of Science and Technology, Government of the People's Republic of Bangladesh. Special thanks to Mr. Arif Hossain Khan, Joint Director, BADC, Rajshahi for supplying Magic Growth Solution.

REFERENCES

1. Alam BMR, 2002. Effect of different levels of urea super granule on the growth and yield of three varieties of boro rice. MS Thesis, Department of Agronomy., Bangladesh Agricultural University, Mymensingh.

2. BBS (Bangladesh Bureau of Statistics), 2012. Monthly Statistical Bulletin of Bangladesh, June 2012. Statistics Division, Ministry of Planning, Government. Of the People's Republic of Bangladesh, Dhaka, p. 54.

3. BRRI (Bangladesh Rice Research Institute), 2010. *Adhunik dhaner chash* (in Bangla). 14thEdn. Bangladesh Rice Res. Inst. Joydebpur, Gazipur. PP. 17-30.

4. Crasswell ET and De Datta SK, 1980. Recent development in research on nitrogen fertilizers for rice. IRRI Research Paper, Series No. 49: 1-11.

5. Gifford RM and Evans LT, 1981. Photosynthesis, carbon partitioning and Yield. Annual Review of Plant Physiology 32: 485-509.

6. Gomez KA and Gomez AA, 1984. Statistical Procedure for Agricultural Research. 2nd Ed. John Wiley and Sons. New York. pp. 643-645.

7. Haque ME, 2002. Effect of time of nitrogen application in two varieties of transplant aman rice. MS thesis, Department of Agronomy., Bangladesh Agricultural University, Mymensingh.

8. Kamal J, Tomy PJ and Rajaappan Nais N, 1991. Effect of sources and levels of nitrogen on the growth, yield and nitrogen use efficiency of wetland rice. Indian Journal of Agronomy 36(1): 40-43.

9. Naznin A, Afroz H, Hoque TS and Mian, MH, 2013. Effects of PU, USG and NPK briquette on nitrogen use efficiency and yield of BR22 rice under reduced water condition. Journal of the Bangladesh Agricultural University, 11: 2015-220.

10. Panir S, 2014. Effect of Magic Growth on the performance of Transplant Aman Rice cv.BRRI dhan37. M.S. Thesis, Department of Agronomy., Bangladesh Agricultural University Mymensingh. pp. 27-39.

11. Quayum A. and Prasad K, 1994. Performance of modified urea materials in rainfed low land rice (*Oryza sativa* L.) Journal and Research of Birsa Agricultural. University, 6: 131-134.

12. Singh, BK. and Singh RP, 1986. Effect of modified urea materials on rainfed low land transplanted rice and their residual effect on succeeding wheat crop. Indian Journal of Agronomy, 31: 198-200.

13. Singh IC and Mahapatra IC, 1989. Economics of Fertilizer use. Fertilizer Marketing News, 5: 1-17.

14. Singh VP, Singh P, Agnihotri A and Singh R B, 1983. Effect of GA$_3$ on plant height of tall, semidwarf and dwarf strains of bread wheat Indian Journal of Plant Physiology, 26 (2): 266-269.

15. UNDP and FAO, 1988. Land resources appraisal of Bangladesh for Agricultural development. Report 2. Agro-ecological regions of Bangladesh. Bangladesh Agricultural Research Council, Dhaka- 1207.pp 212-221.

9

POST-HARVEST FACTORS AFFECTING QUALITY AND SHELF LIFE OF MANGO CV. AMROPALI

Sherajum Monira, M. Abdur Rahim, MAB Khalil Rahad and M. Ashraful Islam*

Department of Horticulture, Faculty of Agriculture, Bangladesh Agricultural University, Mymensingh-2202, Bangladesh

*Corresponding author: M. Ashraful Islam; E-mail: ashrafulmi@bau.edu.bd

ARTICLE INFO	ABSTRACT
Key words Aloe vera, Chitosan, Mango, Postharvest, Quality, Shelf life	A study was conducted to investigate the effect of some postharvest treatments on shelf life and quality of mango. Experiment was conducted at the BAU Germplasm centre, Department of Horticulture, Bangladesh Agricultural University, Mymensingh during May to July, 2013. Two treatments viz. Aloe vera and Chitosan solution were used for this study. The experiment was laid out in Completely Randomized Design (CRD) to observe the post-harvest performance of mango with three replications. The fruits were divided into three grades eg. large, medium and small for the convenience of the experiment. In each group mango were treated with Aloe vera gel and Chitosan solution in three different concentrations of 0.5%, 1% and 1.5%. Amropali cultivar had the highest shelf life (7days) at 1.5% Aloe vera gel treatment to the large size mangoes at room temperature compare to other sizes and treatments. Nearly, similar effect was obtained from 1.5% Aloe vera gel to the medium size fruits than other treatments. 1.5% Chitosan showed the second best result in case of shelf life extension followed by 1% Chitosan solution.

INTRODUCTION

Mango (*Mangifera indica* L.) is a delicious, nutritionally superior and one of the most valuable fruit. Mango is a common fruit throughout the world. Mango, often referred to as the king of tropical fruits, is an important fruit crop cultivated in the tropical regions (Boghrma *et al.,* 2000). There are so many mango varieties in our country in different areas, available in the season (May to July). Mango is grown in almost all districts of Bangladesh, but it is commercially cultivated in the greater districts of Rajshahi, Dinajpur, Rangpur, Kushtia and Jessore. Amropali was an exotic variety from India, although now it has been registered in our country but its commercial production is still poor than other local varieties. Amropali is usually smaller in size but total production is more than other common varieties. Main problems of commercial production of mango are irregular bearing, poor bearing, lower quality of fruits and small duration of market availability. Total production of mango is 992296 metric tons with an average yield of 36.14 tons per hectare in our country (BBS, 2014). Now, main concern is to increase shelf life, maintain better quality. However, a considerable proportion of mango fruit is spoiled each year due to lack of proper storage and marketing infrastructures. Herianus*et al.* (2003) observed the ripening process in mature green mango takes within 9–12 days or 12–14 days after harvest (Manzano*et al.* 1997) with good flavor, texture and color characteristic at ambient conditions at 25°C. In Bangladesh, postharvest loss of mango in supply chain is about 27 %. Hence, adequate measures should be taken to prolong shelf life of mangoes. Proper storage is essential for extending the consumption period of fruits, regulating their supply to the market and also for transportation to long distances. The mature green fruits can be kept at room temperature for about 4 to10 days depending upon the variety (Carrillo *et al.* 2000). Molla*et al. (*2010) also reported that the post-harvest losses of mango in Bangladesh are 51.88% (including agro–food sector) while it is only 5–25% in developed countries (Kader and Mitcham, 2008). Shelf life of fruits could be extended by pre cooling, chemical treatments, low temperature, different botanical extracts and so on. Different botanical extracts were found as to extend the shelf life of Mango (Shindem *et al.*, 2009). The use of Aloe vera gel has drawn interest in the food industry. Aloe vera coating is found to prevent loss of moisture and firmness, control respiration rate and development and maturation, delay oxidative browning, and reduce microorganism proliferation in fruits such as sweet cherry, table grapes and nectarines (Arowora *et al.,* 2013). In experiment, we used two botanical extracts Aloe vera gel and chitosan on different sizes of Mango cv. Amropali. Mango was collected with different pre -harvest treated plants which were graded according to the size (Monira *et al.,* 2015). Our main objectives of the experiment were to observe the performance of shelf life and quality of different size mangoes influenced by different botanical extracts.

MATERIALS AND METHODS

Experimental materials

The experiment was carried out at BAU Germplasm Centre (GPC) under the Fruit Tree Improvement Program (FTIP) during May to July, 2013. The two factor experiment was laid out in Complete Randomized Design (CRD) with three replications. Aloe vera gel was obtained from Aloe vera plant collection of BAU Germplasm Centre (GPC) field. Fruits of uniform size, free from visual blemishes and diseases were employed in this work. Mangoes were randomly divided into 3 grades according to their size, small, medium and large fruits. The mangoes were then treated with Aloe vera gel and Chitosan dilutions, prepared with distilled water, (0.5%, 1.0%and 1.5%).

Effect of fruit size and Post–harvest treatments on shelf life of mango

Two factor experiments were laid out in a Complete Randomized Design (CRD) with three replications.
Factor A (Size of fruit)

        1. L =Large size fruit
        2. M= Medium size fruit
        3. S =Small size fruit

Factor B (Coating Material)

        1. Aloe vera gel solutions: 0.5% , 1% and 1.5%
        2. Chitosan powder solutions: 0.5% , 1% and 1.5%

So, the treatment combinations were 3x3x3x3=54. Fruits were dipped in solution for 2-3 seconds and then placed on card board boxes with tagging. The main benefits of edible active coatings are to maintain the quality and extend shelf-life of fresh fruits and prevent microbial spoilage.

Storing at Atmospheric conditions of store room

The temperature and relative humidity of the storage room were recorded daily during the study period with a digital thermo hygrometer (THERMO, TFA, and Germany). The minimum and maximum temperatures during the study period of the storage room were 10.54 to 33.07°C, respectively. The minimum and maximum relative humidity was 66.4 % and 84.5%, respectively.

Aloe vera gel coating

Aloe veragel was obtained from BAU Germplasm Centre (GPC) field. Fruits of uniform size, free from visual blemishes and diseases were employed in this work. The mangoes were then treated with Aloe veragel dilutions, prepared with distilled water @ (0.5%, 1.0% and 1. 5 %) by dipping for 2-3 seconds. After letting dry, mangoes were placed on card board boxes with tagging. The boxes were stored in room temperature.

Chitosan coating

Chitosan powder is a biological product obtained from grind shell of Arthopoda family. It was imported from Malaysia, as an experimental material, first time used in that experiment in order to identify its potential in maintaining quality and extending shelf-life of fresh fruits through preventing microbial spoilage.

Maturity stage

The mangoes used for the experiment were harvested at maturity stage therefore; the fruits were soon start developing color within few days.

Total soluble solids (% Brix)

Total soluble solids (TSS) content of mango pulp was estimated using Abbe's Refract meter. A drop of mango juice squeezed from the fruit pulp was placed on the prism of the refract meter, and TSS was recorded at % Brix from direct reading of the instrument.

Weight loss

Weight loss was determined considering the fresh weight at harvest using a balance with an accuracy of 0.01 g. Weight loss was calculated from the weight of each mango measured initially before storages and after 3, 6 and 9 days of storage. Firmness was measured visually.

Percent weight loss

The weight loss of mango fruits was determined by using the following equation.

$$\text{Weight loss of fruits} = \frac{\text{Initial weight(g)} - \text{Final weight(g)}}{\text{Initial weight}} \times 100$$

The weight losses were recorded periodically during the storage period.

Disease–pest incidence

Disease and pest incidences of collected fruits were calculated using the following equation

$$= \frac{\text{No.of infected fruits in eachreplication}}{\text{Total no of fruits in each replication}} \times 100$$

The disease–pest incidence was recorded periodically during the storage period.

Shelf life

Shelf life of mango fruits as influenced by different coating substances and temperature of the storage room that was calculated by counting the number of days required to ripen fully with retained optimum marketing and eating qualities.

Collection of data

To assess the effect of different types of solution and physiochemical changes of mango fruits during storage was collected at two days interval during storage period. The shelf life, color development, weight loss or gain (%), rotting (%) were studied during the entire storage period. All the characteristics were recorded until 9th days of storage. Weight loss was determined considering the fresh weight at harvest using a balance with an accuracy of 0.01 g. Weight loss was then calculated from the weight of each mango measured initially before storages and after 3, 6 and 9 days of storage. Firmness was measured visually. Disease–pest incidence was recorded periodically during the storage period.

Statistical analysis

The collected data on various parameters were statistically analyzed using MSTAT statistical package. The means for all the treatments were calculated and analysis of variances (ANOVA) for all parameters was performed by F–test. The significance of difference between the pairs of means was compared by least significant difference (LSD) test at the 1% and 5% levels of probability (Gomez and Gomez, 1984).

RESULT

Quality and shelf life of mango fruits as affected by size and post-harvest treatment

Effect of post-harvest treatments on shelf life, no. of diseased fruits (%), no. of infected fruits (%) and weight loss (%)

Shelf life of mango varied significantly as treated with biological extract at room temperature. Amropali cultivar showed maximum shelf life (5.89days) when it was treated with Aloe vera gel, with comparatively the lowest disease and insect infestations. The extension of shelf life of fruits was one of the major concerns. Results revealed that the largest shelf life (5.89 days) of mango fruits was recorded from the 1.5% Aloe vera gel treated fruits (T_4) with the lowest disease (33.33%) and insect infestations (33.33%). This was followed by 1.5% Chitosan treated (T_7) fruits (5.22 days). The shortest shelf life (3.00days) was observed from the control (T_1), 0.5% Aloe vera (T_2) and 0.5% Chitosan (T_5) treated fruits, with comparatively higher disease and pest incidences. Among other chemical treated fruits, treatment T_3 (1.0% Aloe vera) and T_6 (1.0%Chitosan) were statistically similar regarding this parameter (4.78 and 3.67 days, respectively), with relatively higher disease and pest infestations that was statistically significant at 1%level.

Post-harvest treatments showed effects on total weight loss. Amropali lost the lowest weight 9.61%in T_4 (1.5% Aloe vera) and the highest 13.27% in T_1 (control) at 3rd day of storage. At 9th day of storage the highest weight loss was 15.97% in T_1 (Control) and the lowest weight loss was10.69% in T_4 1.5 % Aloe vera gel treatment. Among the treatments 1.5% Aloe vera gel coating was found to be the best in terms of weight loss followed by 1.5% Chitosan, 1.0% Aloe vera and 1.0 % Chitosan during the storage under room condition.

Effect of size on shelf life, no. of disease fruit (%), no. of infected fruit (%) and weight loss (%)

According to the size of the treatment shelf life, disease and pest infestation show significant difference. Control fruits were distinguished into three size grades eg. Large, medium and small. Here, the large fruits showed the highest shelf life with comparatively lower disease and pest incidences. Shelf life of mango was intensively related with the size of the fruits. Large size fruit showed greater shelf life (4.33days) with the lowest disease (47.62%) and pest (42.87%) incidences on the other hand, the lowest shelf life was found in Small fruits with the highest disease and pest infestations. According to the size of the fruits, weight loss also varied significantly.

Table1. Main effect of post-harvest treatment on shelf life, no. of diseased fruits (%), no. of infected fruits (%) and weight loss (%).

Treatment	Shelf life (Days)	No. of Diseased fruits (%)	No. of Infected Fruits (%)	Weight loss (%) at DAS		
				3	6	9
T_1	3.00	92.59	81.48	13.27	15.13	15.97
T_2	3.00	59.26	55.56	12.28	13.57	14.17
T_3	4.78	40.74	40.74	10.19	11.46	12.39
T_4	5.89	33.33	33.33	9.61	10.09	10.69
T_5	3.00	77.78	66.67	12.81	14.19	14.93
T_6	3.67	55.55	44.44	11.89	12.51	12.96
T_7	5.22	40.74	37.04	9.84	10.68	11.53
LSD (0.05)	0.29	4.03	58.13	0.19	0.16	0.22
LSD(0.01)	0.39	5.38	77.72	0.26	0.21	0.29
Level of significance	**	**	**	**	**	**

T_1-=Control, T_2=0.5% Aloe vera, T_3=1.0% Aloe vera, T_4=1.5% Aloe vera, T_5=0.5% Chitosan,
T_6 =1.0% Chitosan, T_7 =1.5% Chitosan
**=Significant at 1.0% level, DAS = days after storage

Table 2. Main effect of grade/size on shelf life, no. of disease fruit (%), no. of infected fruit (%) & weight loss (%)

Grade/size	Shelf life(Days)	No. of Diseased fruits (%)	No. of Infected fruits (%)	Weight loss (%) at different DAS		
				3	6	9
L	4.33	47.62	42.86	10.51	11.49	12.22
M	4.14	55.56	44.44	11.25	12.31	13.12
S	3.86	68.25	66.67	12.47	13.76	14.36
LSD(0.05)	0.190	2.634	38.037	0.129	0.104	0.142
LSD(0.01)	0.254	3.524	50.892	0.173	0.140	0.190
Level of significance	**	**	**	**	**	**

Here, L= Large, M= Medium, S= Small, DAS = days after storage; **=Significant at 1.0%level

At 3rd day of storage the highest weight loss was found in small fruits (12.47%) and the lowest weight loss in large size fruits 10.51%. On the other hand, at 9th day of storage the highest weight loss (14.36%) was in small size and (12.22%) in large size fruits. Therefore large fruit was found to be the best for storage in respect of weight loss prevention.

Combined effect of size and postharvest treatment on shelf life, no. of diseased fruits (%) and no. of infected fruits (%) and weight loss (%)

The combined effect of chemical treatment at different concentrations and sizes of the fruits had significant effect on shelf life and disease-pest incidence at room temperature. It appears from the table that, Larger fruits of Amropali with the concentration of 1.5% Aloe vera (LT_4) at room temperature, extended maximum shelf life (7.00days) along with the lowest disease and pest incidences. The combined effect of size and post-harvest treatments on weight loss was statistically significant. At 3rd day of storage the highest weight loss (14.80%) in ST_1 and the lowest (8.85%) in LT_4 were observed. On the other hand, at 9th day of storage the highest weight loss (17.22%) in ST^1 and lowest weight loss (9.97%) in LT^4 were found. Therefore

1.5% Aloe vera gel treatment was found to be the most suitable treatments in respect of reducing weight loss. That was followed by (MT4) Medium fruits with 1.5% Aloe vera gel treatment (6.33days). Which were statistically similar with LT7 (5.67days) Large at 1.5% Chitosan in this respect. Among other combinations, control treated fruits, (ST1), (LT1), (LT2), (LT5), (MT2), (MT5), (ST2), (MT1) and (ST5) had the shorter shelf life (3 days) which was also statistically significant at 1% level of probability.

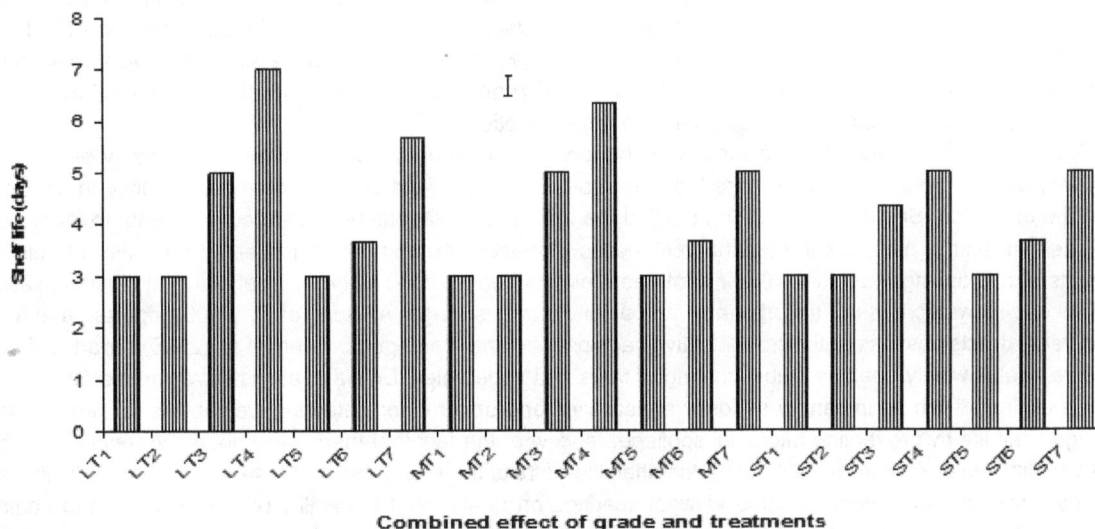

Figure 1.Combined effect of size and postharvest treatment on shelf life. Vertical bar represents 1%

level of probability.

Here,
S= Small, M= Medium, L= Large,
T1-=Control, T2=0.5% Aloe vera, T3=1.0% Aloe vera, T4=1.5%, Aloevera,
T5=0.5% Chitosan, T6 =1.0% Chitosan, T7 =1.5% Chitosan

DISCUSSION

Quality and shelf life of mango fruits from Control treated plants as affected by size and post-harvest treatment

Shelf life extension with different coating materials under room temperature in respect of size of fruits and treatment concentration had significant variation. Highest shelf life (7.00 days) found with Aloe vera gel treatment compared with Chitosan and no coating Control treatment. With the concentration of Aloe vera and Chitosan, shelf life and disease-pest incidence of fruits vary significantly. 1.5% Aloe vera gel treatment was best followed by 1.5% Chitosan. Disease and pest incidence also changes accordingly. Among the other treatments, 1.0% Aloe vera and 1.0% Chitosan was statistically similar. Size of fruits determines the shelf life and post-harvest quality. Large size fruit performed better than medium (7.00days) and small (6.33days) size fruits. As proper size-shape development is necessary for better post-harvest performance in respect of shelf life and disease-pest incidence. Weight loss is an important index of ripening. With the advancement of ripening and conversion of starch to sugar weight decreases. Singh *et al.* (2000) showed the behavior of mango (*Mangiferaindica*) cv. Langra and reported that the treatment of neem oil (10%) showed the minimum physiological weight loss when compared to other treatments and controls. Popy*et al*, (2013) found the similar result in Amropali treating with neem and garlic extract.

Ochiki*et al.,* (2014) stated that mango is a highly perishable fruit and high post-harvest losses occur in Africa. In order to address this problem, 4 concentrations of Aloe vera gel (AG) (0, 25, 50 and 75%) and chitosan (1%) were tested at two temperature levels (room temperature 15-22°C and 13°C) to determine their effect on the postharvest life of mango (var. Ngowe'). The experimental design was a 5 by 2 factorial experiment embedded in a complete randomized design with three replications. It was found that at both temperatures 50 and 75% Aloe vera concentrations significantly increased the shelf life and decrease in titrable acidity. Fruit color and ascorbic acid were also maintained for longer periods in these treatments. Findings of this study demonstrate the potential of using Aloe vera gel at 50% as a coating for improved postharvest shelf life and maintaining quality of mango fruits hence reduced postharvest losses. Hossain et al, (2001) studied the physio-chemical composition of three varieties of mango. The best fruit weight was lowest (221.33 gram) in Amarpali but the variety of Bishawanath had the maximum fresh weight (256.0 gram) and keeping quality (8.75 days). The maximum keeping quality was in Amarpali (12.5 days) mango fruit. The TSS (23.50 percent), total sugar (26.85 percent) and pH of pulp (6. 0) were highest in Amarpali, but Bishawanath indicated highest Vitamin C (14.20 mg /100g) and acidity (titrable) (0.87 %). Amarpali fruit was better in respect of al l characters as compared to other varieties.

Molla*et al,* (2011) studied the postharvest changes in mango and stated that color and quality of mango was very better in treated fruits compared to non-treated fruits. Leaf extract has better performance in case of shelf life extension. Shinde*met al,* (2009) studied the influence of various plant extract treatments to increase the shelf life and to minimize the postharvest losses in mango. Among the fruits treated with different plant extracts and wrapping materials, 10 per cent neem oil has been proved to be most effective in slow increase of TSS and slow decrease of ascorbic acid and acidity during storage. Adetunji *et al,* (2008) reported that the *Aloe vera*extracts possess antimicrobial activity against bacterial pathogens. Dang *et al,* (2008) reported that, *Aloe vera*gel is widely used as edible coating of fruits and vegetables. Edible coatings have various favorable effects on fruits such as imparting a glossy appearance and better color, retarding weight loss, or prolonging storage/shelf life by preventing microbial spoilage. In view of the fact that there were not much recent reports on the utilization of *Aloe vera*in extending the shelf life of fruits therefore, this study was conducted to fulfill the gap by evaluating the effects of *Aloe vera*gel coatings on quality and storability of fruit. Further, the same experiment can be repeated to observe the performance of different varieties of mango.

REFERENCES

1. Arowora ST, 2013. Functional foods: an adventure in food formulation. Food Australia, 53: 428- 432.
2. BBS, 2014. All Crop Summary 2009–10. *Year* Book of Agricultural Statistics of Bangladesh. Bangladesh Bureau of Statistics. Statistics Division, Ministry of Planning, Government of the People Republic of Bangladesh. Agriculture Census/All_Crops_summary_09–10.
3. Boghrma V, RS Sharma and D Puravankara, 2000. Effect of antioxidant principles isolated from mango *(Mangiferaindica*L) seed kernels on oxidative stability of buffalo ghee (butter fat). Journal of Science & Food Agriculture, 80: 522–526.
4. Carrillo LA, F Ramirez–Bustamante, JB Valdez–Torres, R Rojas–Villegas and EM Yahia, 2000. Ripening and quality changes in mango fruit as affected by coating with an edible film. Journal of Food Quality 23: pp.479–486.
5. Dang, Z Singh and EE Shinny, 2008. Edible coatings influence fruit ripening, quality and aroma biosynthesis in mango fruit. Journal of Agriculture and Food Chemistry.56: pp. 1361-1370.
6. Gomez KA and AA Gomez, 1984. Statistical Procedure for Agricultural Research (2nd edition). Willey. International Science of Publication. pp. 28–192.
7. Hossain KH, 2001. Policy for Postharvest Loss Reduction of fruits and Vegetables and Socio– Economic Uplift of the Stakeholders.P.188. Research Project Funded by United SAID and EC, and Jointly implemented by FAO and FPM of the Ministry of Food and Disaster Management (MoFDM).
8. Herianus JD, LZ Singh and SC Tan, 2003. Aroma volatiles production during fruit ripening of 'Kensington Pride' mango. Postharvest Biotechnology, 27: 323 –336.
9. Kader A and B Mitcham, 2008. Optimum Procedures for Ripening Mangoes. In: Fruit Ripening and Ethylene Management: 47-48.University of California.Postharvest Technology Research and Information Center Publication Series number 9:766-897
10. Manzano JE, Y Perez and E Rojas, 1997. Coating waxes on Haden mango fruits (*Mangiferaindica*L.) cultivar for export. ActaHorculture455: 738-746.

11. Molla MM, MN Islam, MA Muqit, KA Ara and MAH Talukder, 2011. Increasing Shelf Life andMaintaining Quality of Mango by Postharvest Treatments and Packaging Technique. Journal ofOrnamental Horticultural Plants, 1: 73-84.

12. Molla MM, MN Islam, TAA Nasrin and MR Karim, 2010. Survey on postharvest practices and losses of mango in selected areas of Bangladesh. Postharvest Management of Horticultural Crops, Annual report, Horticultural Research Centre, Bangladesh Agricultural Research Institute, Gazipur–1701. P.64

13. Monira S, MA Rahim and MA Islam, 2015. Pre-harvest factors affecting yield, quality and shelf life of Mango cv. Amropali. Research in Agriculture, Livestock and Fisheries, 2: 279-286.

14. Ochiki S, G M Robert and W J Ngwela, 2014. Effect of Aloe vera gel coating on postharvest quality and shelf life of mango (Mangifera*indica* L.) fruits Var. 'Ngowe' Department of Crops, Horticulture and Soils Sciences, Egerton University

15. Popy B, 2013. Effect of different chemicals on yield, quality and shelf life of different varieties of mango. MS thesis of Bangladesh Agricultural University, Mymensingh

16. Shindem GS, RR Viradia, SA Patil and DK Kakade, 2009. Effect of post-harvest treatments of natural plant extract and wrapping material on storage behavior of mango (cv. KESAR). International Journal of Agricultural Science, 5: 420-423.

17. Singh DK, S Ram and S Ram, 2000. Effect of paclobutazol application on fruits. Journal of Horticultural Science, 70: 28–59

10

VARIATION IN MORPHOLOGICAL ATTRIBUTES AND YIELD OF TOMATO CULTIVARS

Amit Malaker[1], AKM Zakir Hossain[2], Tahmina Akter[3*] and Md. Shariful Hasan Khan[4]

[1]Department of Seed Science and Technology, [2]Department of Crop Botany, [3]Department of Biochemistry and Molecular Biology and [4]Department of Soil Science; Faculty of Agriculture, Bangladesh Agricultural University, Mymensingh-2202, Bangladesh

*Corresponding author: Tahmina Akter; E-mail: tahminataniabmb@gmail.com

ARTICLE INFO

ABSTRACT

Key words

Tomato,
Genetic variations,
Morphology,
Phenology,
Yield

A pot experiment was carried out in the grill house of the Department of Seed Science and Technology, Bangladesh Agricultural University (BAU), Mymensingh during the period of mid December 2014 to mid April 2015. The experiment conducted with five treatments (five tomato genotypes) and employed in Completely Randomized Design (CRD) with four replications. Of the five genotypes two were obtained Japan (Pasta and Mimi) and the rest three from Bangladesh Agricultural Research Institute (BARI). Genetic variations appeared for morphological and phenological characters and yield attributes in the five tomato genotypes. Taller plants with higher number of branches were observed in the Japanese genotype (Mimi) while number of days required to initiate flower and fruit initiation and fruit maturity were shorter in BARI tomato 12 and BARI tomato 14. Flower and fruit productions plant^{-1} was higher in BARI released tomato cultivars, therefore, the tomato yield was also higher in those cultivars.

INTRODUCTION

Tomato (*Lycopersicon esculentum* Mill.) is one of the most important edible and nutritious vegetable all over the world and it belongs to the family Solanaceae. Because of its wide adaptability, tomato is grown throughout the world. It ranks next to potato and cassava in respect of vegetable production in the world (FAO, 2013). In Bangladesh, it ranks 2nd which is next to potato (BBS, 2010) and top the list of canned vegetables. A large number of tomato varieties are growing in Bangladesh and those are exotic in origin and developed long before. Most of them lost their potentiality due to genetic deterioration, diseases and insect infestations. So, in order to increase the tomato production in Bangladesh, it is very much essential to find out the varieties capable of growing round the year, higher yield and resistant to disease and insect pests. Recently various research organizations have developed a few high yielding, disease and insect resistant varieties but these do not show better performance throughout the year because of photo-sensitiveness and insufficient wider adaptability.

In Bangladesh, tomato has great demand throughout the year especially in early winter and its production is mainly concentrated during the winter season. Recent statistics showed that tomato was grown in 15,790 hectares of land and the total production was approximately 102,000 metric tons in 2010-11. Thus the average yield of tomato was 1.29 t ha^{-1}(BBS. 2010), while it was 69.41 t ha^{-1}in USA, 14.27 t ha^{-1}in India, 26.13 t ha^{-1}in China, 13.25 t ha^{-1}in Indonesia and 59.6 t ha^{-1}in Japan (FAO, 2009). To meet nutritional demand of population, it is highly important to increase the yield of tomato per unit area of land. Increase of production depends on many factors, such as the use of improved varieties, proper management and awareness about improved production technologies. So, using different types of techniques, such as nuclear techniques, fertilizer management, proper spacing, applying plant growth regulator, synthetic mulching, natural mulching and even conventional breeding methods may improve production level and quality under the existing environmental conditions. Inventing morphological characters is required to identify important canopy features and yield attributes in tomato. Such studies are also necessary for varietal improvement. The purposes of this study is to evaluate some morphological attributes and yield of tomato genotypes and to select the better genotypes of tomato in respect of yield performance from five elite local and exotic tomato varieties.

MATERIALS AND METHODS

Experimental site and time

The experiment was conducted at the grill of the Department of Seed Science and Technology, Bangladesh Agricultural University (BAU), Mymensingh, during the period from mid December 2014 to mid April 2015.

Experimental materials

Five tomato genotypes Japanese tomato (Pasta), Japanese tomato (Mimi), BARI tomato 2, BARI tomato 12 and BARI tomato 14 used as experimental materials. The seed of Japanese tomato genotypes were collected from Japan (Kaneko Seed Co. Ltd.) and BARI tomato variety from Horticulture division of Bangladesh Agricultural Research Institute (BARI), Gazipur.

Treatments

Five tomato genotypes considered as five different treatments and four replications were used in the experiment.

Preparation of pot

Pots were used for the experiment (20 inch) and total numbers of pots were 20. The selected pots were filled with soil. Then the soil mixed with organic manure and fertilizer (urea 2.09 g, MoP 0.99 g, cowdung 400 g and compost 50 g pot^{-1})

Statistical analysis

Yield and yield contributing data were collected and analyzed statically following the analysis of variance (ANOVA) technique and the mean differences were adjudged by Duncan's Multiple Range test (DMRT) using the statistical computer package program, MSTAT-C (Gomez and Gomez, 1984).

RESULTS

Percent germination of seeds

The highest germination percentage of seeds (89.30%) was recorded in BARI tomato 14 followed by BARI tomato 12. On the contrary, the lowest germination percentage of seeds (74.70%) was observed in Japanese tomato PASTA which was also statistically different from other genotypes (Figure 1). Variation of seed germination might be due to genotypic effect.

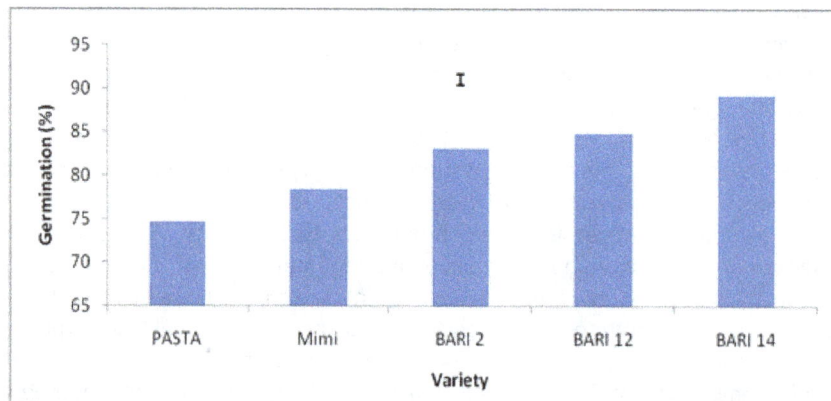

Figure 1. Effect of different genotypes of tomato on germination percentage of seeds
(Vertical bar represents LSD at 0.5 % level of significance)

Plant height (cm)

The highest plant height (80.67cm) was recorded in Mini and the lowest (62.00cm) was recorded in BARI 2 which was statistically identical with PASTA (64.00cm) but statistically different from other genotypes (Table 1).

Number of branches plant^{-1}

The maximum (21.0) number of branches plant^{-1} was found in Japanese tomato Mini which was statistically differed from other genotypes (table 1). The minimum number of branches (13.0) was recorded in Japanese tomato Pasta while same (13.0) number of branches plant^{-1} was also recorded in BARI tomato 14.

Leaf size

The highest leaf size (5.30cm) was recorded in BARI tomato 14 which was statistically similar to BARI tomato 2 (5.20cm^2). On the other hand, the lowest leaf size (2.50cm^2) was obtained from the genotype Japanese tomato Mimi which was statistically differed from other genotypes (Table 2).

Number of leaves branch^{-1}

The highest number of leaves branch^{-1} (15.00) was produced from the both genotypes of BARI tomato 12 and BARI tomato 14 while Japanese tomato Pasta also produced statistically identical highest number of leaves branch^{-1} (13.33) in this study. On the other hand, the lowest number of leaves branch^{-1} (11.00) was found in BARI tomato 2 while it was statistically close to Mimi (13.00) (Table 2).

Days to flower initiation

Days to flower initiation showed highly significant variation among the genotypes. Among the genotypes, Japanese tomato Pasta required more time (38.90 DAT) for initiation of flowering while Japanese tomato Mini also required the statistically similar time (38.60 DAT) for flower initiation. In contrast, the genotype BARI tomato 14 showed the lowest time (31.30 DAT) for flower initiation which was statistically differed from other genotypes of the study (Figure 2).

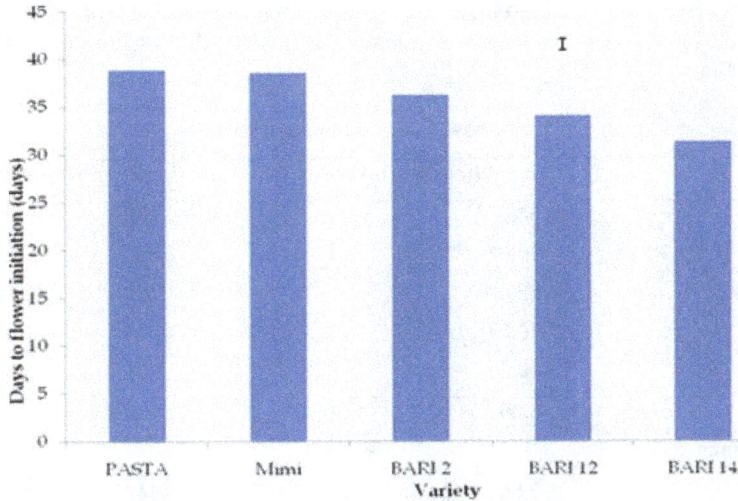

Figure 2. Effect of different genotypes of tomato on days of flower initiation
(Vertical bar represents LSD at 0.5 % level of significance)

Days to fruit initiation

The mean value for the first fruit setting of different genotype ranged from 35.10 to 42.90 DAT (Figure 3). Mean data showed that the BARI tomato 14 provided very earlier (35.10 DAT) fruit setting and the Japanese tomato Pasta showed delayed fruit setting (42.90 DAT) which was statistically close to Japanese tomato Mimi (41.10 DAT).

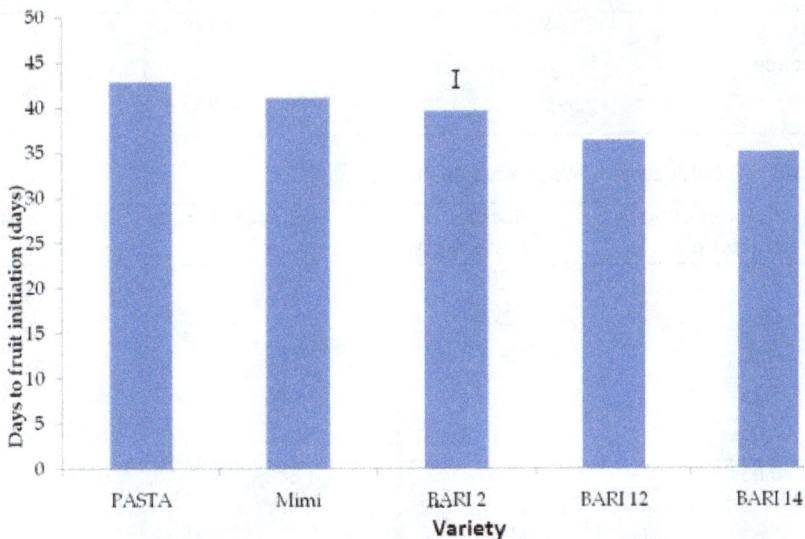

Figure 3. Effect of different genotypes of tomato on days to fruit initiation
(Vertical bar represents LSD at 0.5 % level of significance)

Number of flowers cluster plant[-1]

The highest number of flowers cluster plant[-1] (7.90) was obtained in BARI tomato 14 which was statistically close to BARI tomato 12 (7.25). The lowest number of flowers cluster plant[-1] (5.70) was found in Japanese tomato pasta (Table 3).

Number of fruits cluster plant[-1]

The maximum number of fruits cluster plant[-1] (5.67) was produced in BARI tomato 2 which was statistically similar to BARI tomato 12 and BARI tomato 14 (both same 5.33) (Table 3). The minimum number of fruits cluster[-1] (4.00) was found in Japanese tomato Pasta while Mimi also showed the statistically similar fruits cluster plant[-1] (4.33).

Table 1. Plant height and number of branches in five tomato cultivars

Variety	Plant height (cm)	Number of branches plant[-1]
PASTA	64.00	13.00
Mimi	80.67	21.00
BARI 2	62.00	15.00
BARI 12	70.67	14.33
BARI 14	73.00	13.00
LSD$_{(0.05)}$	3.36	1.475
Level of significance	**	**
CV (%)	2.55	5.147

Table 2. Leaf size and No. of leaf/branch in five tomato cultivars

Variety	Leaf size (cm^2)	Number of leaves branch[-1]
PASTA	3.30	13.33
Mimi	2.50	13.00
BARI 2	5.20	11.00
BARI 12	5.00	15.00
BARI 14	5.30	15.00
LSD$_{(0.05)}$	0.164	2.110
Level of significance	**	*
CV (%)	2.095	8.353

Table 3. Flower and fruit production in five tomato cultivars

Variety	Number of flowers cluster plant[-1]	Number of flowers plant[-1]	Number of fruits plant[-1]	Number of fruit cluster plant[-1]
PASTA	5.70	9.70	7.33	4.00
Mimi	5.90	12.00	9.00	4.33
BARI 2	6.10	16.15	10.66	5.66
BARI 12	7.25	17.35	11.00	5.33
BARI 14	7.90	19.90	12.00	5.33
LSD $_{0.05}$	0.951	2.117	1.646	0.972
Level of significance	**	**	**	*
CV (%)	7.31	9.01	8.759	10.465

Number of flowers plant^{-1}

At harvest, the maximum number of flowers plant^{-1} (19.90) was recorded in BARI tomato 14 and the minimum number of flowers plant^{-1} (9.105) was recorded in Japanese tomato Pasta (Table 3).

Number of fruits plant^{-1}

The maximum number of fruits plant^{-1} (12.00) was recorded in BARI tomato 14 which was statistically similar to BARI tomato 2 (10.66) and BARI tomato 12 (11.00). On the other hand, the minimum number of fruits plant^{-1} (7.33) was obtained in Japanese tomato Pasta which was statistically differed from other genotypes of the study (Table 3).

Weight of individual fruit

The highest weigh of individual fruit (35.44g) was obtained in the genotypes BARI tomato 14 which was statistically identical to BARI tomato 12 (35.21g). The lowest weight of individual fruit (23.17g) was found in Japanese tomato Pasta which was statistically differed from other genotypes (Figure 4).

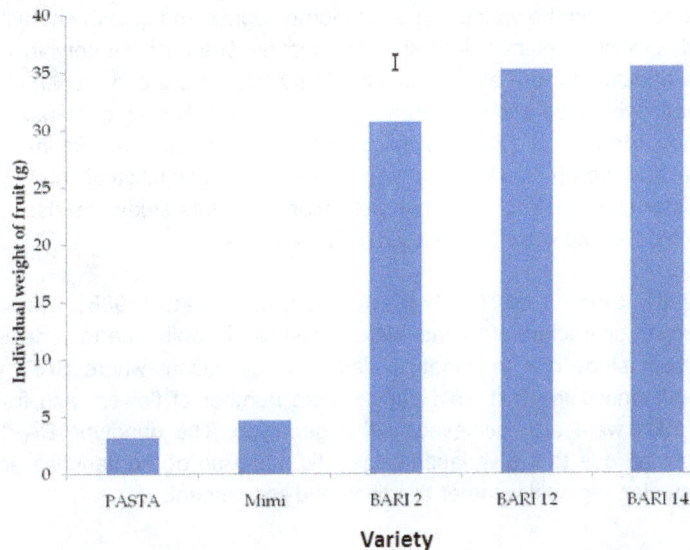

Figure 4. Effect of different genotypes of tomato on individual weight of fruit.
(Vertical bar represents LSD at 0.5 % level of significance)

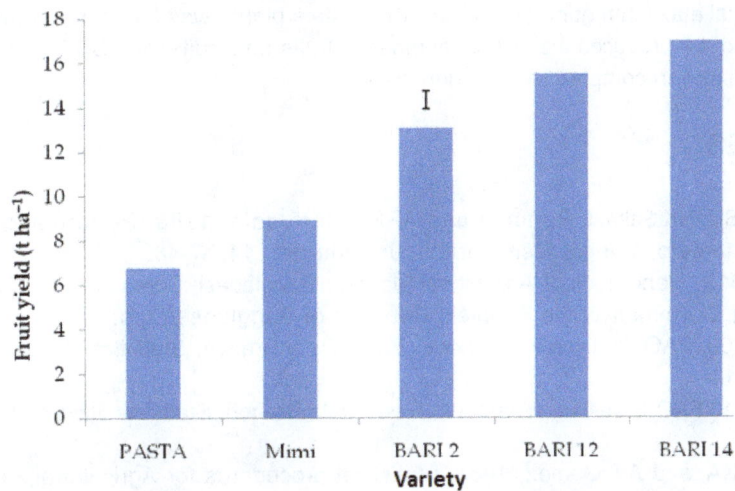

Figure 5. Effect of different genotypes of tomato on yield of fruit per ha.
(Vertical bar represents LSD at 0.5 % level of significance)

Fruit yield (t ha^{-1})

The highest fruit yield (17.011 tha^{-1}) was found in BARI tomato 14 followed by BARI tomato 12 (15.492 tha^{-1}). The lowest fruit yield (6.796 tha^{-1}) was obtained in Japanese tomato Pasta which was statistically different from other genotypes of the study (Figure 5).

DISCUSSION

In the conducted study, genotypic effect was significantly influenced the studied whole traits regarding morpho–physiological, fruit yield and yield attributes where most of the characters were highly influenced by different genotype BARI tomato 14. Germination percentage different significantly due to genotypes while BARI tomato 14 had more significant than other varieties which might be due to the variation in genetic characteristics of the studied varieties. The same observation was also found by the researchers named Kabir (2004), Hussain (2001),Khokhar (2001) and Ahmed et al.(1986). They also found significant variation on seed germination due to the variation in genetic makeup of the studied genotypes. Similar effect due to genotypes was also obtained on the various types of morphological and growth characters of the tomato plant such as plant height, branches plant^{-1}, leaf size and leaves branch^{-1}. Genotypic variation of the studied genotypes was more effective for the above variation in morphological and growth characters while climatic factors and soil nutrients were also effective for the above variation. The agreement of the present study were also similar to the study. Singh et al. (2002), Kabir (2004),Hossain (2003) and karim (2005) who also reported that the morphological and growth characters of tomato were varied significantly due to varieties. Due to more number of flower cluster represents the highest yield Plant^{-1}. In this study we also found that, more flower cluster (BARI tomato 14) had more productive than other varieties.

Phookan et al. (1998), Sing et al. (1997) and Sonone et at. (1986) who also reported that the morphological and growth characters of tomato were varied significantly due to varieties. Fruit yield and yield attributes of the present study due to varieties were also significant where BARI tomato 14 were more effective and highly influenced the fruit yield due to more number of flowers and fruits plant^{-1} and highest weight of individual fruits were also achieved in this genotype. The genotype BARI tomato 14 had more productive which might be due to the variation in genetic make up of the varieties and also the variation in adaptability with the studied regional weather condition and soil nutrients.

CONCLUSION

The genotype BARI tomato14 produced the highest number of flower clusters plant^{-1} and fruit clusters plant^{-1}. Japanese tomato Mimi and BARI tomato12 showed higher plant height and in a total maximum number of primary branches plant^{-1} was found in Japanese tomato Pasta and BARI tomato 14 produced the highest number of fruits and fruit yield. BARI tomato 14 showed the best yield performance in compare to other genotypes.

REFERENCES

1. Ahmed SU, HK Saha, L Rahman and AFM Sharfuddin, 1986. Performance of some advance lines of tomato. Bangladesh Horticulture Journal, 14: 47-48.
2. BBS, 2010. Hand book of Agricultural Statistics. Bangladesh Bureau of Statistics (BBS), Ministry of Planning, Govrnment of the People's Republic of Bangladesh, pp: 14.
3. FAO, 2009. FAO Statistical Yearbook, Basic Data Branch, Statistics Division, FAO, Rome, Italy, 56: 142-144.
4. FAO, 2013.FAO Statistical Yearbook, Basic Data Branch, Statistics Division FAO, Rome, Italy. pp: 54.
5. Gomez KA, and AA Gomez, 1984. Statistical procedures for Agricultural Research. John Wiltey and Sons, New York. pp: 640.
6. Hossain MM, 2003. Comparative morpho-physiological studies of some exotic and local

genotypes of tomato. M.S. Thesis, Department of Crop Botany, Bangladesh Agricultural. University, Mymensingh, pp: 26-28.

7. Hussain SI, KM Khokhar, T Mahmood, MH Laghari, and MM Mahmud, 2001. Yield potential of some exotic and local tomato cultivars grown for summer production. Pakistan Journal of Biological Science, 4: 1215-1216.

8. Kabir MSA, 2004. Morpho-physiological evaluation of elite genotypes of tomato in early summer season. M.S. Thesis, Department of Crop Botany, Bangladesh Agricultural University, Mymensingh. pp: 31-39.

9. Karim MR, 2005. Study of growth and yield performance of late transplanted summer tomato genotypes. M.S. Thesis, Department of Crop Botany, Bangladesh Agricultural University, Mymensingh. pp: 34.

10. Khokhar KM, HI Hussain, T Mahmood, Hidayatullah and MH Laghari, 2002. Winter production of tomatoes under plastic tunnel. Asian Journal of Plant Science, pp: 659-660.

11. Phookan DB, P Talukdar, A Shadeque, and BK Chakravarty, 1998. Genetic variability in tomato (*Lycopersicon esculentum* Mill.) genotypes during summer season under plastic house condition. Indian journal of Agricultural Science, 68: 304-306.

12. Singh AK, SK Bhalla and S Verma, 2002. A note on the variability in tomato *(Lycopersicon esculentum* Mill.).Journal of Agricultural Research, 19: 98-100.

13. Singh UN, A Saha, and AK Panda, 1997. Genetic variability and correlation studies in tomato *(Lycopersicon esculentum* Mill.). Environmental Ecology, 15: 177-181.

14. Sonone AH, SB Patil and DC More, 1986. Variability and heritability in inter-varietal crosses of tomato *(Lycopersicon esculentum* Mill.). Current Research Report, Mohatma Phule Krishi Vidyapeeth, Rahuri, India 2: 125-127.

11

SCREENING OF POTATO GENOTYPES BASED ON GLUCOSE AND ASPARAGINES CONTENT TO MINIMIZE ACRYLAMIDE FORMATION IN POTATO CHIPS AND FRENCH FRIES

Fatema Zahan, Md. Masudul Karim, Tahmina Akter[1] and Md. Alamgir Hossain[*]

Department of Crop Botany and [1]Department of Biochemistry and Molecular Biology, Faculty of Agriculture, Bangladesh Agricultural University, Mymensingh-2202, Bangladesh

[*]**Corresponding author:** Md. Alamgir Hossain; E-mail: alamgir.cbot@bau.edu.bd

ARTICLE INFO

Key words

Glucose,
Asparagines,
Reducing sugar,
Total soluble sugar
and Acrylamide

ABSTRACT

Seven potato genotypes that are available in Bangladesh, were grown at the field laboratory under the Crop Botany Department, Bangladesh Agricultural University in 2014. Reducing sugars and free asparagine were determined at freshly harvested potato tubers and those after storing at 8 ^{0}C for 8 months. There was no significant variation of asparagine content in all genotypes of freshly harvested tubers. But a significant difference was found in reducing sugar content. The lowest was in the samples of the genotypes Cardinal and Rumanapakri, and the highest in Hagrai. The variety Diamant appeared to contain the lowest amount of reducing sugars after 8 months storage. The results showed that freshly harvested Cardinal, Rumanapakri and Diamant after storage produced less amount of acrylamide after frying as potato chips or French fries. It may be concluded that screening potato genotypes primarily on their reducing sugar contents could be useful tool to minimize acrylamide formation in potato chips and French fries. Further investigation is needed to find out the factors affecting reducing sugar and asparagine content in potato tubers.

INTRODUCTION

Heating of glucose with asparagine yields acrylamide according to the Maillard reaction (Mottram et al. 2002) and the rate is strongly increasing with temperature increasing from 120°C to 170°C (Stadler et al. 2002). Acrylamide (plant's origin) is a neurotoxin to human (Tareke et al. 2002) and known as a carcinogen in experimental studies (Mucci and Wilson, 2008), and it is classified as a "probable human carcinogen" by the International Agency for Research on Cancer (IARC, 2002). In 2002, the Swedish National Food Administration detected high concentrations of acrylamide in common heated starch rich foods such as French fries (www.slv.se.). It has evoked great concern for the health effects of acrylamide to the public. Soon after, the World Health Organization published that the average daily intake of acrylamide for the general population was about 1 μg/kg of body weight (bw) (FAO/WHO, 2005). The main precursors for this undesirable substance are reducing sugars and the amino acid asparagine (Stadler et al. 2004, Mottram et al. 2002, Weisshaar et al. 2002). Heating of reducing sugars (glucose and fructose) with asparagine yields acrylamide. Therefore, the accumulation of asparagines and reducing sugars in the harvested organs of plants had implications for food safety. Interestingly, asparagine accumulation can be caused by stresses like salinity, drought, mineral deficiencies, toxic metals and pathogen attack. Crop production by escaping the above mentioned challenges is practically impossible. In addition, asparagine may also be formed following the detoxification of cyanide.

Potato tuber (*Solanum tuberosum* L.) is rich in asparagine content (33-59% of the total free amino acids) (Eppendorfer and Bille, 1996). So, potato is very susceptible to acrylamide formation during heating at temperature above 120 °C for the preparation of several potato products such as potato chips, potato crisps, French fries, roasted and baked potatoes. However, reducing sugar (glucose), another important precursor of acrylamide is limiting in potato tuber at harvest. Interestingly, storage of potatoes at temperature below 8-10°C induces a strong increase in sugar contents. The phenomenon is commonly known as "low-temperature sweetening" (Coffin et al. 1987). Prolonged storage at low temperature markedly increases in glucose, fructose and sucrose. As a consequence, the potential of acrylamide formation at 120°C rose by a factor of 28 (Biedermann et al. 2002, Chuda et al. 2003).

Acrylamide is ubiquitous in the human diet, and more than one-third of the calories we take in each day come from foods with detectable levels of acrylamide (Petersen and Tran, 2005). The levels of acrylamide can exceed 1000 parts per billion in potato products (Lea and Azevedo, 2007). For comparison, the tolerance level for water set by WHO is 1 ppb. Estimated dietary acrylamide intake in populations has been calculated by national food administrations for several countries. For adults, estimated average intakes range from approximately 0.3 to 0.6 μg/kg of body weight (bw)/day. Children and adolescents tend to eat more acrylamide on a per body weight basis. This may be due to a combination of children's higher caloric intake relative to body weight as well as their choice for higher consumption of certain acrylamide-rich foods, such as French fries and potato crisps (Dybing et al. 2005).

To date, modified processing methods has been the main approach used in the mitigation of the acrylamide problem. Decreasing the content of precursors may reduce the amount of acrylamide in the products. Keeping this point in mind, selection of genotypes should be done that contain low levels of asparagine to minimize acrylamide in the processed foods. Therefore, attention is turning to improving the raw material by decreasing the levels of sugars and/or free asparagine and thereby the risk of acrylamide formation. The aim of this study was to identify potato genotypes available in Bangladesh that have a lower potential for acrylamide formation during frying. Being a public health issue (especially child health concern), which is highly related to plant origin and glucose/asparagine ratio in the raw material, the present study is designed to minimize acrylamide problem by selecting potato genotypes. Potato genotypes with low reducing sugar as well as low asparagine content may be a more useful target for crop improvement nutritionally in future. To the best of my knowledge, it will be the first scientific investigation in Bangladesh.

METHODOLOGY

Experimentation

An experiment was carried out at the field laboratory, Department of Crop Botany, Bangladesh Agricultural University, Mymensingh (24° 75' N, 90°50' E and the elevation of the area is approximately 19 m from the average sea level) during November 2014–March 2015. The plant materials are seven potato genotypes, namely Cardinal, Diamant, Rumanapakri, Fata pakri, Gutypakri, Tel pakri and Hagrai. The experiment was laid out in the Randomized Complete Block Design (RCBD) with three replications. The total number of plots was 7×3. The size of the unit plot was 6m × 1m. A spacing of 0.5 m was provided between the plots and 1.0 m spacing was provided among three blocks. The seed tubers were planted at a depth of 5 cm in the experimental plots on 15 November, 2014. A spacing of 60 cm × 30 cm was used and tubers were planted in rows. The soil along the rows of seed tubers were ridged up immediately after planting. All intercultural practices were done when necessary.

Sampling and data recoding

Data were recorded on different morphological, yield components and yield from 5 randomly selected sample plants. Data were recorded on different morphological, yield components and yield from 5 randomly selected sample plants. With the help of digital weight machine 100 g of potato flesh was weighed, dried in an electric oven at 80^0c for 72 hours until the weight become constant. It was then cooled and weighed. Percent moisture content was calculated according to the following formula:-

$$(\%) \text{ Moisture} = \frac{IW - FW}{IW} \times 100$$

Where,

IW= Initial weight of flesh,   FW= Final weight of oven dried flesh

Then percentage of dry matter content of the flesh was calculated from the data obtained during moisture estimation using the following formula.

% Dry matter content = 100 -% moisture content

Estimation of water soluble carbohydrates (WSCs) in potato tuber

The WSCs in tubers were extracted and measured using anthrone method (Yemm and Willis, 1954) as described in Hossain et al. (2009, 2010 and 2011). The fresh tubers were chopped and oven dried and then milled to rough powder. The tuber powder was weighed and extracted once with 80% ethanol at 60°C for 30 minutes followed by 2 successive 15 minutes extractions with distilled water at 80° C. The extracts were combined and evaporated to dryness at 65°C. The dried carbohydrates were resolved in 5 mL distilled water. A fraction of the extract solution (about 1 mL) was taken in a Micro-centrifuge tube (1.5 mL) and charcoal powder was added to it. After mixing the powder and extract solution with a vortex (touch mixture), the solution was centrifuged at 5000 rpm for 5 min to make a clear solution. The clear solution was diluted 200 times with distilled water. Diluted solution (0.06 mL) was mixed with ice-cold anthrone reagent (3 mL). The mixture was heated for 10 min in a boiling-water bath and subsequently cooled with ice water. The absorbance of the reacted solution for standard and samples was measured with a spectrophotometer at 620 nm. The content of WSCs in the sample was calculated as mg WSCs per gram of tuber dry mass using regression equation.

Determination of reducing sugar

Reducing sugar content of potato tuber was determined by dinitrosalicylic acid method as described by Miller (1972).

Preparation of tuber extract

The fresh tubers were chopped and oven dried and then milled to rough powder. The tuber powder was weighed to 1g and dissolved with 70% ethanol and was kept for overnight and then filtered to collect the extract solution. The extracts were then evaporated to remove ethanol in a sand bath. After evaporating ethanol the volume of the extract was made up to 10 ml.

Reagent required

Dinitrosalicylic acid (DNS) reagent (1 g of DNS, 200 mg of crystalline phenol and 50 mg of sodium sulphite were placed simultaneously in a beaker and mixed with 100 ml of 1 % NaOH by stirring). 40% solution of Rochelle salt (It was prepared by dissolving 40 g of sodium potassium tartarate with 100 ml of distilled water in 100 ml volumetric flask).

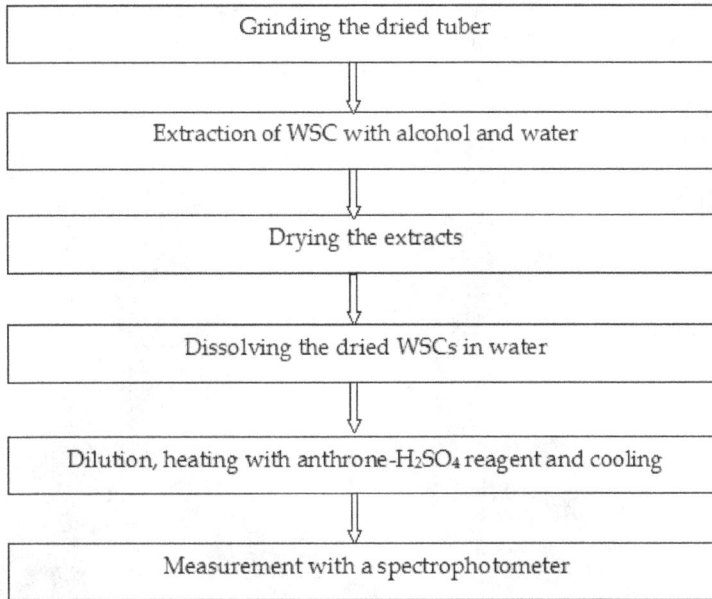

Figure 1. Protocol for measuring tuber WSCs by anthrone method

Estimation of reducing sugar of the potato tuber

Three ml of the extract was pipetted into a test tube and 3 ml of DNS reagent was added to the solution and mixed well. The test tube was heated for 5 minutes in a boiling water bath. After the color had developed, 1 ml of 40% Rochelle salt was added when the contents of the tube was still warm. The test tubes were then cooled under a running tap water. A reagent blank was prepared by taking 3 ml of distilled water and 3 ml of DNS reagent in a tube and treated similarly. The absorbance of the solution was measured at 575 nm in a colorimeter. The amount of reducing sugar was calculated from the standard curve of glucose.

Free Amino Acid Content

Ninhydrin assay (Friedman, 2004) was adapted for free amino acids determination including asparagines. Briefly, the lyophilized (freeze-dried) (approximately 100 mg) sample was deproteinized by stirring in 50 mL of a 0.3 M sulfosalicylic acid solution for 5 min. This was followed by centrifugation at 8200 × g for 5 min. An aliquot of the supernatant (250 µL) was used for free amino acids determination based on a color reaction using a buffered ninhydrin solution and a continuous measurement of the absorbance at 570 and 440 nm (De Wilde et al. 2005).

Statistical analysis

The data obtained for yield contributing characters and yield were statistically analyzed to find out the significance of the differences among the treatments, The mean value of all the characters for seven genotypes were calculated and the analysis of variance was performed by 'F' (variance ratio) test. The significance of difference among treatment means was evaluated by least significant difference (LSD) test at 5% and 1% levels of probability (Gomez and Gomez, 1984).

RESULTS

Morphological characteristics of stem

The variation of stem characteristics of the genotypes under study is shown in Table 1. Stem characters like stem colour, number, hairiness of the genotypes under study varied considerably. The stem color of the genotypes Cardinal and Diamant were green, whereas that of Fata pakri was green to light pinkish. On the other hand, the stem color of Gutipakri was light green, but the nodal region is pinkish. The stem color of the remaining genotypes was green to medium pinkish except Hagrai. Hagrai had pinkish stem. The main stems of the genotypes Cardinal, Gutipakri, Rumanapakri and Diamant were few in number, while hagrai and Tel pakri were many in number. On the other hand, the number of main stem of the variety Fata pakri was least. The stems of genotypes Cardinal and Gutipakri were slightly hairy, whereas Diamant was moderately hairy. The remaining genotypes had very hairy stem (Figure 1).

a. Cardinal   b. Diamant   c. Fatapakri   d. Rumanapakri

e. Tel pakri   f. Gutipakri   g. Hagrai

Figure 1. Photographic view of leaves and stem characteristics of seven potato genotypes

Table 1. Stem characteristics of seven potato genotypes

Potato Genotypes	Stem		
	Color	Number	hairiness
Cardinal	Green	Few	Slightly hairy
Diamant	Green	Few	Moderately hairy
Rumanapakri	Green to medium pinkish	Many	Very hairy
Fata pakri	Green to medium pinkish	Least	Very hairy
Gutipakri	Light green, pinkish at the nodal zone	Few	Slightly hairy
Tel pakri	Green to medium pinkish	Many	Very hairy
Hagrai	Pinkish	Few	Very hairy

Least = <5, Few = 5-9, Many = >9

Tuber characteristics

Table 2 is shown the tuber characteristics of seven potato genotypes under study. Remarkable variation was observed in different characters of tubers. The tubers of the genotypes Cardinal and Diamant were large in size whereas those of the remaining genotypes were medium in size. The tuber of the genotypes Fata pakri was round, while they were oval in Tel pakri. The tubers of the genotypes Hagrai were round irregular, while those of the Gutipakri and Rumanapakri were oval round. The tubers of the variety Cardinal were elongated. The skin of the tubers of Cardinal and Gutipakri were pinkish with creamy patches where eyes were more pinkish but the base of the tubers of the Gutipakri show a characteristic concentration of creamy patches. In the case of Hagrai the skin of tubers were pinkish white. The skin of the tubers Diamant were creamy white with the exception of eyes of distal end of Diamant which was pink in color. In case of Rumanapakri and Fata pakri the skin of the tubers were red (Figure 2). But the eyes of Rumanapakri were deep red while those of Fata pakri were pink with white eyebrows. Skin of tuber of Tel pakri was redish with yellow patches.

The skin of the tubers of the genotypes Tel pakri and Diamant were smooth, whereas those of Cardinal and Fata pakri were rough. On the other hand, smooth and shiny skin was observed in the genotypes Gutipakri and Hagrai and medium smooth skin in Rumanapakri. The average number of eyes on the tuber of the genotypes Rumanapakri and Diamant were moderate, while those of the remaining genotypes were low. Eyes were not uniformly distributed in the genotypes Cardinal, Fata pakri and Diamant. On the other hand, eyes were more or less uniformly distributed in Hagrai and Rumanapakri.

Table 2. Tuber characteristics of 7 potato genotypes

Potato genotypes	Tuber		
	Size	Shape	Skin color
Cardinal	Large	Elongated	Pinkish with creamy patch. Eyes are more pinkish.
Diamant	Large	Round irregular	Creamy white. Eyes of distal end are pink color.
Rumanapakri	Medium	Oval round	Red. Eyes are more red than other position.
Fata pakri	Medium	Round	Red. Eyes are pink and eyebrows are white.
Gutipakri	Medium	Oval round	Pinkish with creamy patch. Pink color cone, at the eye and creamy patches conc. at the base.
Tel pakri	Medium	Oval	Red. Yellow patchs are present at the surrounding of eyes.
Hagrai	Medium	Round irregular	Pinkish white

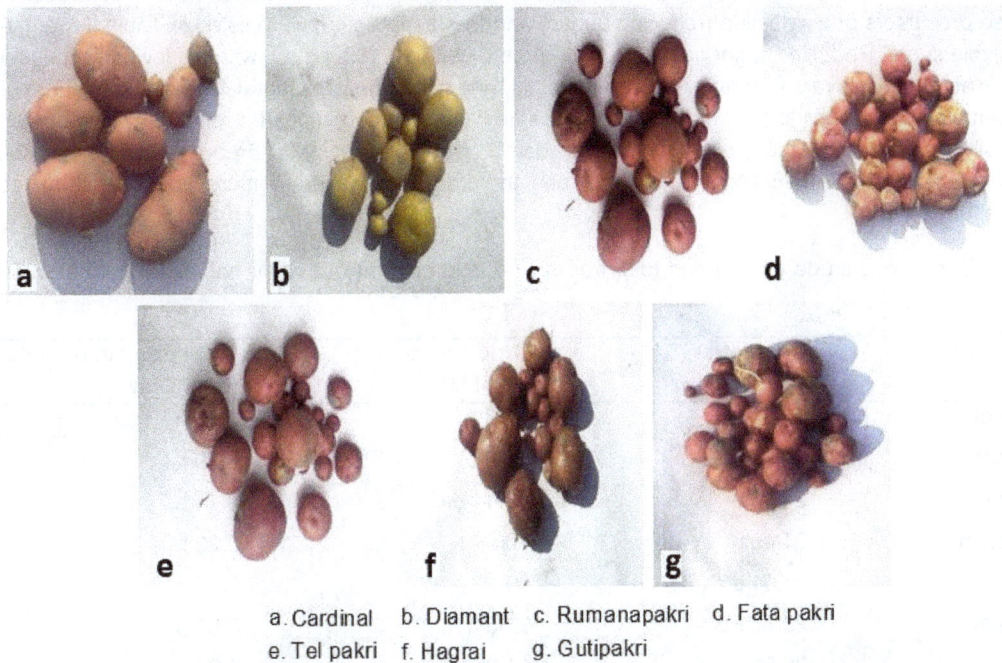

a. Cardinal   b. Diamant   c. Rumanapakri   d. Fata pakri
e. Tel pakri   f. Hagrai   g. Gutipakri

Figure 2. Photographic view of tuber of seven potato genotypes

Yield and yield components

Significant variation was observed among the experimental potato genotypes in the yield of tuber per plot and other yield components (Data not shown). Cardinal produced the highest yield of tubers (19.0 Kg plot^{-1}). The lowest yield of tubers was produced by Hagrai (8.3 Kg plot^{-1}). When the yield per plot was converted into yield per hectare, Cardinal was the highest yielded (26.49 tons ha^{-1}) followed by Diamant (24.36 t ha^{-1}), Gutipakri (19.76 t ha^{-1}), Tel pakri (15.8 t ha^{-1}), and Fata pakri (18.51 t ha^{-1}). The lowest tuber yield (10.23 t ha^{-1}) was obtained from the variety Hagrai (Figure 3).

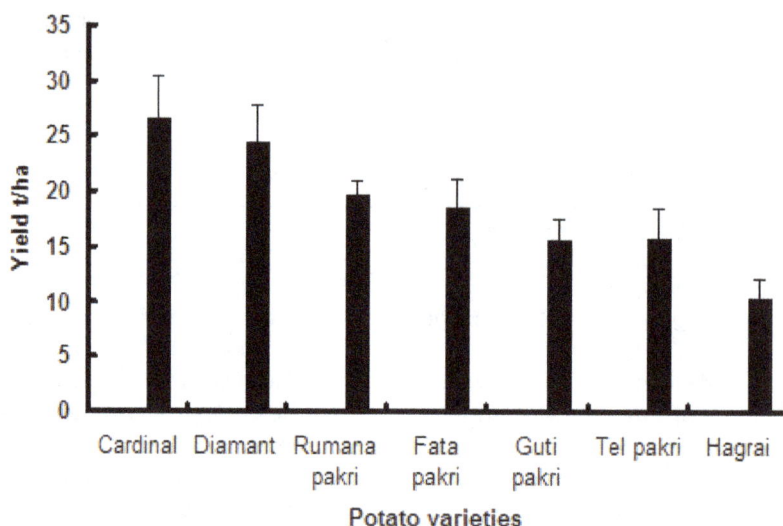

Figure 3. Yield of different potato genotyps available and cultivated in Bangladesh

The DM content, crude protein and total free amino acid content were determined for almost all of the potato samples (Table 3). Dry matter content was significantly varied among the 7 genotypes. Per cent moisture content was significantly greater in Cardinal and Diamant than the others. The concentrations of the assumed precursors of acrylamide (reducing sugars and free asparagine) were listed in Table 4 together with total soluble sugar. Reducing sugar content ranged from 0.094 to 0.482 mg g^{-1}, with the lowest values found in the samples of the cultivars Rumanapakri and Cardinal, and the highest in Hagrai. The highest sugar contents in the samples of the cultivar Tel pakri and the lowest in Diamant and gutipakri. Free asparagine content was statistically insignificant in the freshly harvested tubers of seven genotypes. Free asparagine was found at concentrations between 0.96 and 1.51 mg g^{-1}, and therefore was generally more abundant than reducing sugars after harvest.

Table 3. Dry matter, crude protein and total free amino acid content of freshly harvested potato tubers of 7 genotypes

Genoypes	Dry matter (%)	Crude protein (% of DM)	Total free amino acid (% of DM)
Cardinal	22.51 bc	9.48 c	3.42 a
Diamant	22.17 c	9.59 c	3.09 b
Rumanapakri	24.13 b	10.44 b	3.11 b
Fata pakri	24.49 b	10.54 b	3.23 b
Gutipakri	23.04 bc	10.27 b	2.89 c
Tel pakri	23.77 b	9.85 c	3.01 b
Hagrai	25.51 a	11.20 a	2.98 c
LSD$_{0.01}$	1.679	0.93	0.41

Table 4. Total soluble sugar, reducing sugar and asparagine content of freshly harvested potato tubers of 7 genotypes

Genotypes	Total soluble sugar (mg g^{-1} DM)	Reducing sugar (mg g^{-1} DM)	Asparagine content* (mg g^{-1} DM)
Cardinal	40.13 b	0.100 e	1.29
Diamant	20.57 e	0.141 d	1.48
Rumanapakri	40.74 b	0.094 e	0.96
Fata pakri	25.25 d	0.153 c	1.25
Gutipakri	22.14 e	0.154 c	1.51
Tel pakri	52.85 a	0.341 b	1.02
Hagrai	36.13 c	0.482 a	1.46
LSD$_{0.01}$	18.830	0.118	

In column, dissimilar letter differ significantly, and * indicates non-significant.

Table 5. Total soluble sugar, reducing sugar and asparagine content of potato tubers of 7 genotypes that stored at 8° C over 8 months

Genotypes	Total soluble sugar (mg g^{-1} DM)	Reducing sugar (mg g^{-1} DM)	Asparagine content* (mg g^{-1} DM)
Cardinal	55.36 b	0.462 c	1.35
Diamant	38.04 c	0.201d	1.50
Rumanapakri	68.60 a	0.763 ab	0.96
Fata pakri	31.44 d	0.525 bc	0.79
Gutipakri	42.32 c	0.653 b	0.97
Tel pakri	58.55 b	0.853 a	1.00
Hagrai	67.23 a	0.952 a	0.95
LSD$_{0.01}$	13.24	0.426	

In column, dissimilar letter differ significantly, and * indicates non-significant.

The main acrylamide precursors (reducing sugars and free asparagine) were determined after 8 months storing at at 8°C and data were shown in Table 5. During storage time, asparagine concentrations of all genotypes did not change significantly whereas the reducing sugar concentrations of all genotypes of the tubers increased significantly. The variety Hagrai showed the highest value (0.95 mg g^{-1}) and was significantly different from all other genotypes. The variety Diamant appeared to contain the lowest amount of reducing sugars (0.201 mg g^{-1}). A significant difference of starch breakdown was observed in the studied genotypes, which was reflected by total soluble sugar content. Total soluble sugar content was significantly highest in the variety Rumanapakri, which was the lowest in Fata pakri.

DISCUSSION

Potato tubers contain substantial amounts of the acrylamide precursors free asparagines and reducing sugars (Becalski et al. 2004), which may explain the high concentrations of acrylamide in certain potato products. Potato tuber is rich in asparagine content (33-59% of the total free amino acids) (Eppendorfer and Bille, 1996). In the present study, asparagines content was greater than reducing sugar content in freshly harvested tubers of all genotypes (Table 4). Since asparagines content was identical, the susceptibility to acrylamide formation depends on the content of reducing sugar. Cardinal, Diamant and Rumanapakri could be used for potato chips and French fries just after harvest (Table 4). Amrein et al. (2003) reported that the asparagine content did not correlate with acrylamide formation in french fries while the high correlation between the potentials os acrylamide and the reducing sugars.

Both the total soluble sugar and reducing sugar content increased during longer storage periods because of the degradation of starch in all genotypes under the present investigation (Table 5). This increase can be described as "senescent sweetening" (Burto 1989). Low temperature storage also influenced the content of asparagines in potato tubers. Asparagine oxidation occurred due to oxidative stress (H_2O_2) during storage (Tareke et al. 2009). Actually, low-temperature storage is very detrimental for acrylamide formation (De Wilde et al. 2005). In the present study, Cardinal and Diamant could be used for frying as potato chips and French fry after storage.

From these results, selection of the appropriate variety seems of extreme importance to control acrylamide formation during frying. Because acrylamide formation is strongly correlated with the amount of reducing sugars present in the raw material, it could be useful to screen potato genotypes primarily on their reducing sugar contents to select the genotypes suitable for frying. Moreover, fresh potatoes sold in retail should be labeled clearly that they are suitable for frying, because potatoes used for other applications are often stored at low temperatures to suppress sprouting. The main precursors for this undesirable substance are reducing sugars and the amino acid asparagine (Stadler et al. 2004, Mottram et al. 2002, Weisshaar et al. 2002). However, despite the high content of free asparagine in potatoes (Olsson et al. 2004), the limiting factor for the creation of acrylamide in potato products is the content of reducing sugars (Amrein et al. 2003, Becalski et al. 2004). It is well-known that the storage temperature influences the amount of reducing sugars in potato tubers, but this does not appear to be the case for asparagine (Olsson et al. 2004). It was also reflected in the present study (Table 4 and Table 5).

CONCLUSIONS

Reducing sugar content was significantly varied among the genotypes whilst asparagines content remained unchanged. The results revealed that (a) the content of reducing sugars is more important factor than the content of asparagine for acrylamide formation. So, reducing sugar content should be determined immediately before home and commercial processing of the potatoes; (b) to reduce cold-storage- induced sugar levels, potatoes should be reconditioned after storage.

ACKNOWLEDGEMENT

University Grants Commission (UGC) is gratefully acknowledged to provide financial support for conducting the research.

REFERENCES

1.   Amrein TM, Bachmann S, Noti A, Biedermann M, Barbosa MF, Biedermann-Brem S, Grob K, Keiser A, Realini P, Escher F and Amado R, 2003. Potential of acrylamide formation, sugars, and free asparagine in potatoes: a comparison of cultivars and farming systems. Journal of Agricultural and Food Chemistry, 51: 5556 – 5560.

2.   Becalski A, Lau BPY, Lewis D, Seaman SW, Hayward S, Sahagian M, Ramesh M, Leclerc Y, 2004. Acrylamide in French fries: influence of free amino acids and sugars. Journal of Agricultural and Food Chemistry, 52: 3801–3806.

3.   Biedermann M, Noti A, Biedermann-Brem S, Mozzetti V and Grob K, 2002. Experiments on acrylamide formation and possibilities to decrease the potential of acrylamide formation in potatoes. Mitteilungenausdem Gebiete der Lebensmitteluntersuchung und Hygiene, 93: 668–687.

4.   Burton WG, 1989. The Potato, 3rd edition, Longman Scientific and Technical, Essex, U.K., Chapter 5: Yield and Content of Dry Matter: 2, pp: 156–215.

5.   Chuda Y, Ono H, Yada H, Ohara-Takada A, Matsuura-Endo C and Mori M, 2003.Effects of physiolical changes in potato tubers (*Solanum tuberosum* L.) after low temperature storage on the level of acrylamide formed in potato chips. Bioscience Biotechnology and Biochemistry, 67:1188–90.

6. Coffin RH, Yada RY, Parkin KL, Grodzinski B and Stanley DW, 1987. Effect of low-temperature storage on sugar concentrations and chip color of certain processing potato cultivars and selections. Journal of Food Science, 52: 639–645.

7. De Wilde T, De Meulenaer B, Mestdagh F, Govaert Y, Vandeburie S, Ooghe W, Fraselle S, Demeulemeester K, Van Peteghem C, Calus A, Degroodt J and Verhe R, 2005. The influence of storage practices on acrylamide formation during frying. Journal of Agricultural and Food Chemistry, 56: 6550–7.

8. Dybing E, Farmer PB, Andersen M, Fennell TR, Lalljie SP, Muller DJ, Olin S, Petersen BJ, Schlatter J, Scholz G, Scimeca JA, Slimani N, Tornqvist M, Tuijtelaars S and Verger P, 2005. Human exposure and internal dose assessments of acrylamide in food. Food and Chemical Toxicology, 43: 365–410.

9. Eppendorfer W H &Bille S W, 1996. Free and total amino acid composition of edible parts of beans, kale, spinach, cauliflower and potatoes as influenced by nitrogen fertilisation and phosphorus and potassium deficiency. Journal of the Science of Food and Agriculture, 71: 449–458.

10. FAO/WHO, 2005. Joint FAO/WHO Expert Committee on Food Additives, 64th meeting, FAO, Rome, Italy.

11. Friedman M, 2004. Application of the ninhydrin reaction for the analysis of amino acids, peptides, and proteins to agricultural and biomedical sciences. Journal of Agricultural and Food Chemistry, 52: 385–406.

12. Gomez KA and Gomez AA, 1984. Statistical Procedure for Agricultural Research. Second Edition. A Willey Inter-Science Publication, John Wiley and Sons, New York. p: 680.

13. Hossain MA, Takahashi T, Zhang L, Nakatsukasa M, Kimura K, Kurashige H, Hirata T. and Ariyoshi M, 2009. Physiological mechanisms of poor grain growth in abnormally early ripening wheat grown in west Japan. Plant Production Science, 12: 278–284.

14. Hossain MA, Takahashi T and Araki H, 2012. Mechanisms and Causes of Poor Grain Filling in Wheat. Lambert Academic Publishing, Saarbrucken, Germany.

15. IARC (International Agency for Research on Cancer), 2002. Acrylamide: Monographs on the evaluation of carcinogenic risks to humans: Some industrial chemicals.

16. Lea PJ and Azevedo RA, 2007. Nitrogen use efficiency. II. Amino acid metabolism. Annals of Applied Biology, 151: 269–275.

17. Miller GL, 1972. Use of Dinitro Salicylic acid reagent for determination of reducing sugar. International Journal of Analytical chemistry, 31: 426–428.

18. Mottram DS, Wedzicha BL and Dodson AT, 2002. Acrylamide is formed in the Maillard reaction. Nature, 419: 448–449.

19. Mucci LA and Wilson KM, 2008. Acrylamide intake through diet and human cancer risk. Journal of Agricultural and Food Chemistry, 56: 6013–6019.

20. Olsson K, Svensson R and Roslund CA, 2004. Tuber components affecting acrylamide formation and colour in fried potato: variation by variety, year, and storage temperature and storage time. Journal of the Science of Food and Agriculture, 84: 447–458.

21. Petersen BJ and Tran N, 2005. Exposure to acrylamide: Placing exposure in context. In Chemistry and Safety of Acrylamide in Food. Friedman, M., Mottram, D., Eds.; Springer Press: New York, pp: 63–76.

22. Stadler RH, Blank I, Verga N, Robert F, Hau J, Guy PA, Robert M and Riediker S, 2002. Acrylamide from Maillard reaction porducts. Nature, 419: 449–50.

23. Tareke E, Heinze TM, da Costa GG and Ali S, 2009. Acrylamide formed at physiological temperature as a result of asparagine oxidation. Journal of Agricultural and Food Chemistry, 57: 9730–9733.

24. Tareke E, Rydberg P, Karlsson P, Eriksson S and Tornqvist M, 2000. Acrylamide: a cooking carcinogen. Chemical Research in Toxicology. 13: 517–522.

25. Weisshaar R and Gutsche B, 2002. Formation of acrylamide in heated potato productssmodel experiments pointing to asparagine as precursor. Deutsche Lebensmittel Rundschau, 98: 397–400.

26. www.slv.se, 2002. Acrylamide in foodstuffs, consumption and intake. Swedish National Food Administration.

27. Yemm EW and Willis AJ, 1954. The estimation of carbohydrates in plant extracts by anthrone. Biochemical Journal, 57: 508–514.

EVALUATION OF PRODUCTIVE PERFORMANCE OF BROILER IN RESPONSE TO KOROCH (*Pongamia pinnata*) CAKE FEEDING

Masuma Habib[1], Abu Jafur Md. Ferdaus[2], Md. Touhidul Islam[3], Begum Mansura Hassin[4] and Md. Shawkat Ali[*2]

Graduate Training Institute Bangladesh Agricultural University (BAU), Mymensingh-2202, [2]Department of Poultry Science, BAU, Mymensingh-2202, [3]Nourish Poultry and Hatchery Ltd. Uttara, Dhaka-1230, [4]Department of Livestock Services (DLS), Bangladesh, Dhaka-1215

*Corresponding author: Md. Shawkat Ali, E-mail: mdshawkatali@hotmail.com

---

ARTICLE INFO   **ABSTRACT**

Key words

Productive performance,
Growth depression,
Meat yield characteristics,
Koroch seed cake,
Broiler chickens

The study was conducted for a period of 28 days to investigate the response of broiler to the inclusion of de-oiled koroch (*Pongamia Pinnata*) seed cake in the diet of broiler. A total of 192 day-old broiler chicks were individually weighed and randomly allocated to 4 dietary treatment groups having 4 replications of 12 chicks each, in a completely randomized design. Broilers under treatment 1 received a basal diet containing no koroch seed cake, considered as control; in treatment 2, 3 and 4, broilers were fed on basal diet containing 2%, 4% and 6% koroch seed cake, respectively. All productive performances (live weight, live weight gain, feed consumption and feed efficiency) of broiler fed on koroch seed cake were significantly (P<0.01) depressed compared to the control. The degree of depression was increased with the increasing level of koroch seed cake in the diet at all ages of broiler. The meat yield parameters showed a non-significant (P>0.05) effect except for the percentage of dressing yield, breast, liver and gizzard weight for the broilers in all treatment groups. Broilers fed on diet containing 6% koroch seed cake yielded the lowest dressed weight in the treatment groups. Inclusion of 2% koroch seed cake resulted in higher breast meat yield compared to any other level of koroch cake inclusion in the diet (P<0.01). However, liver and gizzard weight were increased significantly (P<0.05) for incorporation of de-oiled koroch seed cake in the diet at all levels (2%, 4% or 6%). On the basis of these results it is concluded that feeding de-oiled koroch cake had no positive effect on growth response and meat yield characteristics of broiler. Therefore, it is suggested that the koroch seed cake may contain anti-nutritional factor(s), which seemed to be associated with growth depression in broilers. Further research is warranted to alleviate the potential toxic effect of koroch seed cake on broiler performances.

INTRODUCTION

Increasing feed cost and non-availability of major grains (maize, wheat or soybean) are the root causes of increasing the cost of poultry production. There is a constant competition among human, livestock and poultry for grains. Cost and availability of grains fluctuate due to natural calamities. Dependency on costly imported grains and their use increase feed cost that hardly permits profitable poultry rearing. Therefore, it was suggested to formulate cheaper diet using unconventional feeds.

Karanj (*Pongamia pinnata)* is one of the nitrogen fixing trees, which produces huge seeds with high nutritional value. In Bangladesh, it is popularly known as Koroch. It is available in the highway, roads, haor and canal of Sylhet region, Sundarban mangrove forest region of Bangladesh. Huge amount of Koroch seeds in Bangladesh was not tested to introduce as unconventional poultry feed, despite its potential chemical composition. It has been reported that air dried kernels had 19.0% moisture, 27.5% fatty oil, 17.4% protein, 6.6% starch, 7.3% crude fiber, and 2.4% ash (James, 1983). The oil cake, when extracted from the seeds, is used as a feed to be cheaper than soybean meal (James, 1983). However, the koroch seed has been shown to contain tannin (23.2g/kg DM) and trypsin inhibitor (62g/kg protein) (Natanam *et al.*, 1989a). The oil from the koroch seed has a high content of triglycerides and has disagreeable taste and odor due to the presence of bitter flavonoid constituents, pongamin and karanjin (Allen and Allen, 1981). Therefore, the oil extracted cake after processing through oil extraction technique might be feasible to formulate broiler diet. In order for the elucidation of the feeding value of de-oiled koroch seed cake in broiler, this study investigated the effect of different levels of oil extracted koroch cake on productive performance and meat yield characteristics of broilers.

MATERIALS AND METHODS

The study was conducted at Bangladesh Agricultural University Poultry Farm, Mymensingh, Bangladesh. The experimental day-old broilers were purchased from "Nourish Poultry and Hatchery Ltd. Shreepur, Gazipur, Bangladesh. The chicks were reared up to 28 days of age. Koroch seeds were collected from Sunamgonj Hoar area. The oil from seeds was expelled locally from an oil extraction mill. Then the remaining cake was dried and proximate analysis was done for Koroch cake.The proximate composition of the feed ingredients was analyzed as per AOAC (1995) and is furnished in Table 1.

The experiment was carried out with a total number of 192 broilers. They were fed on diet containing either 0%, 2%, 4% or 6% de-oiled Koroch cake. These experimental diets were considered as 4 different treatments. Each treatment was consisting of 4 replications having 12 birds per replication. Broilers in each treatment group were fed *ad libitum* on starter (Table 2) and grower diet (Table 3) for 0-21 days and 22-28 days period, respectively. The recommended medications and vaccines (Table 4) were administered to ensure good health status of the experimental birds.

Birds were weighed at the first day of experiment (initial body weight) and weekly basis for all birds from each replication. Average body weight gain of the broiler in each replication was calculated by deducting initial body weight from the final body weight. The amount of feed consumed by the birds in each replication of each treatment group were calculated for every week by deducting the amount of feed left over from the amount supplied for a particular week. Feed conversion ratio (FCR) was calculated as the unit of feed consumed per unit of body weight gain.

At the end of feeding trial, one broiler having near to pen average weight was taken from each pen for recording meat yield parameters. Broilers were slaughtered and allowed to bleed for 2 minutes and immersed in hot water (semi-scalding; 51-55°C) for 120 seconds in order to loose feathers followed by removal of feathers by hand pinning. Then head, shank, viscera, giblet (heart, liver and gizzard) and abdominal fat were removed for determination of meat yield parameters. Dressed broilers were cut into different parts such as breast, thigh, drumstick and wing. Finally, every cut up parts were weighed and recorded separately for each broiler of all replications.

Table 1. Proximate composition of Koroch seed and cake

	Moisture (%)	CP (%)	Ash (%)	CF (%)	EE (%)
Seed	9.04	22.43	2.41	5.57	26.17
Cake	8.89	24.27	3.12	2.37	14.61

Table 2. Ingredients and nutrient composition (%) of starter diet (0-21 days)

Ingredients	Different level of oil extracted Koroch seed cake			
	0%	2%	4%	6%
Maize	61.00	59.90	59.00	58.00
Rice Polish	2.00	2.00	1.00	1.00
Soybean meal	24.9	24.0	23.65	22.4
Protein Concentrate.	9.00	9.00	9.00	9.00
Koroch cake	0.00	2.00	4.00	6.00
Vegetable oil	0.00	0.00	0.25	0.50
Dicalcium phosphate	2.00	2.00	2.00	2.00
Vit.-min. premix	0.25	0.25	0.25	0.25
DL-Methionine	0.20	0.20	0.20	0.20
L-lysine	0.15	0.15	0.15	0.15
Common Salt	0.50	0.50	0.50	0.50
Nutrient composition				
ME Kcal/kg	2971	2952	2945	2941
Crude protein (%)	22.38	22.37	22.50	22.35
Crude fiber (%)	3.82	3.78	3.68	3.62
Calcium (%)	1.28	1.28	1.28	1.27
AV. Phosphorus (%)	0.71	0.71	0.70	0.70
Lysine (%)	1.30	1.27	1.25	1.22
Methionine (%)	0.55	0.54	0.54	0.53

Table 3. Ingredients and nutrient composition (%) of grower diet (22-28 days)

Ingredients	Different level of oil extracted Koroch seed cake			
	0%	2%	4%	6%
Maize	65.05	63.44	61.62	60.23
Rice Polish	2.00	2.00	2.00	2.00
Soybean meal	20.69	19.91	19.18	18.35
Protein Concentrate.	7.50	7.50	7.50	7.50
Koroch cake	0.00	2.00	4.00	6.00
Vegetable oil	1.66	2.05	2.50	2.82
Dicalcium phosphate	2.00	2.00	2.00	2.00
Vit.-min. premix	0.25	0.25	0.25	0.25
DL-Methionine	0.20	0.20	0.20	0.20
L-lysine	0.15	0.15	0.15	0.15
Common Salt	0.50	0.50	0.50	0.50
Nutrient composition				
ME Kcal/kg	3100	3100	3100	3100
Crude protein (%)	20.0	20.0	20.0	20.0
Crude fiber (%)	3.61	3.56	3.51	3.47
Calcium (%)	1.20	1.19	1.19	1.19
AV. Phosphorus (%)	0.68	0.67	0.67	0.66
Lysine (%)	1.14	1.12	1.09	1.06

Methionine (%)	0.52	0.51	0.50	0.49

Statistical analysis

All recorded and calculated data were statistically analyzed using analysis of variance (ANOVA) technique by a computer using SAS statistical package program in accordance with the principles of Completely Randomized Design (SAS, 2009). Duncan's multiple range test (DMRT) was done to compare variations among diets where ANOVA showed significant differences.

Table 4. Vaccination schedule for the experimental broilers

Age of broilers (day)	Name of Vaccine	Trade Name*	Dose	Route
4	IB+ND	MA5+Clone30	One drop	Ocular
10	IBD	D-78	One drop	Ocular
21	IBD	D-78	One drop	Ocular

IB, Infectious Bronchitis; ND, Newcastle Disease; IBD, Infectious Bursal Disease *Intervet International,
B.V. BOXMEER, The Netherlands.

RESULTS AND DISCUSSION

Growth performance

Overall performances of broiler chicks as affected by inclusion of different dietary levels of koroch seed cake are shown in Table 6. Final body weight, body weight gain, feed consumption and FCR were significantly (P<0.01) influenced by dietary treatments. All performances of broilers fed on koroch seed cake at different levels were significantly (P<0.01) depressed compared to the control. The degree of depression was increased with the increasing inclusion level of koroch seed cake in the diet of broiler. The weekly changing pattern of broiler performances was shown in Table 5.

Table 5. The weekly growth performance of broilers fed on oil extracted Koroch seed cake

Trait	Age (week)	Different level of oil extracted Koroch seed cake				Significant Level
		0%	2%	4%	6%	
Body weight (g)	Day-old	40.11±0.19	40.08±0.07	40.11±0.37	40.04±0.39	NS
	1st	150.42±4.22[a]	132.92±4.22[a]	120.58±6.85[b]	109.50±6.32[c]	**
	2nd	379.54±22.46[a]	317.21±7.78[b]	276.88±21.58[c]	242.58±20.11[d]	**
	3rd	704.38±33.30[a]	595.73±14.92[b]	476.25±17.87[c]	421.04±35.97[d]	**
	4th	1087.9±25.15[a]	940.54±25.57[b]	720.05±26.24[c]	635.63±67.57[d]	**
Body weight gain (g)	1st	110.31±4.36[a]	92.83±5.34[b]	80.48±3.57[c]	69.46±6.24[d]	**
	2nd	229.12±20.17[a]	184.29±8.50[b]	156.29±15.71[c]	133.08±14.40[c]	**
	3rd	324.83±23.95[a]	278.52±18.41[b]	199.38±23.83[c]	178.46±19.98[c]	**
	4th	383.59±12.65[a]	344.81±11.43[b]	243.80±11.27[c]	214.59±33.07[c]	**
Feed intake (g)	1st	101.82±1.40[a]	100.13±1.39[ab]	95.21±3.81[bc]	93.28±4.85[c]	**
	2nd	266.15±10.58[a]	253.75±4.39[ab]	238.13±15.33[bc]	226.96±14.74[c]	**
	3rd	494.27±20.44[a]	456.87±16.34[b]	402.50±23.23[c]	390.63±32.82[c]	**
	4th	774.68±13.54[a]	734.37±13.29[a]	610.21±14.30[b]	570.33±46.95[b]	**
Feed conversion ratio	1st	0.93±0.05a	1.08±0.06[b]	1.19±0.06[c]	1.35±0.06[d]	**
	2nd	1.17±0.15[a]	1.38±0.05[b]	1.53±0.07[b]	1.71±0.11[c]	**
	3rd	1.53±0.05[a]	1.64±0.06[a]	2.03±0.14[b]	2.20±0.08[c]	**
	4th	2.02±0.09[a]	2.13±0.10[b]	2.51±0.13[c]	2.69±0.26[c]	**

[a-d]Mean±SD values with different superscripts within same row differ significantly;
NS=Non-significant; **=significant (p<0.01)

The depression in body weight and feed intake and poor weight gain in the present study on koroch seed cake incorporated diet was in agreement with the findings of Mandal and Banerjee (1974) and Panda et al. (2008). Panda et al. (2008) observed that increasing the inclusion of solvent extracted karanj cake significantly decreased the body weight gain and feed intake, and resulted in poor feed efficiency. On the other hand, Mandal and Banerjee (1979) recommended that de-oiled karanj cake (EE<0.05%) could safely be included at 5% level in the diet of broiler chicken. Similarly, Dhara et al. (1997) reported that de-oiled karanj cake could be included in the diet to a maximum level of 4.45%. The adversed effect of high level of koroch seed cake inclusion on growth and feed intake might be due to the residual oil content of the cake. Although the oil was extracted by normal milling procedure, the oil content of cake was 14.61% which was attributed to the presence of toxic factors left back in the cake. Allen and Allen (1981) stated that the oil of koroch seed has a high content of triglycerides, and has disagreeable taste and odor for bitter flavonoid constituents, pongamin and karanjin.

Table 6. Performances of broilers fed on oil extracted Koroch (Pongamia Pinnata) seed cake during the experimental period

Parameter	Different level of oil extracted Koroch seed cake				Signi. level
	0%	2%	4%	6%	
Initial weight (g)	40.11±0.19	40.08±0.07	40.11±0.37	40.04±0.39	NS
Final weight (g)	1087.96±25.15[a]	940.54±25.57[b]	720.05±26.2[c]	635.63±67.57[d]	**
Body weight gain (g)	1047.85±25.29[a]	900.46±25.53[b]	679.95±26.5[c]	595.58±67.71[d]	**
Feed Intake (g)	1636.92±16.83[a]	1545.13±4.82[b]	1346.04±7.7[c]	1281.2±70.88[d]	**
Feed conversion ratio (FCR)	1.56±0.02[a]	1.72±0.05[b]	1.98±0.03[c]	2.20±0.09[d]	**

[a-d]Mean±SD values with different superscripts within same row differ significantly; Signi. = Significant; NS=Non-significant; **=significant (p<0.01)

Dietary incorporation of koroch seed cake also depressed the feed efficiency in the present study. Increasing the inclusion level from 2% to 6%, resulted in lowered feed efficiency, which is in line with the findings of Dhara et al. (1997). They observed poor feed efficiency by dietary incorporation of 11.20% to 22.40% de-oiled karanj cake in the diet of Japanese quail. But, Natnam et al. (1989b) reported a comparable feed efficiency with control diet in broiler chicks by incorporating solvent extracted karanj cake at 10% level in the diet. Efficiency of feed utilization depends on the level of protein and energy (Mellen et al., 1984) and their ratio (Davidson, 1964) in addition to the presence of incriminated factors (Chand, 1987). Probably the presence of higher levels of residual toxic factors in diets containing koroch seed cake at all levels of incorporation resulted in poor feed efficiency in broiler chicks.

Meat yield characteristics

Dietary effect on organs, tissues weight and body development may affect carcass characteristics of birds. The data on carcass traits like dressing percentage, percent live weight of neck, head, gizzard, heart, liver, shank, breast, drumstick, thigh, wing, skin and abdominal fat have been set out in Table 7. All the parameters determined showed a non-significant treatment effect except for the percentage of dressing yield, breast, liver and gizzard weight. Percent live weight of dressing yield, gizzard and liver were significantly (P<0.05) influenced by dietary treatments. Dressing percentage of birds fed on 6% koroch seed cake was significantly (P<0.05) depressed compared to any other dietary treatment, whereas birds fed on 2% koroch seed cake exhibited similar results compared with their counterparts on the control diet. The percentage of breast meat was also significantly (P<0.01) decreased for 4% and 6% koroch seed cake dietary group compared to either 2% koroch seed cake or control group. However, liver and gizzard weight increased significantly (P<0.05) due to incorporation of koroch seed cake in the diet at all levels.

Table 7. Meat yield characteristics of broilers fed on oil extracted Koroch *(Pongamia Pinnata)* seed cake up to 28 days of age

Parameter	Different level of oil extracted Koroch seed cake				Significance Level
	0%	2%	4%	6%	
Live weight (g)	1004.5[a]	857[b]	836[b]	816[b]	*
Dressing yield (%)	64.97[a]	64.91[a]	62.05[ab]	59.59[b]	*
Blood (%)	5.34	4.32	4.19	3.60	NS
Feather (%)	3.55	5.03	3.96	3.60	NS
Neck (%)	1.99	2.04	2.04	2.02	NS
Head (%)	3.13	3.41	3.12	3.14	NS
Gizzard (%)	1.79[a]	1.82[ab]	2.17[b]	2.39[b]	*
Heart (%)	0.35	0.35	0.41	0.32	NS
Liver (%)	2.93[a]	2.79[ab]	3.16[ab]	3.61[b]	*
Shank (%)	1.94	1.93	1.97	1.89	NS
Breast (%)	13.00[a]	13.60[a]	10.62[b]	9.90[b]	**
Drumstick (%)	5.07	5.02	5.48	6.14	NS
Thigh (%)	7.46	7.04	6.71	6.83	NS
Wing (%)	3.49	3.64	3.22	3.41	NS
Skin (%)	1.94	2.21	2.22	2.16	NS
Abdominal fat (%)	0.89	0.99	0.82	0.62	NS

[a,b] Mean values with different superscripts within same row differ significantly; NS=Non-significant; **=significant ($p<0.01$)

Mandal and Banerjee (1982) reported no difference in organ weights like liver, heart, kidney and spleen of cockerels due to dietary replacement of black til cake with de-oiled karanj cake at 30% level. Dhara *et al.* (1997) also found no significant variation in weight of different commercial cuts (neck, wing, thigh, shank, breast and trunk) and organs (liver, heart and gizzard) due to incorporation of deoiled karanj cake upto 22.40% in the diet of Japanese quail. However, in the present study, dietary inclusion of koroch seed cake significantly decreased the weight of dressing yield and breast meat.

Panda et al. (2008) observed that the dietary inclusion of either solvent extracted karanj cake (SKC) or NaOH treated SKC at 25% level or Ca(OH)$_2$ treated SKC at 12.5% and 25% levels significantly increased the weight of liver leading to liver hypertrophy. Gizzard weight also increased significantly due to incorporation of SKC at both levels. These findings are similar to the findings of the present study where liver and gizzard weight increased significantly due to incorporation of koroch seed cake in broiler diet at all levels.

Panda *et al.* (2008) noticed that the adverse effect on growth and feed intake was found when daily karanj intake was 18mg and above. Similarly, Natanam (1989a) reported that the adverse effect of karanj cake feeding was due to the leftover of karanj oil in the processed cake. Probably the higher daily karanj intake (>18mg) could be the potential reason for showing adverse effect on growth and meat yield characteristics of brolers in the present study.

CONCLUSION

The results revealed that the feeding of koroch seed cake to broilers had no improvement on growth rate and feed conversion efficiency up to 28 days. However, treatment of koroch seed cake might be potentially beneficial to the performance of broiler.

ACKNOWLEDGEMENT

The first author would like to express his cordial thanks to Bangladesh Agricultural University Research System (BAURES) for the full financial support of conducting this research work.

REFERENCES

1. Allen ON and EK Allen, 1981. The Leguminosae. The University of Wisconsin Press. 812 p

2. AOAC, 1995. Association of Official Analytical Chemistry, Official Methods of Analysis. (13th edn.) Washington, DC.

3. Chand S, 1987. Nutritional evaluation of neem seed meal in chicks. Ph.D. Thesis, submitted to the Rohilkhand University, Bareilly.

4. Davidson J, 1964. The effciency of conversion of dietary metabolizable energy into tissue energy in chicken as measured by body analyses. European Association of Animal Production; 3rd symposium on Energy Metabolism, Troon.

5. Dhara TK, N Chakraborty, G Samanta and L Mandal, 1997. Deoiled karanj (*P. glabra* vent) cake in the ration of Japanese quail. Indian Journal of Poultry Science, 32: 132-136.

6. James AD, 1983. Handbook of Energy Crops. Unpublished. http://www.hortpurdue.edu/ duke_ energy/p ongamiapinnata.htm.

7. Mandal L and GC Banerjee, 1974. Studies on the utilization of karanj (*P. glabra*) oil cake in poultry rations. Indian Journal of Poultry Science, September: 141-147.

8. Mandal L and GC Banerjee, 1979. Studies on the utilization of karanja (*Pongamia glabra* vent.) cake in layer diet. Indian Journal of Poultry Science, 14:105-109,

9. Mandal L and GC Banerjee, 1982. Studies on the utilization of karanj (*Pongamia glabra)* cake in poultry rations Effect on growers on blood composition and organ weight of cockerels. Indian Veterinary Journal, 59: 385-390.

10. Mellen WJ, FW Hill and HH Dukes, 1984. Studies on the energy requirement of chickens, 2. Effect and dietary energy level on basal metabolism of growing chickens. Poultry Science, 31: 735-740.

11. Natanam R, R Kadirvel and R Balagopa, 1989a. The effect of kernels of karanj (*P. glabra vent*) on growth and feed efficiency in broiler chicks to 4 weeks of age. Animal Feed Science Technology, 25: 201-206.

12. Natanam R, R Kadirvel and R Ravi, 1989b. The toxic effects of karanj (*P.glabra* bent) oil and cake on growth and feed efficiency in broiler chicks. Animal Feed Science Technology, 27: 95-100.

13. Panda AK, VRB Sastry and AB Mandal, 2008. Growth, nutrition utilization and carcass characteristics in broiler chickens fed raw and alkali processed solvent extracted karanj (*Pongamia glabra*) cake as partial protein supplement. The Journal of Poultry Science, 45:199-205.

14. SAS, 2009. User's Guide: Statistics. Version 9.2, SAS Institute Inc., Cary, NC. USA.

PERFORMANCE OF BROILER FED ON DIET CONTAINING DEOILED KOROCH (*Pongamia Pinnata*) SEED CAKE TREATED WITH NaOH AND HCl

Masuma Habib[1,] Abu Jafur Md. Ferdaus[2], Md. Touhidul Islam[3], Begum Mansura Hassin[4] and Md. Shawkat Ali[*2]

Graduate Training Institute Bangladesh Agricultural University (BAU), Mymensingh-2202, Bangladesh; [2]Department of Poultry Science, BAU, Mymensingh-2202, Bangladesh; [3]Nourish Poultry and Hatchery Ltd. Uttara, Dhaka-1230, [4]Department of Livestock Services (DLS), Dhaka-1215, Bangladesh

*Corresponding author: Md. Shawkat Ali, E-mail: mdshawkatali@hotmail.com

ARTICLE INFO	ABSTRACT
	The study was conducted to investigate whether the productive performances and meat yield characteristics of broiler would be improved by feeding diet containing koroch (*Pongamia Pinnata*) seed cake (KSC) treated with NaOH and HCl. A total number of 160 day-old straight run broiler chicks were fed on 4 iso-energetic and iso-nitrogenous diets containing either basal diet with no KSC or basal diet containing 2% KSC treated with or without 1% NaOH or 1% HCl for a period of 28 days. These diets were considered as 4 different treatments. Each treatment was replicated 4 times, each having 10 birds. Inclusion of 2% KSC treated with NaOH or HCl in the diet resulted in lower live weight, live weight gain, feed intake, feed efficiency of broilers at all ages compared to inclusion of dietary 2% KSC alone ($P<0.01$). However, broilers, irrespective of age, fed on diet containing 2% KSC treated with 1% HCl exhibited better productive performances than those fed on 1% NaOH treated 2% KSC incorporated diet. The amount of feed intake of broilers on either KSC alone or HCl treated KSC was almost similar to that on control diet. The meat yield parameters showed a non-significant treatment effect except for the percentage of dressing yield, liver, gizzard, head and skin weight. Dressing percentage of broilers fed on HCl treated KSC at 2% level was significantly ($P<0.05$) lower compared to any other dietary treatment, whereas broilers in 2% KSC dietary group showed similar results to the control group. Percentage of head and skin weight relative to body weight were higher in treated (NaOH or HCl) KSC dietary group compared to either KSC alone or control group. However, the highest liver and gizzard weights were obtained from the broilers fed on diet containing 2% KSC treated with 1% HCl, followed by diet with 1% NaOH treated KSC, KSC alone diet and the control diet. The results of the present study clearly indicated that neither the 1% NaOH treated nor 1% HCl treated KSC allliviated the depressed productive performances and meat yield characteristics of broilers. It is concluded that the poor productive performances of broilers fed on KSC incorporated diet cannot be improved by treating KSC with either NaOH or HCl, and the oil extracted KSC may contain leftover oil which might have a potential toxic effect on growth and meat yield of broilers.
Key words Productive performance, Growth depression, Meat yield characteristics, Koroch seed cake, Broiler chickens	

INTRODUCTION

Efficient and economic poultry production is only possible when using cheaper locally available feed ingredients, because the feed alone contribute to 70 to 75 per cent of the total cost of poultry production (Panda and Mahapatra, 1989). The researchers are thus compelled to explore the possibilities of feeding unconventional agro-forest based industrial by-products, to meet the nutritional requirement of birds. Since protein sources are expensive, measures are often adapted to partially or completely replace the conventional dietary protein with unconventional protein supplements in order to reduce the cost of production. *Pongamia pinnata* known as koroch in Bangladesh could be used as an unconventional feed. Koroch or karanja, a small handsome evergreen shade tree with glabrous bright green foliage, grows wildly almost in all the districts of Bangladesh (Rahman *et al.*, 2011). Koroch kernels (air dried) had 19.0% moisture, 27.5% fatty oil, 17.4% protein, 6.6% starch, 7.3% crude fiber, and 2.4% ash (James, 1983).

Feeding of de-oiled koroch cake after extraction at high level (>5%) adversely affect the performance (Natanam *et al.*, 1989a; Dhara *et al.*, 1997) due to residual toxin (karanjin) left in the solvent extracted cake. Karanjin can be converted to less toxic intermediates by treatment with alkali (Seshadri and Venkateswarlu, 1943). However, studies have not been made for its proper utilization in poultry feeding after detoxification, except physical treatments with water washing and soaking (Mandal and Banerjee, 1974) and autoclaving (Natanum *et al.*, 1989a). Studies also showed that alkali treatment reduces the karanjin content of solvent extracted koroch seed cake (Panda *et al.*, 2008). Prabhu (2002) stated that hydrochloric acid (HCl) and glacial acetic acid treatments have reduced only 1/3rd of karanjin. Therefore, processing of koroch seed and cake by oven-drying, autoclaving, water extraction, alkali or acid extracted cake may have positive effect to formulate poultry diet. Thus, this study investigated the effect of oil extracted koroch seed cake (KSC), NaOH treated KSC and HCl treated KSC on growth and meat yield of broilers.

MATERIALS AND METHODS

The study was conducted at Bangladesh Agricultural University Poultry Farm, Mymensingh, Bangladesh. The experimental day old broilers were purchased from "Nourish Poultry and Hatchery Ltd." Shreepur, Gazipur, Bangladesh. Koroch seeds were collected from Sunamgonj Hoar area. The oil from seeds was expelled locally from an oil extraction mill. Then the remaining cake was dried and proximate analysis was done for KSC. The proximate compositions of the feed ingredients were analyzed as per AOAC (1995) and are furnished in Table 1. The NaOH and HCl treated KSC was prepared by soaking the cake for 24h in water (1:1, W/V) containing either NaOH or HCl at 1% (v/v). The processed cake was sun dried ground and stored in gunny bags for incorporation into the diet.

The experiment was carried out with 160 broilers fed on diet containing either 0% KSC, 2% KSC 2% KSC treated with 1% NaOH or 2% KSC treated with 1% HCl, considered as 4 different dietary treatments. Each treatment was consisting of 4 replications having 10 birds per replication. The chicks were reared up to 28 days of age. The vaccination schedule that was followed during the experimental period is given in Table 4.
The chicks were individually weighed at the beginning of the experiment (initial body weight) and allocated to different replication pens of each of the 4 treatment groups. Body weight was also measured on a weekly basis for all birds from each replication. The average body weight gain of the broiler in each replication was calculated by deducting initial body weight from the final body weight. The amounts of feed consumed by the birds in each replication of each treatment group were calculated from the amount supplied at each week and the amounts leftover at the end of that week. Feed conversion ratio (FCR) was calculated as the unit of feed consumed per unit of body weight gain.

At the end of feeding trial, one broiler weighing average from each pen was randomly selected for recording meat yield parameters. Broilers were fasted from feed and water for twelve hours prior to slaughtering. After complete bleeding, birds were immersed in hot water (51-55°C) for 120 seconds for proper defeathering manually. After defeathering, the birds were again individually weighed. Then head, shank, viscera, giblet (heart, liver and gizzard) and abdominal fat were removed for determination of meat yield parameters. Dressed broilers were cut into different parts such as breast, thigh, drumstick and wing. Finally, every cut up parts were weighed and recorded separately for each broiler of all replications.

Statistical analysis

All recorded and calculated data were analyzed using analysis of variance (ANOVA) technique by a computer using SAS statistical package program in accordance with the principles of Completely Randomized Design (SAS, 2009). Duncan's multiple range test (DMRT) was done to compare variations among diets where ANOVA showed significant differences.

Table 1. Proximate composition of Koroch seed and cake

	Moisture (%)	CP (%)	Ash (%)	CF (%)	EE (%)
Seed	9.04	22.43	2.41	5.57	26.17
Cake	8.89	24.27	3.12	2.37	14.61

Table 2. Ingredients and nutrient composition (%) of starter diet (0-21 days)

Ingredients	Different level of oil extracted Koroch seed cake			
	0% KSC	2% KSC	2% KSC treated with 1% NaOH	2% KSC treated with 1% HCl
Maize	61.00	59.90	59.90	59.90
Rice Polish	2.00	2.00	2.00	2.00
Soybean meal	24.9	24.0	24.0	24.0
Protein Con.	9.00	9.00	9.00	9.00
Koroch cake	0.00	2.00	2.00	2.00
Vegetable oil	0.00	0.00	0.00	0.00
DCP	2.00	2.00	2.00	2.00
Vit.-min. premix	0.25	0.25	0.25	0.25
DL-Meth.	0.20	0.20	0.20	0.20
L-lysine	0.15	0.15	0.15	0.15
Common Salt	0.50	0.50	0.50	0.50
Nutrient composition				
ME Kcal/kg	2971	2952	2952	2952
Crude protein (%)	22.38	22.37	22.37	22.37
Crude fiber (%)	3.82	3.78	3.78	3.78
Calcium (%)	1.28	1.28	1.28	1.28
AV. Phosphorus (%)	0.71	0.71	0.71	0.71
Lysine (%)	1.30	1.27	1.27	1.27
Methionine (%)	0.55	0.54	0.54	0.54

RESULTS AND DISCUSSION

Growth performance

The overall productive performances of broiler chicks as affected by inclusion of different dietary levels of KSC are shown in Table 6. Final body weight, body weight gain, feed consumption and FCR were significantly (P<0.01) influenced by dietary treatments. All performance of broiler fed KSC alone or treated KSC were significantly (P<0.01) depressed compared to the control. However, the feed intake of birds fed either KSC alone or HCl treated KSC dietary treatment was comparable with the control diet. Dietary incorporation of treated (NaOH or HCl) KSC at 2% resulted in poor feed efficiency, whereas FCR of broilers fed on KSC alone at 2% level was comparable with the control diet. All sorts of performance during the experimental period (1st to 4th week) followed the same trend as described above (Table 5).

The depression in body weight and feed intake and poor weight gain in the present study on koroch seed cake incorporated diet was in agreement with the findings of Mandal and Banerjee (1974) and Panda et al. (2008). Panda et al. (2008) observed that enhancing the inclusion of solvent extracted karanj cake significantly decreased the body weight gain and feed intake and resulted in poor feed efficiency. On the other hand, Mandal and Banerjee (1979) recommended that deoiled karanj cake (EE<0.05%) could safely be included at 5% in the diet of broiler chicken. Similarly, Dhara et al. (1997) reported that deoiled karanj cake could be included in the diet to a maximum level of 4.45%. Higher level resulted in adverse effect on growth and feed intake, which was attributed to the presence of toxic factors left back in the cake after oil extraction.

Dietary incorporation of treated (NaOH or HCl) KSC at 2% depressed the feed efficiency in the present study. Dhara et al. (1997) found poor feed efficiency by dietary incorporation of 11.20% to 22.40% deoiled karanj cake in the diet of Japanese quail. But, Natanam et al. (1989b) reported comparable feed efficiency with control diet in broiler chicks by incorporating solvent extracted karanj cake at 10% level in the diet. Efficiency of feed utilization depends on the level of protein and energy (Mellen et al., 1984) and their ratio (Davidson, 1964) in addition to the presence of incriminated factors (Chand, 1987). Probably the presence of higher levels of residual toxic factors in diets containing koroch seed cake at all levels of incorporation resulted in poor efficiency in broiler chicks.

Table 3. Ingredients and nutrient composition (%) of grower diet (21-28 days)

Ingredients	Different level of oil extracted Koroch seed cake			
	0% KSC	2% KSC	2% KSC treated with 1% NaOH	2% KSC treated with 1% HCl
Maize	65.05	63.44	63.44	63.44
Rice Polish	2.00	2.00	2.00	2.00
Soybean meal	20.69	19.91	19.91	19.91
Protein Con.	7.50	7.50	7.50	7.50
Koroch cake	0.00	2.00	2.00	2.00
Vegetable oil	1.66	2.05	2.05	2.05
DCP	2.00	2.00	2.00	2.00
Vit.-min. premix	0.25	0.25	0.25	0.25
DL-Meth.	0.20	0.20	0.20	0.20
L-lysine	0.15	0.15	0.15	0.15
Common Salt	0.50	0.50	0.50	0.50
Nutrient composition				
ME Kcal/kg	3100	3100	3100	3100
Crude protein (%)	20.0	20.0	20.0	20.0
Crude fiber (%)	3.61	3.56	3.51	3.47
Calcium (%)	1.20	1.19	1.19	1.19
AV. Phosphorus (%)	0.68	0.67	0.67	0.66
Lysine (%)	1.14	1.12	1.09	1.06
Methionine (%)	0.52	0.51	0.50	0.49

Table 4. Vaccination schedule for the experimental broilers

Age of broilers (day)	Name of Vaccine	Trade Name*	Dose	Route of vaccination
4	IB+ND	MA5+Clone30	One drop	Ocular
10	IBD	D-78	One drop	Ocular
21	IBD	D-78	One drop	Ocular

IB, Infectious Bronchitis; ND, Newcastle Disease; IBD, Infectious Bursal Disease *Intervet International, B.V. BOXMEER, The Netherlands

Meat yield characteristics

The data on carcass traits like dressing percentage, percent live weight of neck, head, gizzard, heart, liver, shank, breast, drumstick, thigh, wing, skin and abdominal fat have been shown in Table 7. All the parameters determined showed a non-significant treatment effect except for the percentage of dressing yield, liver, gizzard, head and skin weight (Table 7). Percent live weight of dressing yield, liver, head and skin were significantly (P<0.05) influenced by dietary treatments. Dressing percentage of birds fed on HCl treated KSC at 2% level was significantly (P<0.05) depressed compared to any other dietary treatment, whereas birds fed on 2% KSC dietary group was similar to the control group. Although, broilers in NaOH treated 2% KSC dietary group was comparable with those in control group. Percentage of head and skin weight relative to body weight were higher in treated (NaOH or HCl) KSC dietary group compared to either KSC alone or control group. However, liver weight (P<0.05) and gizzard weight (P<0.05) increased significantly due to incorporation of KSC alone, NaOH treated KSC and HCl treated KSC diet at 2% levels as compared to that of control diet.

Table 5. The weekly growth performance of broilers fed on oil extracted Koroch seed cake

| Trait | Age (week) | Different level of oil extracted Koroch seed cake | | | | Level of Significance |
		0% KSC	2% KSC	2% KSC treated with 1% NaOH	2% KSC treated with 1% HCl	
Body weight (g)	Day-old	44.23±0.48	44.15±0.10	44.00±0.16	44.20±0.16	NS
	1st	132.80±2.45[a]	131.05±3.42[ab]	126.00±8.66[ab]	123.45±3.78[b]	NS
	2nd	311.50±6.27[a]	281.65±14.32[b]	270.33±18.02[b]	265.90±14.47[b]	**
	3rd	635.28±7.01[a]	565.70±22.14[b]	537.20±19.54[b]	559.70±57.40[b]	**
	4th	1096.6±29.51[a]	1026.0±32.12[ab]	874.80±50.80[b]	978.0±106.97[b]	**
Body weight gain (g)	1st	88.58±2.08[a]	86.90±3.45[ab]	82.00±8.79[ab]	79.25±3.73[b]	NS
	2nd	178.70±7.87[a]	150.60±11.47[b]	144.33±12.97[b]	142.25±11.05[b]	**
	3rd	323.78±0.75[a]	284.05±30.13[ab]	266.88±13.20[b]	293.8±48.31[ab]	NS
	4th	461.33±23.92[a]	388.48±14.85[ab]	337.60±33.12[b]	354.1±29.62[ab]	**
Feed intake (g)	1st	95.73±1.95	96.88±4.11	102.48±5.14	101.80±5.46	NS
	2nd	205.15±1.00[ab]	204.80±1.83[b]	207.55±2.24[ab]	207.83±1.53[a]	NS
	3rd	538.75±25.94	525.00±24.49	530.00±8.16	542.50±11.90	NS
	4th	848.63±50.79[a]	776.05±48.24[b]	753.60±25.57[b]	779.60±33.8[b]	**
Feed con -version ratio	1st	1.08±0.04[a]	1.12±0.09[ab]	1.26±0.16[b]	1.29±0.10[b]	*
	2nd	1.15±0.05[a]	1.37±0.11[b]	1.45±0.12[b]	1.47±0.12[b]	**
	3rd	1.66±0.08[a]	1.86±0.15[ab]	1.99±0.08[ab]	1.88±0.25[b]	NS
	4th	1.84±0.12[a]	2.00±0.12[ab]	2.25±0.25[b]	2.22±0.24[b]	*

[a-d] Mean±SD values with different superscripts within same row differ significantly; NS=Non-significant; **=significant (p<0.01)

Table 6. Performances of broilers fed on oil extracted Koroch (Pongamia Pinnata) seed cake during the experimental period

| Parameter | Different level of oil extracted Koroch seed cake | | | | Level of Significance |
	0% KSC	2% KSC	2% KSC treated with 1% NaOH	2% KSC treated with 1% HCl	
Initial weight (g)	44.23±0.48	44.15±0.10	44.0±0.16	44.20±0.16	NS
Final weight (g)	1096.6±29.51[a]	1026.0±32.12[b]	874.8±50.80[b]	978.0±106.97[b]	**
Body weight gain (g)	1052.4±29.85[a]	910.03±29.35[b]	830.8±50.78[b]	869.6±70.7[b]	**
Feed consumption (g)	1688.3±67.94[a]	1602.7±68.78[ab]	1593.6±24.12[b]	1631.7±37.94[ab]	**
Feed conversion ratio	1.61±0.08[a]	1.76±0.08[ab]	1.92±0.11[b]	1.89±0.16[b]	**

[a-d] Mean±SD values with different superscripts within same row differ significantly;
NS=Non-significant; **=significant (p<0.01)

Table 7. Meat yield characteristics of broilers fed on oil extracted Koroch *(Pongamia Pinnata)* seed cake up to 28 days of age

Parameter	Different level of oil extracted Koroch seed cake				Level of Significance
	0% KSC	2% KSC	2% KSC treated with 1% NaOH	2% KSC treated with 1% HCl	
Live weight (g)	1125.5[a]	1026.0[ab]	945.5[b]	978.0[b]	*
Dressing yield (%)	62.80[a]	64.91[a]	62.05[ab]	59.59[b]	*
Blood (%)	4.09	4.60	6.11	4.89	NS
Feather (%)	4.66	3.75	3.52	2.88	NS
Neck (%)	3.38	4.15	3.80	3.65	NS
Head (%)	2.21[b]	2.49[ab]	2.68[a]	2.76[a]	*
Gizzard (%)	1.46[b]	2.24[a]	2.23[a]	2.25[a]	**
Heart (%)	0.49	0.63	0.54	0.46	NS
Liver (%)	1.99[b]	3.12[a]	2.78[a]	3.03[a]	*
Shank (%)	3.55	3.79	3.78	3.79	NS
Breast (%)	11.03	10.46	9.88	10.58	NS
Drumstick (%)	7.38	8.30	7.71	7.90	NS
Thigh (%)	7.89	8.85	8.43	8.67	NS
Wing (%)	6.41	6.33	6.56	6.72	NS
Skin (%)	1.77[b]	1.71[a]	2.35[a]	2.21[a]	*
Abdominal fat (%)	0.36	0.34	0.27	0.46	NS

[a,b] Mean values with different superscripts within same row differ significantly; NS=Non-significant; **=significant (p<0.01)

Mandal and Banerjee (1982) reported no difference on organ weights like liver, heart, kidney and spleen of cockerels due to dietary replacement of black til cake with deoiled karanj cake at 30% level. Dhara *et al.* (1997) also found no significant variation in weight of different commercial cuts (neck, wing, thigh, shank, breast and trunk) and organs (giblet, liver, heart and gizzard) due to incorporation of deoiled karanj cake up to 22.40% in the diet of Japanese quail. However, in the present study, dietary inclusion of KSC alone, NaOH treated KSC and HCl treated KSC diet at 2% levels significantly reduced the weight of dressing yield. Panda *et al.* (2008) observed the dietary inclusion of either solvent extracted karanj cake (SKC) or NaOH treated SKC at 25% level and $Ca(OH)_2$ treated SKC at 12.5% and 25% levels significantly increased the weight of liver leading to liver hypertrophy. Gizzard weight also increased significantly due to incorporation of SKC at both the levels. These findings are similar to the findings of the present study where liver and gizzard weight increased significantly due to incorporation of KSC alone, NaOH treated KSC and HCl treated KSC diet at 2% levels. Higher percentage of head and skin weight in treated (NaOH or HCl) KSC dietary group compared to KSC alone and control group was observed in the present study. Body surface area might be the causes of weight variation of head and skin in different dietary treatment groups. Panda et al. (2008) noticed that the adverse effect on growth and feed intake was found when daily karanjin intake was18mg and above. Similarly Natanam (1989a) reported the adverse effect of karanj cake feeding is due to the left over karanjin in the processed cake. Probably the higher daily karanjin intake (>18 mg) could be the potential reason for showing adverse effect on growth and meat yield characteristics in the present study.

CONCLUSION

The results revealed that the treatment of koroch seed cake with either NaOH or HCl did not improve performance of broilers. The oil content of koroch seed cake may leftover even after NaOH or HCl treatment which might have an antagonistic effect on metabolism of broilers. However, further research to investigate the efficacy of other alternative processing techniqute for the complete removal of leftover oil content in association to the improvement of broiler growth performance and meat yield characteristics is advocated.

ACKNOWLEDGEMENT

Cordial thanks are due to the Bangladesh Agricultural University Research System (BAURES) for its financial support to the successful completion of this research work.

REFERENCES

1. AOAC, 1995. Association of Official Analytical Chemistry, Official Methods of Analysis. (13th edn.) Washington, DC.
2. Chand S, 1987. Nutritional evaluation of neem seed meal in chicks. Ph.D. Thesis, submitted to the Rohilkhand University, Bareilly.
3. Davidson J, 1964. The effciency of conversion of dietary metabolizable energy into tissue energy in chicken as measured by body analyses. European Association of Animal Production; 3rd symposium on Energy Metabolism, Troon.
4. Dhara TK, N Chakraborty, G Samanta and L Mandal, 1997. Deoiled karanj (*P. glabra* vent) cake in the ration of Japanese quail. Indian Journal of Poultrt Science, 32: 132-136.
5. James AD, 1983. Handbook of Energy Crops. Unpublished. http://www.hortpurdue.edu/ duke_energy/p ongamiapinnata.htm.
6. Mandal L and GC Banerjee, 1974. Studies on the utilization of karanj (*P. glabra*) oil cake in poultry rations. Indian J. Poul. Sci. September: 141-147.
7. Mandal L and GC Banerjee, 1979. Studies on the utilization of karanja (*Pongamia glabra* vent.) cake in layer diet. Indian Journal of Poultry Science, 14:105-109,
8. Mandal L and GC Banerjee, 1982. Studies on the utilization of karanj (*Pongamia glabra)* cake in poultry rations Eff ect on growers on blood composition and organ weight of cockerels. Indian Veterinary Journal, 59: 385-390.
9. Mellen WJ, FW Hill and HH Dukes, 1984. Studies on the energy requirement of chickens, 2. E ffect and dietary energy level on basal metabolism of growing chickens. Poultry Science, 31: 735-740.
10. Natanam R, R Kadirvel and R Balagopa, 1989a. The effect of kernels of Karan (*P. glabra* vent) on growth and feed efficiency in broiler chicks to 4 weeks of age. Animal Feed Science Technology, 25: 201-206.
11. Natanam R, R Kadirvel and R Ravi, 1989b. The toxic effects of karanj (*P.glabra* bent) oil and cake on growth and feed efficiency in broiler chicks. Animal Feed Science Technology, 27: 95-100.
12. Panda AK, VRB Sastry and AB Mandal, 2008. Growth, nutrition utilization and carcass characteristics in broiler chickens fed raw and alkali processeed solvent extracted karanj (*Pongamia glabra*) cake as partial protein supplement. The Journal of Poultry Science, 45:199-205.
13. Panda B and SC Mahapatra .1989: Poultry Production, ICAR, New Delhi, PP190.
14. Prabhu TM, 2002. Clinico nutritional studies in lambs fed raw and detoxified karanj (*P. glabra* vent) meal as protein supplement. PhD thesis submitted to Indian Veterinary Research Institute, Izatnagar.
15. Rahman MS, MB Islam, MA Rouf, MA Jalil, and MZ Haque, 2011. Extraction of Alkaloids and Oil from Karanja (*Pongamia pinnata*) Seed. Journal of Scientific Research, 3: 669-675.
16. SAS, 2009. User's Guide: Statistics. Version 9.2, SAS Institute Inc., Cary, NC. USA.
17. Seshadri TR and V Venkateshwarlu, 1943. Synthetic experiments in the Benzo-Pyrone series. Proceedings of the Indian Academy of Sciences, 17(A): 16-19.

CLIMATE CHANGE EFFECTS AND ADAPTATION MEASURES FOR CROP PRODUCTION IN SOUTH WEST COAST OF BANGLADESH

Rajib Jodder[1], Mohammad Asadul Haque[1*], Tapan Kumar[1], M Jahiruddin[2], M. Zulfikar Rahman[3] and Derek Clarke[4]

[1]Department of Soil Science, Patuakhali Science and Technology University, Patuakhali-8602, Bangladesh; [2]Department of Soil Science and [3]Department of Agricultural Extension Education, Bangladesh Agricultural University, Mymensingh-2202, Bangladesh; [4]Department of Civil and Environmental Engineering, University of Southampton, UK

*Corresponding author: Mohammad Asadul Haque; E-mail: masadulh@yahoo.com

---

ARTICLE INFO   **ABSTRACT**

Key words

Adaptive measure,
Climate change,
Water scarcity,
Salinity

A survey was conducted to determine the effect of climate change on crop production and water quality in 12 villages of Deluty and Garaikhali unions under Paikgacha upazila of Khulna district, Bangladesh. Total of 100 farmers were interviewed using a pre-tested questionnaire. The climatic hazards as reported on the study area are salinity, cyclone, drought, hailstorm, river erosion and waterlogging, of them salinity is the most dominant hazard. Due to salinity the cropping system has undergone changes. Many crops and varieties have been either extinct or their cultivation has come down. Both soil and water are severely affected by salinity. Most of the farmers (90%) use pond water for irrigation and the majority farmers use pond and rain waters for drinking purpose. Results of the present study serves as a good basis for in-depth study to achieve successful crop production in the south west coastal area of Bangladesh.

INTRODUCTION

Bangladesh is often cited as one of the most vulnerable countries to climate change in the world (MoA and FAO, 2013). The vulnerability to climate change is high due to a number of hydro-geological and socio-economic factors (Ahmed, 2004, 2006). The country often experiences natural disasters as an effect of climate change, particularly in coastal areas. Crop agriculture is often constrained by different hazards and disasters such as floods, droughts, soil and water salinity, cyclones and storm surges (MoEF, 2009).Salinity is a great constraint to growing crops, especially in rabi season (dry months) when water and soil salinity arises and reaches to the peak in March-April before monsoon starts (Haque et al., 2008, 2014).

The main crop grown in the saline areas is local transplanted *aman* rice which has low yield potential. Haque (2006) reported that most of the coastal areas are located over medium highlands, where flooding depth ranges from 0.3-0.9 meter. This category of land is suitable for the minimum two crops and three crops could be possible if some suitable interventions are done. Kim et al., (2016) described that the decrease in crop yields with the increase in the salinity of irrigation water was caused by disturbances in physiological and biochemical activities under saline conditions.

Salinity causes unfavorable environment and hydrological situation that restrict the normal crop growth. The factors which contribute significantly to the development of saline soil are tidal flooding during wet season (June-October), direct inundation by saline water, and upward or lateral movement of saline ground water during dry season (November-May) (Karim et al.,1990).The prevailing salinity intrusion due to climate change has severely affecting the crop productivity in the saline regions of Bangladesh (Haque, et al., 2015). Although people of the south-west region of the country are mostly dependent on crop farming, most of the farmers do not know how to address soil salinity by modern techniques for better crop production. The present study was done to find out the effects of climate change with special focus in salinity on crop production and to identify the adaptation measures used by the farmers to cope with salinity.

MATERIALS AND METHODS

Selection of the study area

Recent rapid industrialization in the developed countries and deforestation increased global warming. Due to global warming sea level is raised and ultimately low laying area of the world is inundated by saline sea water. Due to deltaic geography of Bangladesh, the country is most vulnerable to climate change. Among the thirteen coastal districts, Khulna is most affected by climate change. Under this district Paikgacha upazila is one of the worst affected upazilas by natural disasters such as salinization, cyclone, drought, hailstorm, river erosion and waterlogging. This situation is a big threat to successful crop production. To achieve the objectives a survey work was done at 12 villages of Paikgacha upazila under Khulna district. The villages were Fulbari, Bigordana, Gobipagla, Horinkhola, Hatbari, Senerber, Kalinagar, Telikhali, Noldanga, Darunmallik, Kumkhali and Bainbariya covering two unions - Deluty and Garaikhali.

Population and sampling

Total 100 farmers across the villages were randomly selected. In order to estimate the existing hazards with their severity and adverse effects on crop production, the respondents were asked some common questions.

Data collection and analysis

Before data collection, an interview schedule was prepared keeping in view the objectives of the study. Then it was pre-tested among the population who were not included in the sample. After necessary corrections and modifications, a final questionnaire was prepared and multiplied. Data were collected through face to face interview with the respondents. The data generated from this experiment were entered in Microsoft Excel Worksheet, checked, organized and processed for further analysis. Frequency and percentage for different variables were estimated with help of SPSS 17 computer software.

RESULTS AND DISCUSSION

Socio-economic status of the farmers

The socio-economic status in the present study includes age, education and land ownership of the farmers. It is summarized and presented in Table1. It was found that 14% of the respondents belonged to young aged group (15-30 years), followed by 33% middle aged group (31-45 years), 36% 46-60 years aged group,15% 61-75 years aged group and only 2% old aged group (76-90 Year). About 47% of the farmers received secondary education followed by 30% primary education, 8% higher secondary education, 6% illiterate, 4% can sign only, 3% above higher secondary education and 2% can read and sign only. In case of household size,46% households were small (1-4), 49% medium (5-8), 5% large and 1% was very large.

Table 1 also shows the land holding status of the farmers. Most of them (74%) were marginal farmers(<1 ha), 21% small farmers (1-2 ha) and 5% semi-medium farmers (2-4 ha). None of the farmers belonged to medium and large categories. Table 1 also reveals that 46% farmers earned 5-8 thousand taka per month followed by 22% less than 5 thousand taka, 14% 8-11 thousand taka, 12% more than 14 thousand taka and 6% earned 11-14 taka per month.

Table 1. Socio-economic status of the farmers

Age category (yr)	Frequency	No. of household members	Frequency	Farmer category (land ownership)	Frequency
15-30	14%	1-4	46%	Marginal (<1 ha)	74%
31-45	33%	5-8	49%	Small (1-2 ha)	21%
46-60	36%	8-12	4%	Semi-medium (2-4 ha)	5%
61-75	15%	above	1%	Medium (4-10 ha)	0%
76-90	2%			Large (>10 ha)	0%
Level of education	Frequency	Monthly income (Tk × 10³)	Frequency		
Below primary	30%	<5	22%		
Below SSC	47%	5-8	46%		
Below HSC	8%	8-11	14%		
HSC and above	3%	11-14	6%		
Illiterate	6%	>14	12%		
Can sign only	4%				
Can read and sign	2%				

Temporal variability of crop cultivation

Cultivation of crops markedly varied with advancement of years. Variation was recorded from past 15 years to the present year (Table 2). The most common crop in the area is T. *aman* rice followed by bitter gourd. *Aus*rice cultivation ceases and *Boro* cultivation is minimum. In case of *Aus* rice, watermelon, sweet potato, jute, dhaincha, melon and cucumber cultivation, a decreasing trend was found. Ten percent of the farmers were engaged in *Aus* rice cultivation in 10-15 years ago but now *Aus* rice is no more cultivated. In this salt affected area transplanted *Aman* rice (July-November) is the dominant crop and farmers mainly cultivate it. But a decreasing trend also found in case of *Aman* rice due to salinity. Rahman *et al.*(2015) reported that risk of cyclone occurrence is high in April - May (pre-monsoon) and October - November (post-monsoon). Local farmers cultivate pulse, T-*aus*, T-*aman* and some minor vegetables during pre and post monsoon and these crops are highly vulnerable to cyclone and storm surges.

Presently the farmers have started to use drought and salinity tolerant crops and varieties as adaptive measures. These crops include bitter gourd, sweat gourd, okra, sesame, ridge gourd and bottle gourd. Cultivation of bitter gourd has remarkably increased.

Table 2. Temporal variability in cultivation of major crops in Paikgachaupazila, Khulna

0-5 years	5-10 years	10-15 years
Aus rice- Nil	Aus rice-2%	Aus rice-10%
Aman rice-90%	Aman rice-100%	Aman rice-100%
Boro rice-4%	Boro rice-8%	Boro rice-1%
Bitter gourd-43%	Bitter gourd-21%	Bitter gourd-2%
Sweat gourd-20%	Sweat gourd-10%	Sweet gourd-17%
Sweet potato-10%	Sweet potato-5%	Sweet potato-15%
Sesame-35%	Sesame -3%	Sesame -15%
Pulse- 28%	Pulse-1%	Pulse-Nil
Water melon-15%	Water melon-6%	Water melon-32%
Okra-14%	Okra-7%	Okra-13%
Ridge gourd-12%	Ridge gourd-2%	Ridge gourd-1%
Bottle gourd-9%	Bottle gourd-3%	Bottle gourd-7%
Amaranth-7%	Amaranth-5%	Amaranth-14%
Potato-13%	Potato-14%	Potato-Nil
Brinjal-8%	Brinjal-8%	Brinjal-Nil
Yard long bean-5%	Yard long bean-5%	Yard long bean-Nil
Banana-2%	Banana-12%	Banana-19%
Jute-2%	Jute-9%	Jute-23%
Dhaincha-Nil	Dhaincha-8%	Dhaincha-11%
Tomato-11%	Tomato-25%	Tomato-Nil
Melon-1%	Melon-1%	Melon-6%
Chili-3%	Chili-8%	Chili-3
Groundnut-2%	Groundnut-Nil	Groundnut-Nil
Wax gourd-1%	Wax gourd-1%	Wax gourd-Nil
Radish-18%	Radish-Nil	Radish-Nil
Snake gourd-2%	Snake gourd-Nil	Snake gourd-Nil
Taro-1%	Taro-Nil	Taro-Nil
Cucumber-Nil	Cucumber-Nil	Cucumber- 5%
Winter vegetables-8%	Winter vegetables-9%	Winter vegetables-11%

Changes in cropping pattern

Due to ingression of salinity, cyclone, drought, continuous rainfall and other climatic factors, cropping pattern in the study area undergoes changes and *rabi* crops are mostly affected. In the recent years, occurrence of late monsoon rain has affected *aus* rice, jute and dhaincha cultivation. In *kharif* II season cultivation of T. *aman* rice is also hampered by climatic hazards which are clear from the information that in last year 90% of the farmers were engaged in T. *aman* rice cultivation whereas in last 0-5 and 5-10 years it was 100%. Rahman *et al.* (2015) reported that that cultivation of *rabi* crops viz. sunflower, chili and wheat decreases. However, development of saline and drought tolerant varieties, coupled with modern soil & crop management technologies have positively impacted on crop production via increasing cropping area as well as cropping intensity. The changes in cropping pattern are shown in the Table 3.

Table 3. Temporal variability in cropping patterns in the study area

Time	Cropping season		
	Rabi	Kharif I	Kharif II
0-5 years	Bitter gourd-28%, Sweet gourd- 11%, Sweet potato- 4%, Melon-1%, Ground nut-1%, Water melon-8%, Teel-33%, Pulse-24%,*Boro*rice-2%, Okra-5%, Bottle gourd-5%, Radish-18%, Potato-25%, Tomato-17%, Brinjal-21%, Winter vegetables-18%	Bottle gourd-3% Bitter gourd-17% Ridge gourd-6%, Jute-3% Yard long bean-2% Snake gourd-2%	T. *Aman* rice- 90%
5-10 years	*Boro* rice-4%, Bitter gourd-35%, Bottle gourd-4%, Sweat gourd-20% Groundnut-2%, Sweet potato-10%, Sesame-35%, Pulse-28%, Water melon-15%, Okra-14%, Amaranth-7%, Radish-18%, Potato-13%, Brinjal-8%, Banana-2%, Melon-1%, Tomato-11%, Wax gourd-1%, Chili-3%, Winter vegetables-8%	Bitter gourd-13% Ridge gourd-12% Bottle gourd-9% Yard long bean-5% Snake gourd-2% Jute-2%	T. *Aman* rice-100%
10-15 years	*Boro* rice-8%, Bitter gourd-13% Sweat gourd-10%, Sweet potato-5%, Teel-3%, Pulse-1%, Water melon-6%, Okra-7%, Amaranth-5%, Potato-14%, Brinjal-8%, Banana-1%, Melon-1%, Tomato-25%, Chili-8% Winter vegetables-9%	*Aus* rice-2% Ridge gourd-2% Bottle gourd-3% Bitter gourd- 8% Yard long bean-5% Jute-9% Dhaincha-8% Wax gourd-1%	T. *Aman* rice-100%

Major climatic hazards

While investigating farmer's perception, different farmer emphasize on different problems that constrain crop production, it is shown in Table 4. It is clear that salinity, cyclone and monsoon storm is the most impacted hazards in crop agriculture. Next to them, drought, continuous rain,hailstorm, river erosion and waterlogging are the common limitations for crop production in the area. As per World Bank report (2001), 14, 32, and 88 cm sea level rise will occur in 2030, 2050 and 2100, respectively which may inundate about 8, 10 and 16% of total land of Bangladesh. Sea level is rising by about 3 mm/year. Cyclone, floods and tidal surges are common disasters in the coastal regions. Table 4 also indicates that the frequency of salinity, drought and continuous rain are increasing with time. About 95% respondents opine that in the previous year their crops were heavily affected by salinity.

Table 4. Perception of respondent (%) on temporal variability of climatic hazards in the study area

Names of major hazards	Year-1	1-5 years	5-10 years	10-15 years
Salinity	95	87	82	85
Cyclone	28	38	48	48
Drought	10	8	7	6
Monsoon storm	18	10	9	10
Continuous rain	8	5	3	3
Hailstorm	4	2	2	5
River erosion	8	12	9	1
Waterlog	3	4	2	3

Adaptation measures to cope with the salinity

Table 5 demonstrates the existing adaptation options against the climate stress in the study area. It appears that salinity, cyclone, monsoon storm, excessive rain, waterlog, river erosion, droughts and hailstorms are the major climate stresses across the regions. Different types of adaption options are found from the study areas against the climate stresses to minimize the loss and damage. Adaptation practices varied depending on the technical and financial capacity of the farmers. Most of the farmers (65%) prefer homestead cultivation, appropriate fertilization (56%) and mulching (28%) as adaptive measures to suppress salinity.

Table 5. Adaptive measures used by the farmers to cope with salinity

Adaptive measures	% of respondent
Cultivation of salt tolerant rice variety	14
Cultivation of salt tolerant crop variety	3
Appropriate fertilization of crops	56
Mulching	28
Homestead cultivation	65
Frequent irrigation	4
Leveling of land	1
Ridge cropping	15

Causes of salinity

There are two main causes of salinity, one is natural causes and another is human induced causes. In case of natural causes, the majority farmers (70%) emphasized on tidal flooding. They also mentioned sea level rise (37%), increasing temperature (23%), increase of saline intrusion (32%), reduced dry season flow in the Shoilmari River (3%) and cyclone (11%) as a natural causes of increasing salinity in their locality. On the other hand, according to farmer's perception extensive shrimp cultivation, construction of Farakka barrage, faulty management of coastal polders and faulty management of sluice gate are human induced causes of increasing salinity (Table 6).

Table 6. Percent respondent perception about causes of increasing salinity in the study area

Natural causes	% of respondent	Human induced causes	% of respondent
Sea level rise	37	Extensive shrimp cultivation	31
Increasing temperature	23	Construction of Farakka barrage	3
Increase of saline intrusion	32	Faulty management of coastal polders	30
Tidal flooding	70	Faulty management of sluice gate	6
Reduced dry season flow in the Shoilmari river	3		
Cyclone	11		
River erosion	8		

Salinity effects on farming enterprises

Crop agriculture is highly affected by salinity and farmers are the direct victims. Farming enterprises affected by salinity is summarized and presented in Figure 1. It was found that 95% of the farmer thought that their crops are affected by salinity followed by homestead by 32%, fisheries by 18% and livestock by 3%.

Sources of sweet water for irrigation

Data analysis shows that in the current study area there are very limited sources of sweet water for irrigation. About 70% farmers view that there is unavailability of sweet water for irrigation, they have accessibility to use only four types of sweet water sources for irrigation namely pond water, tube well, rain water and canal water.

Farmers widely use pond water as irrigation water and the rest of the sources have very limited use. About 90% of the farmers use pond water while tube well by 1%, rain water by 1% and canal water by 2%.

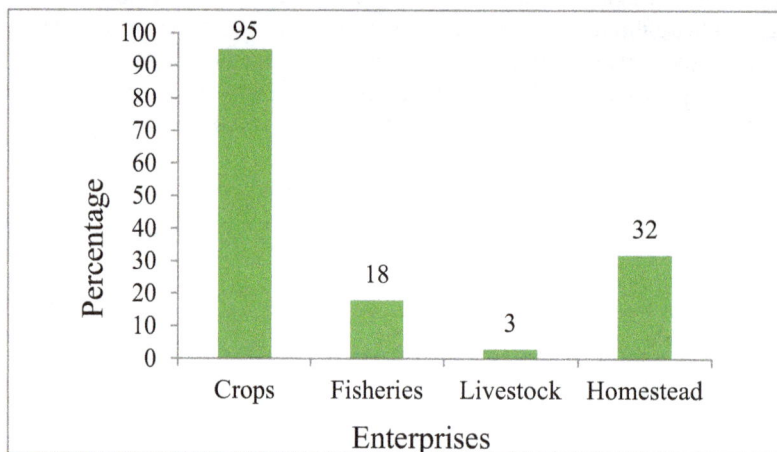

Figure 1. Enterprises affected by salinity

Distance of fresh water source for irrigation

Figure 2 portrays the distance of fresh water source from farmer's crop field which they use for irrigation purpose. Distance of fresh water source is classified into five categories: 0-100 m, 101-200 m, 201-500 m, 501-800 m and 800 m- above. About 24% of the farmers reported that their crop field is within 100 m away from the nearest irrigation water source followed by 201-500 m by 23%, above 800m by 20%, 101-200 m by 15% and 501-800 m by 14%.

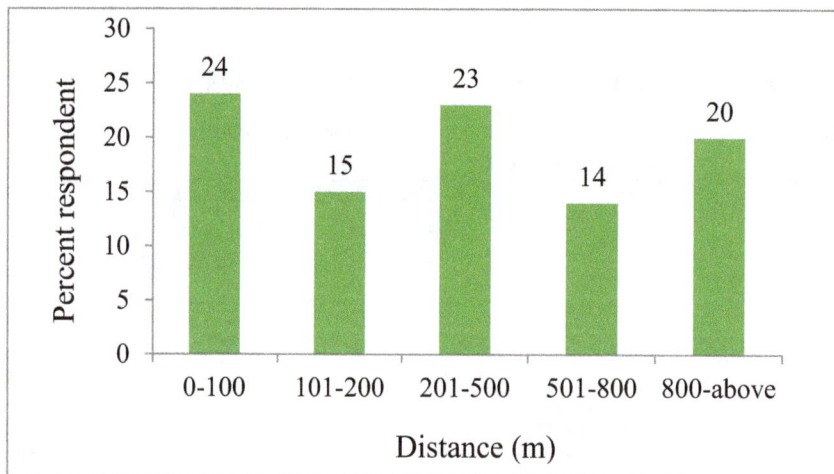

Figure 2. Distance of fresh water source for irrigation from farmer's crop field

Adaptation measures in collecting drinking water

Table 7 is designed to identify the adaptation measures followed in collecting drinking water to cope with salinity both at individual and community levels. Different farmers follow different methods but in few cases they have high similarities. They collect water from far distance, harvest rainwater, dug well, conserve pond water, install tube-well and purchase water as methods of individual effort. Among these methods most of the farmers follow collecting water from far distance and harvesting rainwater. On the other hand in case of community basis effort farmers emphasize on use of pond and filter water (34%) followed by conservation of pond water (29%), digging of pond (11%) and installation of deep tube-well (4%).

Table 7. Adaptation measures followed by farmers to cope with salinity

Individual level	Percentage	Community level	Percentage
Collect water from far distance	64	Conservation of Pond water	29
Rainwater harvest	57	Installation of deep tube-well	4
Dug well	2	Use of pond sand filter	34
Conservation of pond water	26	Digging of pond	11
Installation of tube-well	18		
Purchase water	2		

Organizations involved in salinity mitigation

Figure 3 shows that government, some international and some local organizations have programmes to mitigate salinity in their locality, and however they did not mention any private sector or any other organizations. About 37% of the farmers tell that government organization do this job where 13% by international organization and 16% by local organization.

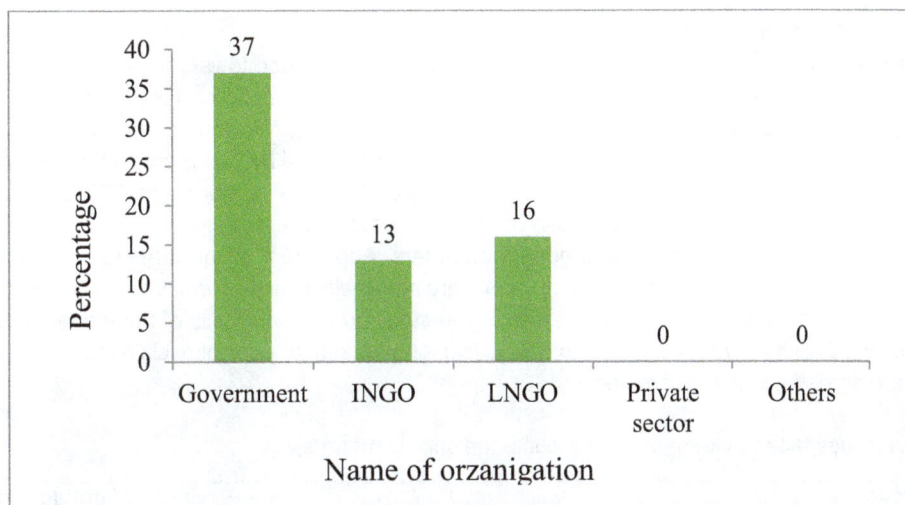

Figure 3. Organizations involved in mitigation of salinity

In study there were seven indicators of adaptive measure to assess the degree of measures that taken by different organization (Table 8) through asking the farmers. It revealed that almost all of the adaptive measures were taken by different organizations. In case of government organization farmers emphasized on Department of Agricultural Extension (DAE), Water Development Board (WDB), Local Government Engineering Department (LGED) and Soil Resource Development Institute (SRDI). In case of international organization they emphasized on Blue Gold and in case of local non-government organization they suggest two organizations- Sofol and Soliderist.

Table 8. Adaptation measures taken by different organizations

Name of the program	Organizations involved in the program
Performing and monitoring soil test- 10%	DAE- 9%, Blue gold- 3%
Establishment of embankment of suitable size and length- 31%	WDB- 31%
Provision of sluice gate on the embankment- 41%	WDB- 33%, LGED- 3%, Soliderist- 7%
Prescription of the amount of fertilizer- 8%	DAE- 8%
Installation of hand driven deep tube-well- 3%	Union porishad-3%
Extension of information about salt tolerant crop varieties- 9%	Blue gold- 7%, DAE- 5%
Arranging training program on salt tolerant crop cultivation- 17%	Blue gold- 13%, DAE- 7%, Sofol- 7%

Preference of cultivating crop varieties against salinity

The most favored crop varieties by farmers in the study area are displayed in Table 9. It shows that in case of rice variety farmers mostly choose modern varieties and among the modern varieties farmers preferred mostly BR10 (51%) and BR23 (50%). Farmers also cultivate some local rice varieties in small scale and most of the farmers (20%) preferred Jotabalam. In case of bitter gourd farmers preferred Meghna-2 variety.

Table 9. Name of the crop varieties used by the farmers to reduce risk of salinity

Name of the crop	Name of the variety				
	BRRI dhan30- 10%	BR23- 50%	BRRI dhan41- 2%	BR11- 6%	BR10- 51%
	Kachra- 12%	Ghunshi- 9%	Hargoza- 6%	Koijur- 1%	Benapol-6%
Rice	Jotabalam- 20%	Hogla- 2%	Kumrogoir-2%	Chappshail-3%	
Bitter gourd	Meghna2- 65%				

Challenges faced to overcome the salinity problem

There are five specific types of problems and based on the attitude toward the problems farmers have been classified into three categories (Table 10). For each of the statements farmers were asked to indicate whether the problem is less or medium or high to them. In case of no governmental as well as organizational support about 40% of the farmer's attitude towards high problem followed by less problem by 30% and medium problem by 28%. However majority of the farmers (39%) attitudes towards less problem, 28% attitudes medium problem and 33% attitudes high problem in case of very poor economic condition. No saline free water source near their crop field is the major problem among the farmers. About 55% of the farmers are agree with this statement, 36% are agree with medium problem and 9% are agree with less problem. Very poor linkage between community and institution is not the major problem in the study area. About 38% of the farmers are agreeing with less problem, 27% by medium problem and 35% by high problem. In case of social/political unrest most of them (44%) are agreeing with high problem.

Table 10. Challenges faced by farmers to overcome the salinity matters

Problems faced	Extent of problem (%)		
	Less	Medium	High
No governmental as well as organizational support	30	28	42
Very poor economic condition	39	28	33
No saline free water source near their crop field	9	36	55
Very poor linkage between community and institution	38	27	35
Social/political unrest	18	38	44

CONCLUSIONS

Agriculture of the coastal area is highly sensitive to climate change. Salinity intrusion was the most significant hazard causing a huge yield reduction. Next to salinity, frequent cyclone has significant influence on crop yield reduction. Due to climate change effect cultivation of some crops such as jute, dhaincha, *aus* rice, sweat potato, water melon is going to be extinct. For risk reduction in crop production farmers avoided aus and boro rice cultivation. They prefer aman season to grow rice. BR10 and BR23 was very popular T aman variety in the study area. In dry season bitter gourd cultivation is gaining popularity. Homestead gardening and appropriate fertilizer management was the promising adaptation technology to cope with salinity.

ACKNOWLEDGEMENT

The authors gratefully acknowledge the British Council funded INSPIRE R4 project "Climate Change Adaptations" for financial support to conduct this study.

REFERENCES

1. Ahmed AU 2004. Adaptation to climate change in Bangladesh: learning by doing. UNFCCC Workshop on Adaptation, Bonn, 18 June.
2. Ahmed AU 2006. Bangladesh: Climate Change Impacts and Vulnerability - a Synthesis. (Dhaka: GoB, MoEF, Department of Environment, Climate Change Cell, July).
3. Haque MA, Jahiruddin M, Hoque MA, Rahman MZ, Clarke D 2014. Temporal variability of soil and water salinity and its effect on crop at Kalapara upazila. Journal of Environnemental Science & Natural Resources, 7: 111-114.
4. Haque MA, Jharna DE, Hoque MF, Uddin MN, Saleque MA 2008. Soil solution electrical conductivity and basic cations composition in the rhizosphere of lowland rice in coastal soils. Bangladesh Journal of Agricultural Research, 33: 243-250.
5. Haque SA 2006. Salinity problems and crop production in coastal regions of Bangladesh. Pakistan Journal of Botany, 38: 1359-1365
6. Hossain ML, Hossain MK, Salam MA and Rubaiyat A 2012.Seasonal variation of soil salinity in coastal areas of Bangladesh. International Journal of Environmental Science, Management and Engineering Research, 1: 172-178
7. Karim Z, Hussain SG and Ahmed M. 1990. Coastal salinity problems and crop intensification In the coastal regions of Bangladesh, publication no. 33, BARC, Farmgate, Dhaka.pp-49.
8. Kim H, Jeong H, Jeon J and Bae S 2016. Effects of Irrigation with Saline Water on CropGrowth and Yield in Greenhouse Cultivation.
9. MoA and FAO 2013. Master Plan for Agricultural Development in the Southern Region of Bangladesh. Ministry of Agriculture. GoB.
10. MOEF (Ministry of Environment and Forest, Government of the Peoples Republic of Bangladesh, National Adaptation Programme of Action. Final Report. 2005, UNFCCC
11. Rahman MS, Biswas AKMAA, Rahman S, Islam MT, Zaman AKMM, Amin MN, Shamsuzzoha M, Shahin M, Rahim MA and Touhiduzzaman M 2015. Climatic Hazards and Impacts on Agricultural Practices in Southern Part of Bangladesh. Journal of Health and Environmental Research, 1: 1-11
12. Seinn SMU, Ahmad MM, Thapa GB and Shrestha RP 2015. Farmers Adaptation to Rainfall Variability and Salinity through Agronomic Practices in Lower Ayeyarwady Delta, Myanmar. Journal of Earth Science and Climatic Change, 2157-7617
13. World Bank, 2001. Bangladesh: Climate Change & Sustainable Development. Report No. 21104 BD, Dhaka.

AGRICULTURAL WASTE MANAGEMENT PRACTICES IN TRISHAL UPAZILLA, MYMENSINGH

Tangina Akhter[1*], Md. Ali Ashraf[2], Md. Monirul Hassan[3], Farzana Akhter[4] and Azmira Nasrin Riza[5]

[1&5]Bangladesh Agricultural Development Corporation (BADC), Dhaka, Bangladesh; [2&3]Department of Farm Structure and Environmental Engineering, Bangladesh Agricultural University, Mymensingh-2202, Bangladesh; [4]Bangladesh Agricultural Research Institute (BARI), Gazipur, Bangladesh

*Corresponding author: Tangina Akhter; E-mail: rainy045@gmail.com

---

ARTICLE INFO

Key words

Waste,
Waste management,
Straw,
Composting,
Biogas

ABSTRACT

A study was conducted to assess the present status of agricultural waste management by farmers in Trishal upazila of Mymensingh district, Bangladesh. During April to May 2015 and data were collected from the sample of 70 farmers and 5 farms. A structured interview schedule was used for collection of data. The study explored the relationship between the four selected type farming (independent variable) of farmers with their generation of agricultural waste (dependent variable). In this study the highest amount waste (straw and husk production) is closely related with the size of cropland. Straw production less than or equal 1000 kg is 36.62%, straw production less than or equal 10000 kg is 54.92%, straw production less than or equal 20000 kg is 5.63% and straw production less than or equal 30000 is 2.81% and husk production less than or equal 1000 kg is 35.71%, husk production less than or equal 10000 kg is 55.71%, husk production less than or equal 15000 kg is 8.57%. So as the dairy and poultry waste is also relate with the number of cows and birds. Average amount of dung is 8.87 kg per day and average amount of used litter was 46.36 kg per 800 bird production. For management biogas was suggested by 12.5 percent respondent. Composting and fish culture were suggested individually by 4.17 and 8.3 percent respectively. Due to manage agricultural waste efficiently it is necessary to initiating program to introduce the economic benefits of waste management and start training programs for farmers.

INTRODUCTION

Waste management is all the activities and actions required to manage waste from its inception to its final disposal. This includes amongst other things, collection, transport, treatment and disposal of waste together with monitoring and regulation. It also encompasses the legal and regulatory framework that relates to waste management encompassing guidance on recycling etc. The term normally relates to all kinds of waste, whether generated during the extraction of raw materials, the processing of raw materials into intermediate and final products, the consumption of final products, or other human activities, including municipal (residential, institutional, commercial), agricultural, and social (health care, household hazardous wastes, sewage sludge). Waste management is intended to reduce adverse effects of waste on health, the environment or aesthetics. Waste management practices are not uniform among countries (developed and developing nations); regions (urban and rural area) etc. The legal definition of "agricultural waste" is: - "waste from premises used for agriculture within the meaning of the Agriculture Act 1947." Waste is a pejorative term for unwanted materials. The term can be described as subjective and inaccurate because waste to one person is not waste to another (Wikipedia, 2013). Commercial poultry industry is growing rapidly in Bangladesh and annual growth rate of chicken population is 5.3 percent (GoB, 2010). Anaerobic digestion process is the most efficient process for biogas production from poultry waste because carbon dioxide (greenhouse gas) never produced in this process (Parmanik, 2000).

Agricultural waste typically associated with animals includes but is not limited to manure, bedding and litter, wasted feed, runoff from feedlots and holding areas, and wastewater from buildings like dairy parlors. Best management practices (BMPs) such as rotational grazing , and pasture renovation to maintain adequate vegetative cover ,riparian buffers and structures built to trap or retain waste should be utilized in order to prevent contamination of both surface waters and groundwater. Wastes are those substances or objects which fall out of the commercial cycle or chain of utility (EIB, 1995). Waste is defined as any substance which constitutes a scrap material or other unwanted surplus substance arising from the application of any process. Waste is defined as any substance which constitutes a scrap material or other unwanted surplus substances coming up from the application of any process. Hazardous waste is defined as any waste or combination of wastes, which could cause or significantly contribute to adverse effects in the health and safety of humans or the environment if improperly managed (EPA, 1990). Waste management includes three steps: transport, treatment and disposal of waste; control, monitoring and regulation of the production, collection, transport, treatment and disposal of waste; and prevention of waste production through in-process modification, reuse and recycling. Supported by Bangladesh Council of Scientific & Industrial Research (BCSIR), he used a bio gasification device to convert 500kg of poultry waste from 9000 birds, per day to generate 7.5 kw of power (Rahman and Zubayer, 2002). A sustainable poultry waste electricity plant established in Faridpur. GTZ Bangladesh has installed a flagship project at Raj Poultry Farm which is situated in Faridpur district. The farm has 15000 birds from which it can produce 105 m^3 biogas per day. The farm has 3 X 35 m^3 or total 105 m^3 biogas plant. GTZ installed 2 X 5 kw i.e. total 10 kW generators to produce electricity (Zaman S.A. et al, 2007).

Litter should be stacked 6 to 8 feet high for 3 to 5 weeks depending on environmental temperature before feeding. Stacking allows the litter to build up heat, thus killing pathogens and improves the palatability to cattle (Hossain et al., 2005). Dried poultry manure has been used as an animal feed for ruminants (Thomas et al., 1972; Alam et al., 2008).There is an increasing rate of waste generation in Bangladesh and it is projected to reach 47, 064 tons per day by 2025. The Waste Generation Rate (kg/cap/day) is expected to increase to 0.6 in 2025. A significant percentage of the population has zero access to proper waste disposal services, which will in effect lead to the problem of waste mismanagement. Bangladesh has minimal waste collection coverage which forces majority of the waste to be dumped in open lands. These wastes are not disposed of properly, where general waste is often mixed with hazardous waste such as hospital waste. In a report on solid waste management in Asia, the data showed that, in Dhaka, only about 42% of generated waste is collected and dumped at landfill sites, and the rest are left uncollected. As much as 400 tons are dumped on the roadside and in open space. Recycling of pesticide waste is not viable due to product quality requirements and the environmental risks involved.

MATERIALS AND METHODS

Study area

Trishal upazila with an area of 338.98 sq km, located in between 24°28′ and 24°41′ north latitudes and in between 90°18′ and 90°32′ east longitudes is bounded by Mymensingh sadar upazila on the north, Bhaluka and Gaffargoan upazilas on the south, Ishwargonj, Nandail and Gaffargaon upazilas on the east, Fulbariaupazila on the west. Main rivers are Old Brahmaputra, Sutia and Banar (Banglapedia, 2013). The study was conducted in different areas randomly. The study involves 75 farmers, various farms situated in different areas of Trishal upazilla. The study considers all type of farmers having cropland, fishery, dairy and poultry etc.

Data collection and analysis

The area of cropland, fishery, dairy and poultry farms of this area were selected at random and the farms are constituted the population for this study. The population constituted 70 farmers and 5 farms. In order to collect relevant information, a semi-structured interview schedule was prepared to collect data. The schedule was carefully designed keeping the objectives of the study in view. Before finalizing the schedule, it was pretested first judging the suitability of schedule to respondents. Necessary correction, modification and alterations were done accordingly.

Data were collected through personal interview during March to April 2015. The researcher explained the purpose of the study and requested necessary help and co-operation in collecting data from the respondents. In order to minimize the response error questions were asked in simple Bangla. After completion of each interview, it was checked to be sure that information had been recorded properly. After completion of the field survey, the information obtained from all the respondents were coded, compiled and tabulated. The responses to the questions in the interview schedule were transferred to a master sheet to facilitate tabulation for statistical analysis. Statistical means such as number and percentage distribution, mean, graph and correlation were calculated and finally analysis of variance were performed to find out the differences between selected variables of the rural areas. The correlation-regression and analysis of variance between dependent and independent variables were carried out to find the relationship and to measure the strength (Gomez and Gomez, 1984).

RESULTS

Age: Age of the respondents ranged from 16 to above 60 years. The respondents were classified into five categories were 16-25, 26-35, 36-45, 46-60 and above 60 years respectively. The highest proportion (33.33%) was in (26 to 35) year range and the lowest (5.33%) respondent's age were above 60 years. In the study area, 13.33, 21.33 and 26.67 percent were in the range of 16-25, 36-45 and 46-60 years respectively (Table 1).

Education level: The level of education undergoes 7 categories. These were can sign only, can read only, primary, secondary, higher secondary, graduate, post graduate. 25% people could sign only and 6% people had reading ability. About 13.33% of respondents have primary education and there were also 13.33% of respondent under secondary level. The percentage under higher secondary level was also same as secondary. There were some graduates who were involved with poultry firming and their percentage was 6.66%. It is very important that there was no respondent who completed post graduate (Table 2). Education broadens outlook of individuals and leads them to explore new ideas for better litter management. The literacy rate in this country is 56.9% (BBS, 2013). Thus, it seemed that rate of literacy of the respondents in the study area was higher than the national context since 100% of individuals had different kind of formal education.

Farming experience: The duration of firming ranged from 8 month to 22 years. Based on their duration of firming, the respondents were classified into three categories. These were less than 5 years, 5 to 10 years and more than 10 years. The duration of firming of 33.33% respondent was less than 5 years. About 26.67% farmers were in the range of 5-10 years while 40% farmer's farming experience was more than 10 years (Table 3).

Table 1. Distribution of respondents according to their age

Age range (years)	Respondents	
	Number	Percentage
16-25	10	13.33
26-35	25	33.33
36-45	16	21.33
46-60	20	26.66
Above 60	04	5.33
Total	75	100

Table 2. Distribution of education level of respondent in study area

Level of education	Respondents	
	number	percentage
Can sign only	15	20
Can read only	25	33.33
Primary	10	13.33
Secondary	10	13.33
Higher secondary	10	13.33
Graduate	5	6.66
Post Graduate	0	0

Table 3. Distribution of firming duration of farmer in the study area

Farming experience	Respondents		Observed range
	Number	Percentage	
Less than 5 years	25	33.33	
5 to 10 years	20	26.67	8 months to 22 years
More than 10 years	30	40	
Total	75	100	

Table 4. Litter management knowledge of farmer in study area

Do you know about litter management?	Respondents	
	Number	Percent
Yes	24	66.66
No	12	33.33

Size of crop land: Total size of crop land is 34125 decimal and the size of cropland varies from the farmers to farmers. Here we find the lowest size of cropland is 71 decimal and the highest size was 3000 decimal. A graph was shown in figure 1 as the size of cropland of the farmers in the selected area. And we get the size less than or equal 100 decimal is 28.98%, size less than or equal 500 decimal is 50.72% and the size less than or equal 1000 decimal is 15.94% and the size less than or equal 2000 decimal is 2.89% and the size less than or equal 3000 decimal is 1.44%.

Table 5. Distribution of animal waste in the study area

Using type of animal waste	Respondents	
	Number	Percent
Bio gas	03	12.5
Composting	01	4.17
Fish culture	02	8.3
Bio gas and Composting	03	12.5
Composting and Crop field	01	4.17
Fish culture and crop field	03	12.5
Bio gas, Composting and Fish culture	02	8.3
Bio gas, Fish culture and Crop field	03	12.5
Composting, Fish culture and Crop field	01	4.17
Bio gas, Composting and Crop field	02	8.3
Bio gas, Composting, Fish culture and Crop field	03	12.5
Total	**24**	**100**

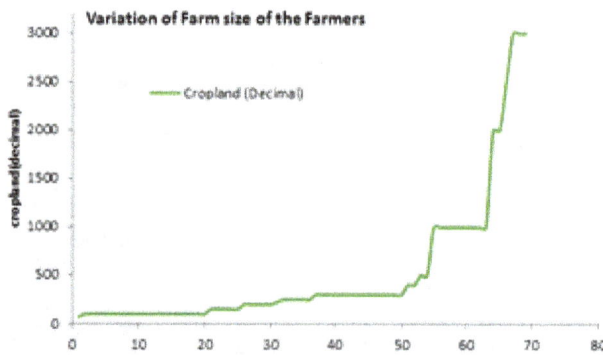

Figure 1. Relationship between the farm size and the farmers

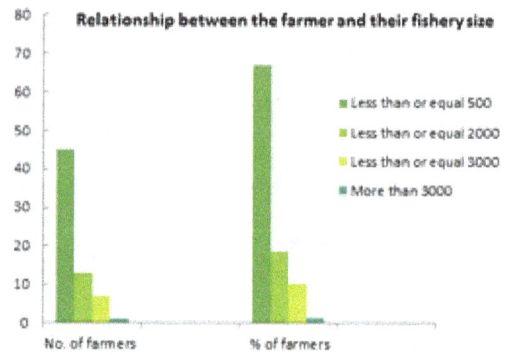

Figure 2. Relationship between the number of farmers and their fishery

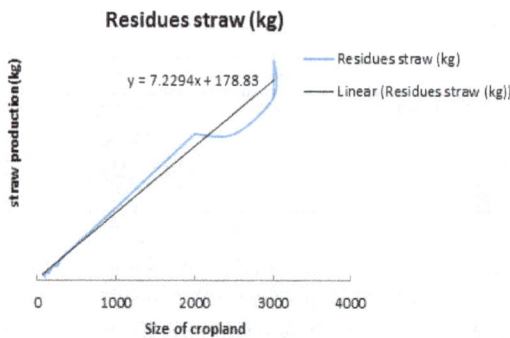

Figure 3. Relationship between cropland and straw

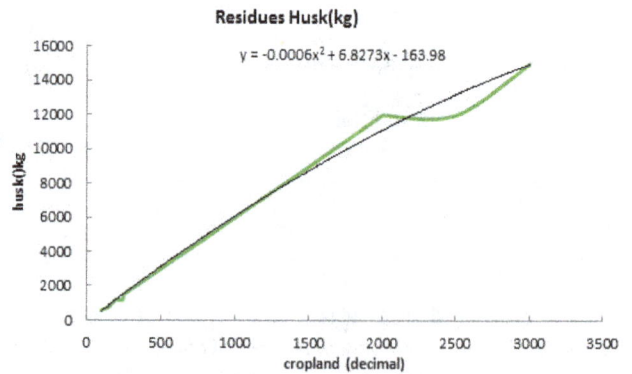

Figure 4. Relationship between cropland and husk

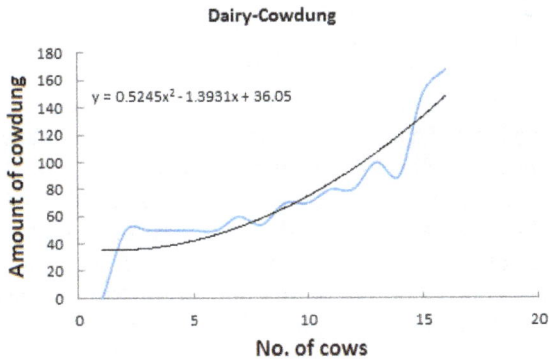

Figure 5. Relationship between dairy and cow dung

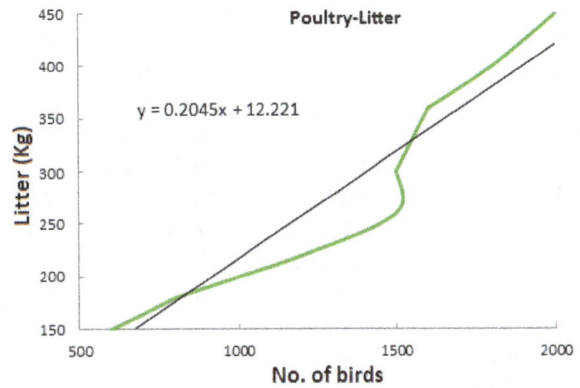

Figure 6. Relationship between the poultry and litter

Amount of fishery: Total size of fishery is 11978 decimal and the lowest size is 50 decimal and the highest size is 4000 decimal. The size of fishery less than or equal 500 decimal is 67.16%,size of fishery less than or equal 2000 decimal is18.84 %,size of fishery less than or equal 3000 decimal is 10.14% and the size of fishery more than 3000 decimal is 1.14% (Figure 2).

Relationship between cropland and straw: The production of straw is closely related with the size of cropland. The total size at the survey area is 35496 decimal and the total straw production at this site is 270250 kg. Here we find that the more size of cropland produce more straw than the less size cropland. We also find that every 100 decimal of cropland produce around 800kg of straw. Straw production less than or equal 1000 kg is 36.62%, straw production less than or equal 10000 kg is 54.92%, straw production less than or equal 20000 kg is 5.63% and straw production less than or equal 30000 is 2.81% (Figure 3).

Relationship between cropland and husk: The production of husk is related with the size of cropland. The total size at the survey area is 35496 decimal and the total husk production at this site is 20900 kg. Here we find that the more size of cropland produce more straw than the less size cropland. We also find that every 100 decimal of cropland produce around 600kg of husk. Husk production less than or equal 1000 kg is 35.71%, husk production less than or equal 10000 kg is 55.71%, husk production less than or equal 15000 kg is 8.57% (Figure 4). The amount of cow dung is related with the number of cows. The number of cows in survey was 132 and the total production of cow dung was 1172 kg, average amount of dung is 8.87kg per day. Here we get the more number of cows produce the more amount of dung. A graph is given below to show the relationship between the dairy and cow dung (Figure 5).

Relationship between poultry and litter: The amount of litter is related with the number of birds. The number of birds in survey was 38600 and the total litter used as one production (800 birds) was 8320 kg, average amount of used litter was 46.36 kg per 800 bird production. Here we get the more number of cows produce the more amount of litter (Figure 6).

Litter management knowledge: About 48% percent people had litter management knowledge and rest 52% percent people had no knowledge about litter management (Table 4).

The respondent used animal waste in composting, biogas, fish culture and crop field (Table 5).

DISCUSSION

Cropland: The lowest size of cropland is 71 decimal and the highest size was 3000 decimal. And the size less than or equal 100 decimal is 28.98%, size less than or equal 500 decimal is 50.72% and the size less than or equal 1000 decimal is 15.94% and the size less than or equal 2000 decimal is 2.89% and the size less than or equal 3000 decimal is 1.44%. Wastes from the cropland are straw and husk. The production of straw is closely related with the size of cropland. The total size at the survey area is 35496 decimal and the total straw production at this site is 270250 kg. Here we find that the more size of cropland produce more straw than the less size cropland. We also find that every 100 decimal of cropland produce around 800kg of straw. Straw production less than or equal 1000 kg is 36.62%, straw production less than or equal 10000 kg is 54.92%, straw production less than or equal 20000 kg is 5.63% and straw production less than or equal 30000 is 2.81%. And the production of husk is related with the size of cropland. The total size at the survey area is 35496 decimal. And the total husk production at this site is 20900kg. Here we find that the more size of cropland produce more straw than the less size cropland. We also find that every 100 decimal of cropland produce around 600kg of husk. Husk production less than or equal 1000 kg is 35.71%, husk production less than or equal 10000 kg is 55.71%, husk production Less than or equal 15000 kg is 8.57%. Farmers generally use straw for dairy feed and fuel. Most of the farmers use this for feeding their own dairy and cattle and they sell the excess straw to the farmers having dairy and cattle or to the dairy farms. And husk is mainly use as fuel in Trishal upazilla. Almost every house used this waste for cooking. So the wastes from cropland are used to feeding (dairy and cattle) and cooking (fuel).

Fishery: Total sizeof fishery is 11978 decimal and the lowest size is 50 decimal and the highest size is 4000 decimal. the size of fishery less than or equal 500 decimal is 67.16%, size of fishery less than or equal 2000 decimal is18.84 %, size of fishery less than or equal 3000 decimal is 10.14% and the size of fishery more than 3000 decimal is 1.14%.Waste generate from the fisheries is mainly pond bottom sediment and it is widely used at this upazilla as binding the pond sidewall. Very few of them used this for gardening as it increase the fertility of soil and the other farmers keep remains it at the pond.

Dairy: The amount of cow dung is related with the number of cows. The number of cows in survey was 132 and the total production of cow dung was 1172 kg, average amount of dung is 8.87kg per day. Here we get the number of cows produce the more amount of dung and the dung is mainly used as bio fuel and composting and the widely use this as natural fertilizer. Few of the farmers are interested to produce biogas because it is expensive to build a biogas plant.

Poultry: The amount of litter is related with the number of birds. The number of birds in survey was 38600 and the total litter used as one production (800 birds) was 8320 kg, average amount of used litter was 46.36 kg per 800 bird production. Here we get the more number of cows produce the more amount of litter, About 66% percent people had litter management knowledge and rest 33.33% percent people had no knowledge about litter management

Management of animal waste: Biogas was suggested by 12.5 percent respondent. Composting and fish culture were suggested individually by 4.17 and 8.3 percent respectively. About 12.5 percent respondent suggested both biogas and composting and 4.17percent was for both composting and crop field where fish culture and crop 32 field both was for 12.5 percent. About 8.3 percent respondent preferred biogas, composting and fish culture. About 12.5 percent respondent preferred biogas, fish culture and crop field. Composting, fish culture and crop field were suggested by 4.17 percent respondent where biogas, composting and crop field were suggested by 8.3 percent respondent. Only 12.5 percent respondent suggested all the methods.

CONCLUSION

Critically, a number of potential barriers to the options at the higher end of the waste hierarchy (that is reduction and recovery) exist. These include: low farmer awareness and motivation; limited cost-effective techniques for on-farm waste recovery; high logistics costs for off-farm recovery and poor markets, high reprocessing costs and limited facilities. So, it is necessary to initiating program to introduce the economic benefits of waste management and start training programs for farmer awareness.

REFERENCES

1.   Alam MS. Khan MJ.Akber MA. Kamruzzaman M, 2008. Broiler litter and layer manure in the diet of growing bull calves. Dhaka. Bangladesh. pp. 62-67.
2.   Banglapedia, 2013. Version 2013.Asiatic society of Bangladesh. Dhaka. Bangladesh.
3.   EIB 1995. Environmental information bulletin July 1995. Rome. Italy. pp. 15.

4. GOB 2010: Government of Bangladesh, Report of the task forces on Bangladesh development strategies of the 2010's. Univ. Press Ltd.,Dhaka, Bangladesh, pp. 2-4.
5. Hossain M, Ijaz M, Tamim M 2005. Energy and sustainable development in Bangladesh. pp. 54.
6. Rahman MT and Zubayer AHM, 2002. Energy Self Reliance in Rural Bangladesh Poultry Waste to Energy. An Unique Bio Gasification Initiative. pp. 01.
7. Thomas JW, Yu Y, Tinnimitt T, Zindel HC, 1972. Dehydrated poultry waste as a feed for milking cows and growing sheep. Journal of dairy Science. pp. 1261.
8. Wikipedia 2013. www.wikipidia.org/waste.html
9. Zaman SA, Klein DW, Boie W, 2007. The Potential of Electricity Generation from Poultry Waste in Bangladesh. A Case Study of Gazipur District. Hamburg. Germany. pp 01.

16

CONTROL OF SEED BORNE FUNGI ON TOMATO SEEDS AND THEIR MANAGEMENT BY BOTANICAL EXTRACTS

Imam Mehedi, Afia Sultana[*] and Md. Amanut Ullah Raju

Department of Plant Pathology, Faculty of Agriculture, Bangladesh Agricultural University, Mymensingh-2202, Bangladesh

*Corresponding author: Afia Sultana; E-mail: af.sultana87@gmail.com

ARTICLE INFO	ABSTRACT

Key words

Seed borne fungi,
Tomato,
Botanical extract,
Control

Seed health test was done in laboratory to determine the status of seed borne fungi on seeds of five tomato (*Lycopersicon esculentum*) varieties viz. Manik, Ratan, Roma VF, Kopotakkho and Monirumpuri and their possible control by using plant extracts. A total of 4 genera of 3 species of fungi were recorded where *Fusarium oxysporum* was the most prevalent and predominant seed borne fungus (25.60%). Other three fungal species are *Aspergillus* sp. of *Aspergillus flavus* (21.70%) and *Aspergillus niger* (11.11%) and *Cladosporium* sp. (13.49%). Tomato seeds were treated with different plant leaf extracts namely Mahogany, Mehendi and Allamanda with different doses viz. 1:1, 1:2 and 1:3 to control the seed borne fungi. In treated seeds, germination was ranged from 72-82% which was 68% in controlled condition. Among the doses of three botanicals, Mahogany, Mehendi and Allamonda extract @ (1:1 w/v) showed significant performance in controlling seed borne fungi and germination of tomato seeds. Among the three botanicals, Mahogany @ (1:1 w/v) was found the best treatment regarding percent reduction of seed borne infection.

INTRODUCTION

Tomato (*Lycopersicon esculentum*) belongs to the genus *Lycopersicon* is considered as the most important food crops in Bangladesh. Present world production is about 100 million tones fresh fruit produced on 3.7 million hectares (www.growtomatoes.com). The yield of tomato is very low in Bangladesh compared to other countries of the world. In Bangladesh, 251.00 tons of tomato was produced in 2013 which is 0.2% of the world share (www.factfish.com). The major constraints of tomato production in Bangladesh are pests, diseases, weeds, lack of quality seeds, postharvest losses, environmental factors etc. Among these constrains, seed borne disease is very crucial. The seed borne pathogens mostly fungi play a vital role in disease development which harms the seed both quantity and quality during pre-emergence to harvest. The seed borne fungal pathogen may because seeds fail to germination or transmit disease from seed to seedling and/or from seedling to growing plant (Islam and Borthakur, 2012). Fungal pathogens may be externally or internally seedborne, extra or intra-embryal or associated with the seeds as contaminants (Singh and Mathur, 2004). Management of plant diseases is important for most crops and it is particularly critical for the production of high quality seed. The control of seed borne pathogens is the first step in any agricultural crop production and protection programmed. Attempts have been made to reduce seed borne infection by chemical treatment of the seeds and some successes have been reported. Though, chemical controls of seed borne pathogens have been very successful, chemical pesticides have the additional potential disadvantages of accumulation in the ecosystem and of induction of pesticide resistance in pathogens (Okigbo, 2004). There is also the problem of lack of expertise in the safe handling of chemical pesticides amongst most of the farmers. Moreover, the costly chemicals are being imported from abroad and farmers have to purchase with high price. In recent years, much attention has been given to nonchemical systems for seed treatment to protect seeds against many plant pathogens (Nwachukwu and Umechuruba, 2001). Eventually a big amount of foreign currency goes out of the country every year due to control of the pathogen. It is therefore, necessary to search for seed quality control measures that are cost effective, ecologically sound and environmentally safe to eliminate or reduce incidence of pathogens of economic importance to increase both seed germination and yield of plant crops. Antifungal activity of different plant extracts have been reported earlier by several investigators against a number of plant pathogens (Hassan *et al.*, 2005; Yang and Clausa, 2007). However, information on management of seed borne fungal pathogens using botanicals on the major vegetable crops especially on tomato is lacking in Bangladesh.

Therefore, there is a great need for recording fungi associated with tomato seeds through easy, quick, reliable and economic seed health testing techniques for proper detection of seed borne pathogens in the crop. The botanicals are easily available and low cost compared to the chemical fungicides. In the view of above facts, the present research was designed to assess the presence of pathogenic fungi on tomato seeds and the possibility of controlling these pathogens using botanical extracts.

MATERIALS AND METHODS

The experiment was conducted at the Seed Pathology Center (SPC) and Department of Plant Pathology, Bangladesh Agricultural University (BAU), Mymensingh.

Collection of seed samples

Seeds of five tomato seed varieties viz. Manik, Ratan, Roma VF, Kapatakkho and Monirampuri were collected from commercial seed shops of Mymensingh sadar upazilla. The seeds were kept in brown paper bags and stored in the refrigerator at 5^0C, till the seeds were used for the subsequent studies.

Seed health test

Seed health test was done by blotter method following the rules of International Seed Testing Association (ISTA, 1996). In this method, three pieces of filter paper (Whatman no.1) were soaked in sterilized water and placed at the bottom of 9 cm diameter glass petridish. Four hundred seeds from each sample were taken randomly and then placed on the moist filter paper at the rate of 25 seeds per plate. The petridishes with the seeds were then incubated at $25\pm2°$C. Seeds produced both plumule and radical after incubation were considered as sprouted seeds. The result was expressed as percentage.

Identification of seed borne fungi on tomato seeds

Incubated seeds were observed under stereomicroscope at 16x and 25x magnification. The incidence of seed borne fungi was detected by observing their growth characters on the incubated seeds on blotter paper following keys outline by Khan (1998). Temporary slides were prepared from the fungal colony and observed under compound microscope. The fungi were identified with the help of keys suggested by Ellis (1976) and Neergaard (1979).

Preparation of plant extracts

For the investigation antifungal effect of plant extract, plant samples such as Mahogany seeds, Mehendi and Allamanda leaves were collected from Bangladesh Agricultural University campus. The collected plant parts were washed carefully in running tap water, dried and weighed by electric balance. Mahogany seed, Mehendi leaf and Allamanda leaf extracts were prepared by grinding in mortar and pastle. Then 1ml, 2ml and 3ml of distilled water were added, respectively with 1 gram of plant material to prepare plant extracts having 1:1, 1:2 and 1:3 doses (weight/volume). The crushed materials were filtered through cheese cloth.

Treatments

A total of 10 treatments (nine botanicals and one control) were used. Those are as follows,

T_0 = Control
T_1 = Mahogany @ (1:1) w/v
T_2 = Mahogany @ (1:2) w/v
T_3 = Mahogany @ (1:3) w/v
T_4 = Mehendi @ (1:1) w/v
T_5 = Mehendi @ (1:2) w/v
T_6 = Mehendi @ (1:3) w/v
T_7 = Allamanda @ (1:1) w/v
T_8 = Allamanda @ (1:2) w/v
T_9 = Allamanda @ (1:3) w/v

Seed treatment with plant extracts

Tomato seeds were treated in each dilution (1:1) w/v, (1:2) w/v and (1:3) w/v of each three plant extracts - Mahogany, Mehendi and Allamanda. Seed samples of tomato were dipped in each extract contained one petridishes at different dilution for one hour. Then the plant extract was drained out from the petridish. The treated seeds were shade dried on blotting papers for one hour. A set of control was maintained by dipping the seeds in tap water. After incubation, germination and seed borne fungi were observed.

Data analyses

The experiment was designed under Completely Randomized Design (CRD). The recorded data on various parameters under the present study were statistically analyzed using MSTAT-C statistical-package program. The level of significance and analysis of variance along with the Least Significance Difference (LSD) were done according to Gomez and Gomez (1984).

RESULT

Pathogenic incidence of seed borne fungi on tomato seeds

Pathogenic incidences of seed borne fungi were observed in the tomato seeds during incubation period. Four species of fungi representing three genera were identified from those incubated tomato seeds and these three predominant fungal genera were identified *Aspergillus* sp., *Fusarium oxysporum* and *Cladosporium* sp. The spores of *Aspergillus* sp., micro conidia of Fusarium *oxysporum* and conidia of *Cladosporium* sp. were observed under sereiobinocular microscope (40 x) by making slide from seed (Figure 1). The highest pathogenic incidence of *Aspergillus flavus* (21.70%) and *Aspergillus niger* (11.10%) were observed in the tomato seed varieties of Manik followed by Roma VF (20.00%) and (9.90%), in Monirumpuri (20.00%) and (10.10%) and others. The one of the most predominant fungal species *F. oxysporum* was recorded in the variety Monirumpuri (25.60%) followed by Manik (23.90%), Roma VF (22.40%) and others. The other fungal pathogen *Cladosporium* sp. was found on the tomato seed varieties Ratan (17.70%) followed by Roma VF (14.40%), Manik (13.30%), Kopotakkho (11.23%) and Monirumpuri (10.80%) (Figure 3-7).

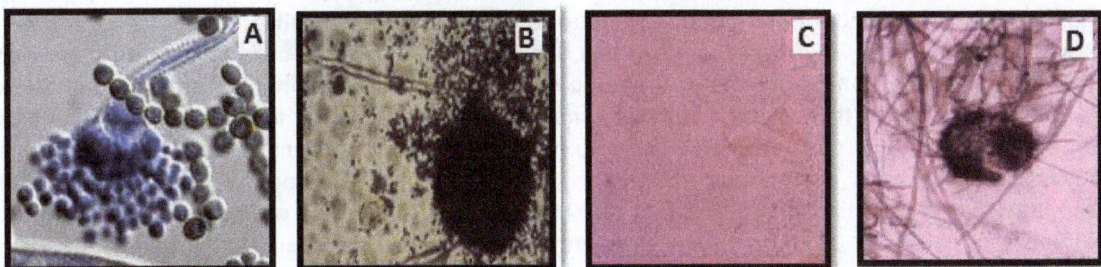

Figure 1. Different structures of seed borne pathogens observed under microscope (40x): **A.** Spore and sporangia of *Aspergillus flavus;* **B.** Spore and sporangia of *Aspergillus niger;* **C.** Micro conidia of *Fusarium oxysporum;* **D.** Conidia of *Cladosporium* sp.

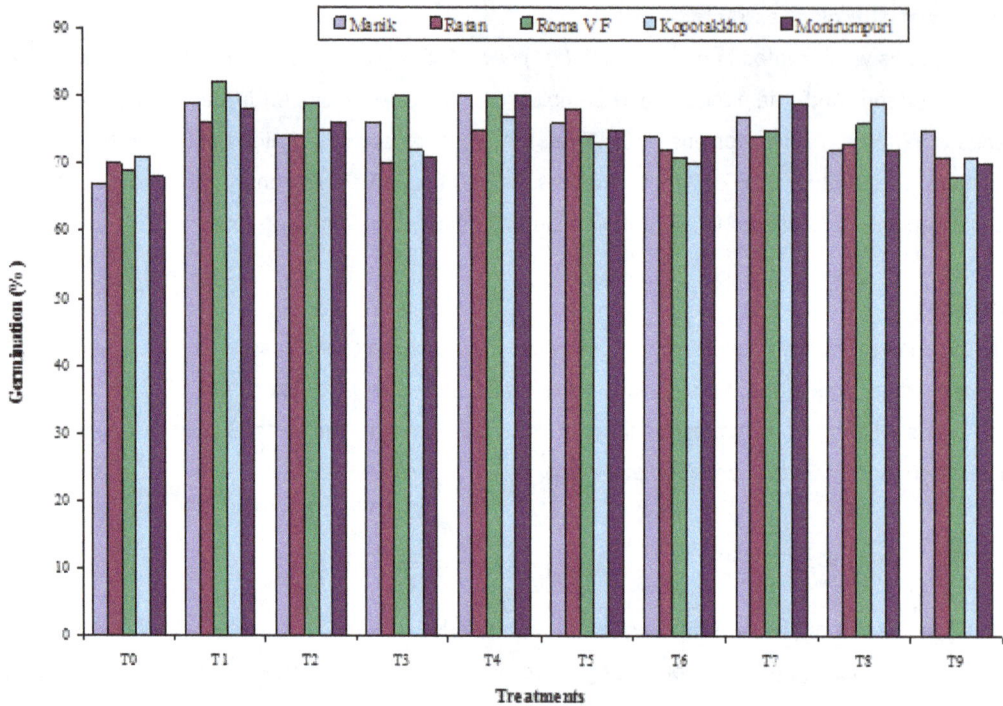

Figure 2. Percent germination of tomato seeds of different varieties with treatments

Effect of seed borne fungi on germination of tomato seeds

Collected five tomato seed samples showed significant differences in terms of seed infection by different seed-borne fungal pathogens in controlled condition and treatment. The highest germination was recorded in Kopotakkho (71%) while the lowest in the variety Manik (67%). The second highest germination was observed in the seed sample Ratan (70%) followed by Monirumpuri (68%) and Roma VF (69%) was statistically similar.

Discussion

Efficacy of plant extracts on incidence of seed borne pathogens

Three plants extracts of Mahogany, Mehendi and Allamanda with three dilutions @ 1:1 w/v, 1:2 w/v and 1:3 w/v for each were used in this experiment. All the selected components showed significant effect on controlling the pathogenic incidence and increasing the seed germination. The results obtained in the present study were shown in Figure 3-7. In the tomato seed variety Manik, the best result in terms of reduction of seed borne pathogens were obtained when seed treated with 1:1 w/v dilution of each plant extract (Mahogany, Mehendi and Allamanda). Seed treated with Mahogany extract 1:1 dilution yielded only 4.40% *Aspergillus* sp., 1.60% of *Fusarium oxysporum* and 1.90% *Cladosporium* sp. Moderate results in terms of reduction of seed borne pathogens were obtained when seed was treated with 1:2 w/v dilution of plant extract (Mahogany, Mehendi and Allamanda).The lowest result of pathogen reduction was found when seeds were treated with 1:3 w/v dilution of plant extracts (Mahogany, Mehendi and Allamanda).

Regarding plant extract, the best result was found when seed was treated with Mahogany extract where the infections were ranged from 2.10% to 4.00% followed by Mehendi extract 3.00% to 4.20%. The least result was found when seed was treated with Allamanda extract where infections were recorded from 3.50% to 5.10%.

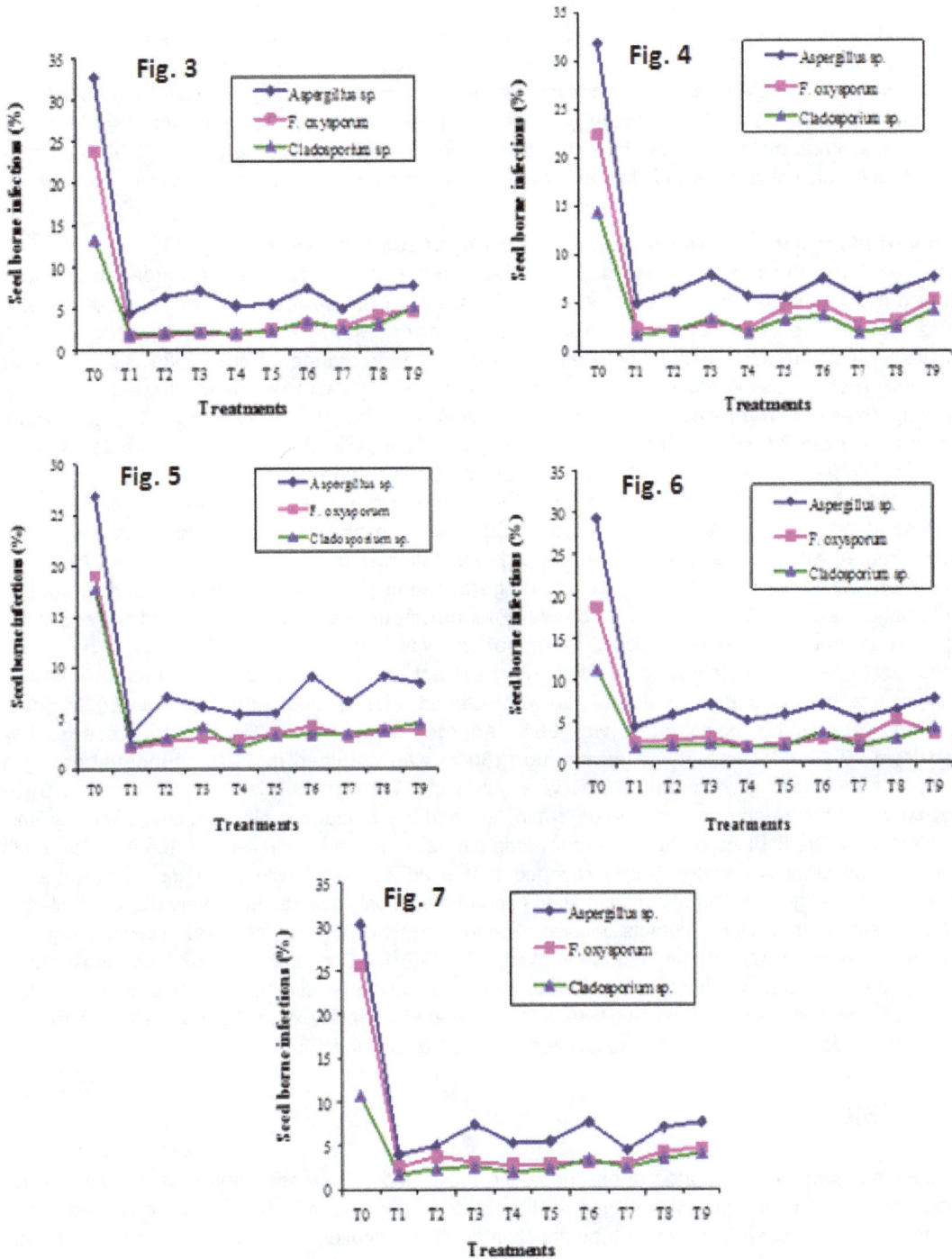

Figure 3. Effect of plant extracts on tomato seed variety Manik; **Figure 4.** Effect of plant extracts on tomato seed variety Ratan; **Figure 5.** Effect of plant extracts on tomato seed variety Ruma V F; **Figure 6.** Effect of plant extracts on tomato seed variety Monirumpuri; **Figure 7.** Effect of plant extracts on tomato seed variety Kopotakkho

The findings of the present investigation agreed with the findings of other previous results. Islam (2012) treated the tomato seeds with Garlic, Neem and Allamonda extracts with1:1 dilution. Among the plant extracts, garlic extract 1:1 was found the best treatment for reduction of seed borne infection. Akter (2008) reported that garlic extract @ (1:1) and chirota @ (1:2) showed best preference to reduce *Collectotrihum dematium, Macrophomina Phaseolina, Cercospora sp. Aspergillers niger, Aspergiller flavus* and *Penicllium* sp. Among them garlic clove *(Allium sativum)* extract was observed to be the most effective followed by allamanda, neem and marigold extract in reducing seed borne fungi. Begum and Momin (2007) successfully demonstrated the antifungal activity of plant extracts in disease control of seed borne fungal pathogen.

Efficacy of plant extracts on the germination (%) of tomato seeds

The results of the present study showed that, almost all the treatments significantly showed positive performance in compared to control after treating the seeds with different botanicals. It is revealed from the Figure 2 that, the tomato seed variety Manik showed the lowest germination which was recorded 67% in control while in treated seeds, ranged from 72-80 %. The highest germination (80%) was obtained in T_4 (Mehendi extract @ 1:1 w/v) where the lowest germination (72%) was obtained in T_8 (Allamanada extract @ 1:2 w/v). In case of variety Ratan, the germination was recorded from 71-78% in treated seeds while 70% was in the control (Figure 2). The seed treated with T_5 (Mehendi extract @ 1:2 w/v) showed the highest germination (78%) while the lowest germination (71%) was obtained from T_9 (Allamonda extract @ 1:3 w/v). In case of Roma VF, the seed germination (%) was recorded as 68-82%. The highest germination (82%) was found the seed treated with T_1 (Mahogany extract @ 1:1 w/v) followed by T_3 (Mahogany extract @ 1:3 w/v) and T_4 (Mehendi extract @ 1:1 w/v). The lowest germination (%) was found in the seed treated with T_9 (Allamanda extract @ 1:3 w/v). In case of Monirumpui, germination was 71% recorded in control treatment. In treated seeds, germination was ranged from 70-80 %. The highest germination (80%) was obtained in T_1 (Mahogany extract @ 1:1 w/v) and T_7 (Allamanda extract @ 1:1 w/v) where the lowest germination (70%) was obtained in T_6 (Mehendi extract @ 1:3 w/v). In case of the variety Kapotakkho, germination was 68% recorded in control while in treated seeds, it was ranged from 70-80 %. The highest germination (80%) was obtained in T_4 (Mehendi extract @ 1:1 w/v) where the lowest germination (70%) was obtained in T_9 (Allamanda extract @ 1:3 w/v). Several studies have demonstrated increased seed germination and vigour index in different crop after application of plant extracts. The findings of the present investigation are very much similar with the findings of other previous investigators. Howlader (2003) reported that seed treatment with Allamanda leaf extract (l:l) effectively increased germination of eggplant seeds and tremendously decreased nursery diseases. Hasan *et al.,* (2005) stated that plant extracts increase seed germination and improves seedling vigour or development. Islam (2009) reported that garlic plant extract @ 1:1 Concentration was found most effective among garlic clove, biskathali, allamanda, tobacco, neem increasing germination percentage and controlling seed borne fungal pathogens of mustard. Hossen 2015 stated that, Garlic extract (1:1) was effective treatment for seed germination and controlling the associated seed borne fungi in chilli.

CONCLUSION

Fusarium oxysporum is reported to be one of the most predominant seed borne as well as soil borne pathogen which can cause reduction in germination by triggering root rot and wilt of tomato. Seed borne pathogen infected the seed from seedling germination to yield formation and transmitted to the seed again. Emphasis should be undertaken for further research with more representative seed samples of different varieties of tomato collected from various tomato growing areas in order to unveil the extract picture of seed borne pathogens and their best control measures by plant extracts. Incorporating Mahogany, Mehendi and Allamanda extracts in the IPM system of Bangladesh can save huge amount of foreign currency and keep the environment pollution free.

REFERENCES

1. Akter N, 2008. Effect of plant extract in the management of seed borne fungal disease of okra. MS Thesis, Department of Plant Pathology, Bangladesh Agricultural Uniiversity, Mymensingh. pp: 36 & 74.

2. Begum HA and A Momin, 2007. Comparison between two detection techniques of seed borne pathogen in cucurbits in Bangladesh. Pakistan Journal of Scientific and Industrial Research, 43: 244-248.

3. Ellis MB, 1976. Dematiaceous Hyphomycetes. Kew, England: Commonwealth Mycological Institute.

4. Gomez KA and AA Gomez, 1984. Statistical procedure for agricultural research. 2nd Edition. John Willey and Sons, New York, pp: 640.

5. Hassan MM, SP Chowdhury, S Alam, B Hossain and MS Alam, 2005. Antifungal effects of plant extracts on seed borne fungi of wheat seed regarding seed germination, seedling health and vigour index. Pakistan Journal of Biological Science, 8: 1284–1289.

6. Hossen MT, 2015. Management of seed borne fungi in chilli. MS Thesis, Department of Plant Pathology, Bangladesh Agricultural University, Mymensingh.

7. Howlader AN, 2003. Effect of seed selection and seed treatment on the development of phomopsis blight and fruit rot of Eggplant. MS Thesis, Department of Plant Pathology, Bangladesh Agricultural University, Mymensingh, pp. 40-68.

8. Islam SME, 2012. Health and quality of tomato seeds collected from Abhoynagor Upazilla of Jessore district and their management by plant extracts. MS Thesis, Department of Plant Pathology, Bangladesh Agricultural University, Mymensingh.

9. Islam MF, 2009. Health and quality of mustard seeds collected from Fulbaria Upazilla of Mymensingh district and control of its fungi by plant extracts. MS Thesis, Department of Plant Pathology, Bangladesh Agricultural University, Mymensingh, pp. 40.

10. Islam NF and SK Borthakur, 2012. Screening of mycota associated with *Aijung* rice seed and their effects on seed germination and seedling vigour. Plant Pathological Quarantine, 2: 75–85.

11. ISTA (International Seed Testing Association), 1996. International Rules of Testing Association. In: Proceedings of International Seed Testing Association pp: 180.

12. Khan MQ and SI Ahmed, 1998. Seed borne microflora of vegetable seed lots in Northern areas of Pakistan. Pakistan Journal of Science and Industrial Research, 41: 47-49.

13. Neergaard P, 1979. Introduction to methods of seed-health testing. Seed Science and Technology, 7: 601-635.

14. Nwachukwu EO and CI Umechuruba, 2001. Antifungal activities of some leave extracts on seed borne fungi of African yam bean seeds. Journal of Applied and Natural Sciences, 1: 29-31.

15. Okigbo RN, 2004. A review of biological control methods for post-harvest yams (*Dioscorea* sp.) in storage in south eastern Nigeria. KMITL Journal of Science, 4: 207-215.

16. Singh D and SB Mathur, 2004. Location of fungal hyphae in seeds. In: Singh D, Mathur SB, editors. Histopathology of Seed-Borne Infections. Boca Ratan, FL, USA: CRC Press, pp: 101–168.

17. Yang VW and CA Clausen, 2007. Antifungal effect of essential oils on southern yellow pine. International Biodeterior Biodegrade, 59: 302–306.

18. www.growtomatoes.com, Vegetable Gardening Hydroponically: Complete Guide for the Home Gardener and Commercial Vegetable Grower: An e-book by Dr. J. Benton Jones of University of Georgia, USA.

19. www.factfish.com, Bangladesh: Tomatoes, production quantity (tons) collected from FAOSTAT (The statistics division of Food and Agriculture Organization of the United Nations).

17

COMPARATIVE STUDY ON HOST PREFERENCE AND DAMAGE POTENTIALITY OF RED PUMPKIN BEETLE, *Aulacophora foveicollis* AND EPILACHNA BEETLES, *Epilachna dodecastigma*

Md. Mahbubur Rahman[1]*and Mohammad Mahir Uddin[2]

Department of Entomology, Faculty of Agriculture, Bangladesh Agricultural University, Mymensingh-2202, Bangladesh

*Corresponding author: Md. Mahbubur Rahman; E-mail: rahmanmm_ent@bau.edu.bd

ARTICLE INFO	ABSTRACT

Key words

Ridge gourd,
Bitter gourd,
Snake gourd,
Leaf infestation

Experiments were carried out in the field and in the laboratory of the Department of Entomology, Bangladesh Agricultural University during the period from February to August, 2013. Three cucurbitaceous vegetables viz. bitter gourd, ridge gourd and snake gourd were used as test crops. Considering to percent leaf infestation and leaf area consumptions, red pumpkin beetle was found mostly harmful to snake gourd (22.62% and 8.84%, respectively) but least harmful to bitter gourd (3.00% and 1.25%, respectively). On the other hand, Epilachna beetle was found mostly damaging to bitter gourd (46.00% and 21.67% respectively) and least damaging to ridge gourd (11.20% and 5.00%, respectively). Similar to the field experiments, red pumpkin beetle consumed more leaf areas of snake gourd (up to 43.36%) and Epilachna beetle (both adult and grubs) consumed more leaf areas of bitter gourd (up to 91.46%) in the laboratory experiments. Bitter gourd (1.42%) and ridge gourd (0.78% to 41.27%) were least preferable to the red pumpkin beetle and Epilachna beetle, respectively.

INTRODUCTION

Bangladesh is a country with sub-tropical climate, where cucurbits are most important vegetables principally grown in summer season. High incidence of insect pests is considered as the major and common constraint in the successful production of cucurbitaceous vegetables in Bangladesh. Cucurbits are severely attacked by a number of insect pests such as red pumpkin beetle, Epilachna beetle, cucurbit fruit fly, aphids, cucumber moth etc. Among them, Red pumpkin beetle and Epilachna beetle are two major pest infesting during the whole cultivating periods and may cause damage up to 80% of the host plants (Rajagopal and Trivedi, 1989).

Red pumpkin beetle is a polyphagous pest in nature feed voraciously on leaves, flower buds and flowers that may cause 30-100% losses in the field (Rashid et al., 2014). Both larval and adult stages of Red pumpkin beetle are injurious to the crop and cause severe damage to almost all cucurbits from the seedling stage (Rahaman and Prodhan, 2007). The adult beetles feed on the leaves making irregular holes and also attack the flowers and flower buds and larvae feed on root tissues (Guruswamy et al., 1995), stems and fruits touching to the ground. On the other hand, both grubs and adults of Epilachna beetle are damaging to the host plants. The grubs feed on the lower epidermal layer of leaves making concentric markings whereas the adults feed irregularly upon the upper surface of leaves by scraping resulting net-like appearance. Later on, the infested leaves become brown in colour, entirely dry up and defoliated (Pradhan et al., 1990). Consequently, vegetative growth and development of the plants are greatly harmed by Epilachna beetle causing significant reduction of their yields (Imura et al., 1978; Srivastava et al., 1998; Ghosh et al., 2001).

The most commonly used control method of Red pumpkin beetle and Epilachna beetle in Bangladesh is the application of chemical insecticides (Anonymous, 1994). But indiscriminate use of broad spectrum insecticides has not only complicated the management strategies, but has also created several adverse effects such as pest resistance, secondary pest outbreak (Hagen and Franz, 1973), health hazards (Bhaduri et al.,1989) and environmental pollution (Kavadia et al.,1984; Desmarchelier, 1985; Fishwick, 1988) etc. On the other hand, adequate knowledge on the feeding choice, damage potentiality of insects as well as host susceptibility can be effective tools in the development of sustainable management strategy of these insects. Moreover, the host preference and food consumptions of Red pumpkin beetle and Epilachna beetle can be useful for interpreting between them. However, comparative research findings of host preference and damage potentiality of Red pumpkin beetle and Epilachna beetle are not adequately available. Considering the above points, the present research was planned to ascertain comparative host preference, food consumption and damage potentially Red pumpkin beetle and Epilachna beetle.

MATERIALS AND METHODS

Experiments were conducted in field laboratory and in the laboratory, of the Department of Entomology, Bangladesh Agricultural University, Mymensingh during February to August 2013. In the field experiment, the land was ploughed with a power tiller and kept open to sunlight for few days. The land was then gradually ploughed and cross-ploughed for several times with a power tiller to obtain desirable tilth. All ploughing operations were followed by laddering for breaking up the clods and leveling the surface of the soil. All weeds and stubbles were removed from the field and finally, the unit plots were prepared as 10cm raised beds along with the addition of the manures and basal doses of fertilizers. The whole experimental field was divided into 9 equal plots (2m×2m) and one pit was prepared in the middle of each plot. Three cucurbitaceous vegetables such as bitter gourd, snake gourd and ridge gourd were used in the experiments with three replications of each. Before sowing, seeds were soaked overnight for proper germination. Three seeds were sown in each pit and one healthy seedling per pit was maintained through thinning at 7 days after germination. Each plant was supported by bamboo platform (bamboo *matchan*) for easy creeping and preventing from lodging. The seedlings were maintained following all recommended horticultural practices. Percent leaf infestation and leaf area consumption by both beetles and Epilachna grubs were used as a parameter for incidence and damage potentiality. Data were recorded at 10 days' interval started from one month age of the plants.

Comparative food consumption of Red pumpkin beetle, Epilachna beetle and Epilachna grubs were studied in the laboratory. Five Red pumpkin beetle, five Epilachna beetle and five Epilachna grubs, were released in separate petridishes containing three leaves (one leaf of each selected vegetables namely bitter gourd, snake gourd and ridge gourd). The set was replicated three times. The cut end of leaf petiole was covered with water-soaked cotton pad to prevent withering. Data on consumed leaf area by beetles and grubs were taken at daily basis and fresh leaves were replaced. Percent leaf area consumed (both in the field and in the laboratory) were calculated by using the following formula-

$$\% \text{ Leaf area consumed} = \frac{\text{Consumed leaf area}}{\text{Total leaf area}} \times 100$$

Where,

Consumed leaf area was measured by using millimeter graph

Total leaf area = L×B×K (Kalra and Dhiman, 1977)

L= length of leaf

B= Breadth of leaf and

K= Kemp's constant

Data obtained from different experiments were analyzed using a statistical package program SPSS20. The mean values were ranked by Duncan's Multiple Range Test (DMRT) at 5% level of probability.

RESULTS AND DISCUSSION

Percent leaf infestation in the field

The percent leaf infestation varied significantly among different vegetables at different counting (Table 1). At first counting, red pumpkin beetle caused the most leaf infestations of snake gourd (22.62%) where the minimum infestations were found for bitter gourd (9.43%) although ridge gourd (20.74%) and snake gourd were found statistically identical. On the other hand, Epilachna beetle infested maximum leaf of bitter gourd (46%) and a minimum of ridge gourd (11.2%). A similar phenomenon was found in the second counting where Red pumpkin beetle infested maximum leaves of snake gourd (23%) and a minimum of bitter gourd (3%). Again, Epilachna beetle infested maximum leaves of bitter gourd (39.91%) and a minimum of ridge gourd (14.79%). At third counting, Red pumpkin beetle infested more leaves of snake gourd (28.28%) than the leaves of ridge gourd and the least percentage of infestations were recorded on bitter gourd (5.76%). On the other hand, Epilachna beetle infested the least (18.48%) and the most (28.28%) percentage of leaves of snake gourd and bitter gourd respectively. At all three-consequent counting, snake gourd was most preferable and bitter gourd was found least preferable to the Red pumpkin beetle which agrees the observation of Hasan et al. (2012). In contrast, bitter gourd was found mostly preferable and the snake gourd was found least preferable to the Epilachna beetle.

Comparative leaf area consumption of Red Pumpkin Beetle and Epilachna Beetle in the field

Percent leaf area consumptions of beetles varied significantly on three cucurbits at different counting (Table 2). At first counting, Red pumpkin beetle consumed maximum leaf area of snake gourd (8.84%) where the minimum leaf area consumption by Red pumpkin beetle was recorded for bitter gourd leaves (5.50%) but the leaf area consumption in ridge gourd (6.84%) and snake gourd were statistically identical. On the other hand, maximum leaf area was consumed by Epilachna beetle was found in bitter gourd (11.17%) and minimum in ridge gourd (6.67%) although ridge gourd and snake gourd (14.3%) were statistically alike. Similarly, maximum leaf area was consumed by Red pumpkin beetle from snake gourd (6.93%) and the least was from bitter gourd (1.25%) where leaf area consumption in ridge gourd was found in between bitter gourd and snake gourd. Again, Epilachna beetle consumed the most leaf areas of bitter gourd (15.10%) followed by snake gourd (8.01%) and a minimumof ridge gourd (5.00%). Similar results were found at third counting, the maximum leaf area was consumed (21.67%) by Epilachna beetle and minimum leaf area consumed (3.50%) by Red pumpkin beetle from bitter gourd but the minimum (5.00%) leaf area consumed by Epilachna beetle from ridge gourd and maximum area consumed by Red pumpkin beetle (6.27%) from snake gourd. At all three-consequent counting, Red pumpkin beetle consumed maximum leaf area of snake gourd but a minimum of bitter gourd and this result is also supported by the findings of Khan et al. (2011). Likewise, Rajak (2001) found bitter gourd as a non-preferable vegetable to the Red pumpkin beetle. On the contrary, Epilachna beetle consumed maximum and minimum leaf area from bitter gourd and ridge gourd, respectively.

Table1. Comparative percent leaf infestation by Red Pumpkin Beetle and Epilachna Beetle on different cucurbits in the field

Parameters	At 1st counting		At 2nd counting		At 3rd counting	
	Red pumpkin beetle	Epilachna beetle	Red pumpkin beetle	Epilachna beetle	Red pumpkin beetle	Epilachna beetle
Bitter gourd	9.43d	46.00a	3.00d	39.91a	5.76b	28.28a
Ridge Gourd	20.74b	11.20cd	20.70bc	14.79c	17.55a	20.93a
Snake Gourd	22.62b	14.30c	23.00b	17.93bc	18.28a	18.48a
LSD$_{0.05}$	3.867		6.27		11.19	

Different letters at same counting are significantly different.

Table 2. Comparative percent leaf area consumption by Red Pumpkin Beetle and Epilachna Beetle on different cucurbits in the field

	At 1st counting		At 2nd counting		At 3rd counting	
	Red pumpkin beetle	Epilachna beetle	Red pumpkin beetle	Epilachna beetle	Red pumpkin beetle	Epilachna beetle
Bitter gourd	5.50c	11.17a	1.25d	15.10a	3.50d	21.67a
Ridge Gourd	6.84bc	6.67bc	6.5bc	5.00c	5.84c	5.00cd
Snake Gourd	8.84b	8.07b	6.93b	8.02b	6.27c	9.55b
LSD$_{0.05}$	2.027		1.617		2.11	

Different letters at same counting are significantly different

Percent leaf area consumption by Red pumpkin beetle, Epilachna beetle and Epilachna grubs on different vegetables in the laboratory

The percent leaf area consumption by both insects varied significantly on different vegetables at successive counting in the laboratory. At first counting, the maximum consumed leaf was recorded for bitter gourd by Epilachna beetle (70.12%) and by Epilachna grubs (60.55%) but Red pumpkin beetle consumed maximum leaf area (43.36%) of snake gourd (Table 4). On the other hand, the small areas of leaves were consumed by Red pumpkin beetle, Epilachna beetle and Epilachna grub from bitter gourd (2.67%), snake gourd (7.38%) and from ridge gourd (1.51%) respectively. In the second counting, maximum leaf area consumption was recorded by Epilachna grub (91.46%) followed by Epilachna beetle (61.61%) in bitter gourd however Red pumpkin beetle consumed maximum leaf area of snake gourd (28.11%) leaves. Again, the minimum leaf area consumed by Epilachna grub (2.00%) and Epilachna beetle (4.17%) in ridge gourd followed by snake gourd but Red pumpkin beetle consumed bitter gourd (2.56%) leaves the least. At third counting, more leaf areas were consumed by Epilachna grub (20.76%), Epilachna beetle (38.50%) from bitter gourd leaves and minimum was consumed by Red pumpkin beetle (1.42%) from the same. However, maximum leaf area consumed by Red pumpkin beetle was found on snake gourd (15.15%) leaves but the minimum leaf area consumed by Epilachna grub and beetles from the ridge gourd (0.78%) and snake gourd (4.61%) leaves respectively. Therefore, at all three-successive counting, maximum leaf area was consumed by Epilachna beetle and Epilachna grubs from bitter gourd leaves and by Red pumpkin beetle from snake gourd leaves. On the other hand, the least leaf area was consumed by Red pumpkin beetle, Epilachna grub and beetles from the leaves of bitter gourd, ridge gourd and snake gourd respectively.

Table 3. Comparative percent leaf area consumption by Red pumpkin beetle, Epilachna beetle and Epilachna grub on different vegetables in the laboratory

	At 1st counting			At 2nd counting			At 3rd counting		
	RPB	EB	EB grub	RPB	EB	EB grub	RPB	EB	EB grub
Bitter gourd	2.67c	70.12a	60.55ab	2.56d	61.61b	91.46a	1.42c	38.50a	20.76a
Ridge Gourd	9.68c	41.27b	1.51c	6.15d	4.17d	2.00d	2.96c	15.15bc	0.78c
Snake Gourd	43.36b	7.38c	3.36c	28.11c	17.52cd	14.48cd	15.15bc	4.61bc	15.46bc
$LSD_{0.05}$	22.91			15.84			14.97		

Different letters at same counting are significantly different; RPB= Red pumpkin beetle, EB= Epilachna beetle

Therefore, snake gourd followed by ridge gourd and bitter gourd were found mostly preferable vegetable to the Red pumpkin beetle which is supported by the observations of Singh et al. (2000); Sharma et al. (1999); Thapa and Neupane (1992). Kamal et al. (2014) also reported the bitter gourd as the lowest preferable to the Red pumpkin beetle. On the other hand, bitter gourd followed by snake gourd and ridge gourd were found mostly preferable vegetable to Epilachna beetle as well as to Epilachna grub which agrees the finding of Hossain et al. (2009) although it was contradictory to the finding of Rahman (2002). However, Bitter gourd was the most susceptible to pest infestation followed by snake gourd and ridge gourd. On the other hand, Epilachna beetle and grub were identified as damaging to the cucurbitaceous leaves than Red pumpkin beetle in both field and laboratory conditions.

ACKNOWLEDGEMENT

We would like to express our deepest sense of gratefulness to Department of Entomology, Bangladesh Agricultural University for providing space and facilities for conducting research in the departmental laboratory and in the field laboratory. We would like to give thanks to the official and laboratory staffs of the Department for their help and support.

REFERENCES

1. Anonymous, 1994. Annual Research Report (1993-94), Entomology Division, Bangladesh Agricultural Research Institute, Joydebpur, Gazipur 132p.
2. Bhaduri M, DP Gupta, S Ram, 1989. Effect of vegetable oils on the ovipositional behaviour of *Callosobruchus maculatus* (Fab.). Proc. 2nd Int. Symp. On Bruchids and Legumes (ISLB-2) held at Okayama, Japan; 6-9 September pp. 81-84.
3. Desmarchelier YM, 1985. Bolivian of pesticide residues on stored grain, Aciar Prof. Series, Australian Centre for International Agricultural Research, 14: 19-29.
4. Fishwick RB, 1988. Pesticide residues in grain arising from post-harvest treatments. Aspects Applied Biology, 17: 37-46.
5. Ghosh SK, SK Senapati, 2001. Biology and seasonal fluctuation of *Henos Epilachna vigintioctocto punctata* Fabr. On brinjal under terai region of West Bengal. Indian Journal of Agricultural Research, 35: 149-154.
6. Guruswamy T, SR Desai, M Veerangouda, 1995. Host preference of pumpkin beetle, *Aulacophora foveicollis* (Lucas) (Coleoptera: Chrysomelidae). Karnataka Journal of Agricultural Sciences, 8: 246-248.

7. Hagen KS, JM Franz, 1973. A history of biological control. In: Smith RF, TE Mittler, CN Smith, (eds.). History of Entomology. Annual Reviews Inc, Palmetto, California, pp. 433-467

8. Hasan MK, MM Uddin, MA Haque, 2012. Host suitability of red pumpkin beetle, *Aulacophora foveicollis* (Lucas) among different cucurbitaceous hosts. International Research Journal of Applied Life Sciences, 1: 91-100.

9. Hossain MS, AB Khan, MA Haque, MA Mannan, CK Dash, 2009. Effect of different host plants on growth and development of Epilachna beetle. Bangladesh Journal of Agricultural Research, 34: 403-410.

10. Imura O, S Ninomiya, 1978. Quantitative measurement of leaf area consumption by *Epilachna vigintioctopunctata* (Fabricius) (Coleoptera: Coccinellidae) using image processing. Applied Entomology and Zoology, 33: 491-495.

11. Kalra GS, SD Dhima, 1977. Determination of leaf area of wheat plants by rapid methods. Journal of Indian Botanical Society, 56: 261-264.

12. Kamal MM, MM Uddin, M Shahjahan, MM Rahman, MJ Alam, MS Islam, MY Rafii, MA Latif, 2014. Incidence and Host Preference of Red Pumpkin Beetle, *Aulacophora foveicollis* (Lucas) on Cucurbitaceous Vegetables. Life Science Journal, 11: 459-466.

13. Kavadia VS, BL Pareek, KP Sharma, 1984. Residues of Malathion and Carbaryl in stored sorghum. Grain Technique Bulletin, 22: 247-250.

14. Khan MMH, MZ Alam, MM Rahman, 2011. Host preference of red pumpkin beetles in a choice test under net case condition. Bangladesh Journal of Zoology, 39: 231-234.

15. Pradhan S, MG Jotwani, S Prakash, 1990. Comparative toxicity of insecticides to the grub and adult of *Epilachna vigintioctopunctata* Fab. (Coleoptera: Coccinellidae). Indian Journal of Entomology, 24: 223.

16. Rahaman MA, MDH Prodhan, 2007. Effects of net barrier and synthetic pesticides on red pumpkin beetle and yield of cucumber. International Journal of Sustainable Crop Production, 2: 30-34.

17. Rahman MM, 2002. Studies on the biology, feeding behavior and food preferences of Epilachna beetle, *Epilachna dodecastigma* Muls. on different host plants. M.S. thesis Deptment of Entomology, Bangladesh Agricultural University, 50p.

18. Rajagopal D, TP Trivedi, 1989. Status, biology and management of Epilachna beetle, *Epilachna vigintioctopunctata* (Fab.) (Coleoptera: Coccinellidae) on potato in India. Tropical Pest Management, 35: 410-413.

19. Rajak DC, 2001. Host range and food preference of the red pumpkin beetle, *Aulacophora foveicollis* (Lucas) (Chrysomelidae: Coleoptera). Agricultural Science Digest, 21: 179-181.

20. Rashid MA, MA Khan, MJ Arif, N Javed, 2014. Red Pumpkin Beetle, *Aulacophora foveicollis* Lucas; A Review of Host Susceptibility and Management Practices. Academic Journal of Entomology, 7: 38-54, 2014.

21. Sharma SS, JP Bhanot, VK Karla, 1999. Host preference, extent of damage and control of red pumpkin beetle, *Raphidopalpa foveicollis* (Lucas). Journal of Insect Science, 12: 168-170.

22. Singh SV, M Alok, RS Bisen, YP Malik, A Misra, 2000. Host preference of red pumpkin beetle, *Aulacophora foveicollis* and melon fruit fly, *Dacus cucurbitae*. Indian Journal of Entomology, 62: 242-246.

23. Srivastava KP, D Butani, 1998. Pest management in vegetable. Research Periodical and Book Publishing House, 197-225.

24. Thapa RB, FP Neupane, 1992. Incidence, host preference and control of the red pumpkin beetle, *Aulacophora foveicollis* (Lucas) on Cucurbitae. Journal of the Institute of Agriculture and Animal Science, 13: 71-77.

PATHOLOGICAL CONDITIONS OF AVIAN COCCIDIOSIS IN THE SMALL SCALE COMMERCIAL BROILER FARMS IN DINAJPUR DISTRICT

Md. Manik Hossain[1], Md. Shahadat Hossain[1*], Md. Tareq Mussa[2], SM Harunur Rashid[3] and Md. Nazrul Islam[1]

[1]Department of Pathology and Parasitology and [2]Department of Anatomy and Histology, Faculty of Veterinary and Animal Science, Jhenidah Government Veterinary College, Jhenidah, Bangladesh; [3]Department of Pathology and Parasitology, Faculty of Veterinary and Animal Science, Hajee Mohammad Danesh Science and Technology University, Dinajpur, Bangladesh

*Corresponding author: Md. Shahadat Hossain; E- mail: shahadatvet@gmail.com

ARTICLE INFO

Key words
Avian coccidiosis,
Commercial broiler
farms,
Coccidiostats,
Dinajpur district

ABSTRACT

The study was designed to investigate the pathological conditions of avian coccidiosis in the small scale commercial broiler farms at different region in Dinajpur district during July, 2012 to December, 2012. A thorough clinical and necropsy examination was done and the characteristics clinical signs and gross lesions were recorded. Different organs mainly caecum and other parts of intestine were collected, preserved and processed for histopathological examination. Intestinal content was also examined for detection of oocyst. Total 234 diseased and dead birds (from 50 farms) were examined out of which 20 (8.54%) birds were found to be positive for coccidiosis. The clinical signs of the affected birds were bloody diarrhea, anemia, reduction of feed and water intake, drooping wings. At necropsy, enlargement and discoloration of caecum with numerous hemorrhagic spots, blood mixed and reddish to brown intestinal contents in the intestinal lumen, hemorrhage on the intestinal wall and mucosa were found. Histopathological examination revels distortion of normal architecture of intestine and desquamation of lining epithelia, formation of tissue debris on the intestinal mucosa and necrotic cells infiltration in the lamina propria and submucosa, degeneration of epithelial cells, glands and intestinal villi. So, outbreaks of coccidiosis in the commercial poultry flocks in Dinajpur district is lower due to farmers are intensely aware of coccidiosis now and they usually use coccidiostats routinely.

INTRODUCTION

Bangladesh is one of the most densely populated countries in the world where 152.5 million people (P.D, 2011) and 31.5 percent people live under malnutrition (Brad, 2010). The average quantity of protein uptake by people is insufficient. Poultry production is an easy and efficient way of producing animal protein. More profit could be earned by producing poultry with less capital investment. The poultry population of Bangladesh has increased from around 71 million in 1986 to around 188 million in 2006, an increase of about 164 percent in 20 years (FAO, 2008). Increasing demand and economic aspect has created a lot of interest among the people to raise poultry either through backyard or intensive commercial farming system. But poultry farming in Bangladesh faces various kinds of hindrance among them coccidiosis is one of the most serious problems for poultry development. Although commercial poultry production has increased manifold during last decade but at the same time, coccidiosis which was primarily a sporadic disease in 1976 has become a diseases of high occurrence in 1986 (FAO/WHO/OIE, 1976, 1986). The coccidia of the genus Eimeria are an obligatory intracellular parasite which has a complex life cycle. Eimeria is distributed worldwide (Macpherson, 1978). Temperature and moisture are two important factors in the epizootic of coccidiosis and faulty waterers have been identified as one source of excess moisture (Davies and Joyner, 1955). The optimum temperature for rapid sporulation of oocyst of different species of Eimeria has been reported to be from 28 to 30^0 (Edgar, 1955). The hot and humid environment of poultry houses in Bangladesh provides an ideal condition for the sporulation of the oocyst of coccidia. The practice of changing litter after each broiler crop apparently removes most of the oocysts, but is not effective in domination of the parasites (Long, 1978). Mondal (1978) made a preliminary report on the occurrence of *Eimeria tenella*, *Elmeria necatrix* and *Eimeria maxima* as by the fecal examination of chicks from Bangladesh Agricultural University Poultry Farm. Karim and Trees (1990) reported the occurrence of *Eimeria acervulina* and Eimeria brunetti in poultry in Bangladesh for the first time. But accurate figure of economic losses due to coccidiosis is not available in Bangladesh.

The parasite appears in the epithelial cells of digestive tract and its associated glands. Coccidiosis has also become a subject of growing interest because it causes significant economic loss in the poultry industry throughout the world. Considerable studies are being conducted to determine its economic importance and associated epizootiological factors and method of control of the disease. Senevlranta (1969) reported that 90 to 100 percent mortality in chicken to be associated with coccidiosis in India. The mortality in young birds is predominant features. In adult also poor growth rate or loss of egg production is observed (Lerine, 1961). The true picture as to the incidence and pathology of coccidiosis in chicken has not been worked out yet in the study region. Until some basic information regarding this disease occurrence and problem is available, therefore it is very difficult to encourage commercial farming in the country. Considering the fact in mind, the present study was undertaken to determine the prevalence of avian coccidiosis with their clinical and histopathological findings in Dinajpur district of Bangladesh.

MATERIALS AND METHODS

Experimental chickens
A total of 234 diseased and dead birds were examined from 50 farms in the small scale commercial broiler farms at different regions of Dinajpur district in Bangladesh. Among the examined bids, only 20 were found to be positive for coccidiosis. A detail flock history in relation to the incidence of diseases including housing, location of farms, source of birds, age, population of the birds per flock, rearing system, litter material, feeding and watering system, biosecurity, previous history on coccidia outbreaks, intervals between the batches, rearing of one more batches in the same farm at the same time etc., were also recorded. The birds affected with coccidiosis were submitted to the pathology laboratory for the diagnosis and treatment and other processing.

Clinical examination of affected birds
The general health condition and age of the chicken were recorded. The clinical signs were recorded during the physical visit of the affected flocks and the farmer's complaints about the affected birds were also considered.

Necropsy findings of suspected birds

The necropsy was done on the suspected dead and diseased birds taken from different upazilla of Dinajpur district. At necropsy, gross morbid changes were observed and recorded carefully by systemic dissection. The collected samples were preserved at 10% formalin for the histopathological study. Gross morbid lesions of different organs were registered during the course of necropsy of the birds.

Histopathological examination

During necropsy, various organs having gross lesions were collected, preserved at 10% formalin, processed for the histopathological study. Formalin fixed samples of the small intestine, large intestine and caeca from the diseased and dead chicken were processed for paraffin embedding, sectioning and staining with haematoxylin and eosin according to standard method (Luna, 1968) for histopathological study.

Examination of faeces

Faecal samples were collected directly from the affected flocks. Interstinal content was collected during the postmortem examination of the birds. The slides were examined under microscope for detection oocysts in low and high magnification.

Photography

All images related to the present study were taken directly from microscope using different objectives manipulation of zooming system of a digital camera (Canon, 1XY, 16.1 Mega pixels, Japan). The images were provided following minute modification for the better illustration of the study.

RESULTS

Pathological investigation of avian coccidiosis encountered in small scale commercial broiler farms in Dinajpur district was studied and different clinical, parasitological, necropsy and microscopic conditions were recorded during the study period.

Clinical findings

The study was conducted in different small scale commercial Broiler farm in different upazilla in Dinajpur district. Total 50 farms were visited. Different species of Eimeria were found to be prevailed in those farms. Total 234 diseased and dead birds were examined out of which 20 birds were found to be positive for coccidiosis i.e. the incidence of coccidiosis was recorded as 8.54 % in relation to age and breed where highest proportion was recorded in birol upozila (13 %) and lowest proportion in phulbari, Kaharol and Birampur upozila (0.00 %). The age groups of 5-6 weeks were mostly affected (38.4 %) where 0-4 weeks age group less affected (12.8%).

Clinical signs were recorded as bloody diarrhea, considered to be a most important clinical sign in the examined chicken, followed by anaemic carcass (Figure 3), attachment of faeces around vent (Figure 1), blood mix with food (Figure 2). The prevalence of various coccidial disorders is shown in Table 1. Proportional mortality rate of coccidiosis in different age group was shown in Table 2.

Histopathological study

In the present study, distortion of normal architecture and desquamation of lining epithelia of intestine were found (Figure 4). Formation of tissue debris on the intestinal ucosa and necrotic cells infiltration in the lamina propria and submucosa (Figure 5) were markedly observed. Degeneration of epithelial cells, glands, intestinal villi and infiltration of inflammatory cell in the musculature (Figure 6) were also found. The villi of the mucosa were destroyed and disorganized and there was no continuation in the lining epithelial cells of villi (Figure 6).

Table 1. Prevalence of coccidiosis at different commercial broiler farms Dinajpur district is graphically shown

Location of the firm (Upazilla)	No. of farm visited	No. of birds Necropsy done	No. of Affected farms	Percentage (%)
Sadar	14	56	6	10
Birol	9	37	5	13
Prabotipur	6	30	3	10
Chairirbondor	5	32	1	3
Kaharol	2	10	0	0
Birgong	3	22	2	10
Khansama	3	15	1	6
Satabgong	5	19	2	10
Phulbari	2	10	0	0
Birampur	1	3	0	0
Total	50	234	20	8.54

Table 2: Proportional mortality rate of coccidiosis in different age groups are graphically shown

Age group	No. of farm affected	Percentage (%)
0-4 weeks	2	12.8
5-6 weeks	8	38.4
7-8 weeks	6	30.4
> 8 week	4	18.4

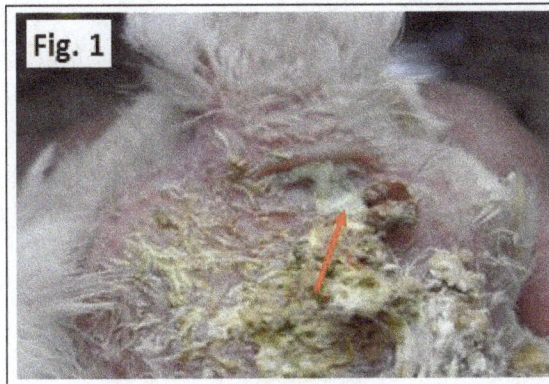

Figure 1. Attachment of feces around vent (red arrow) of chicken;

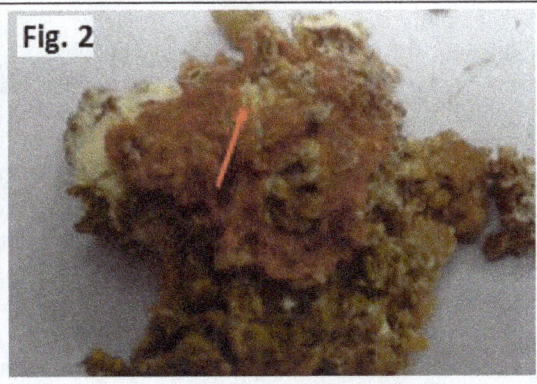

Figure 2. Feed mixed with blood (red arrow);

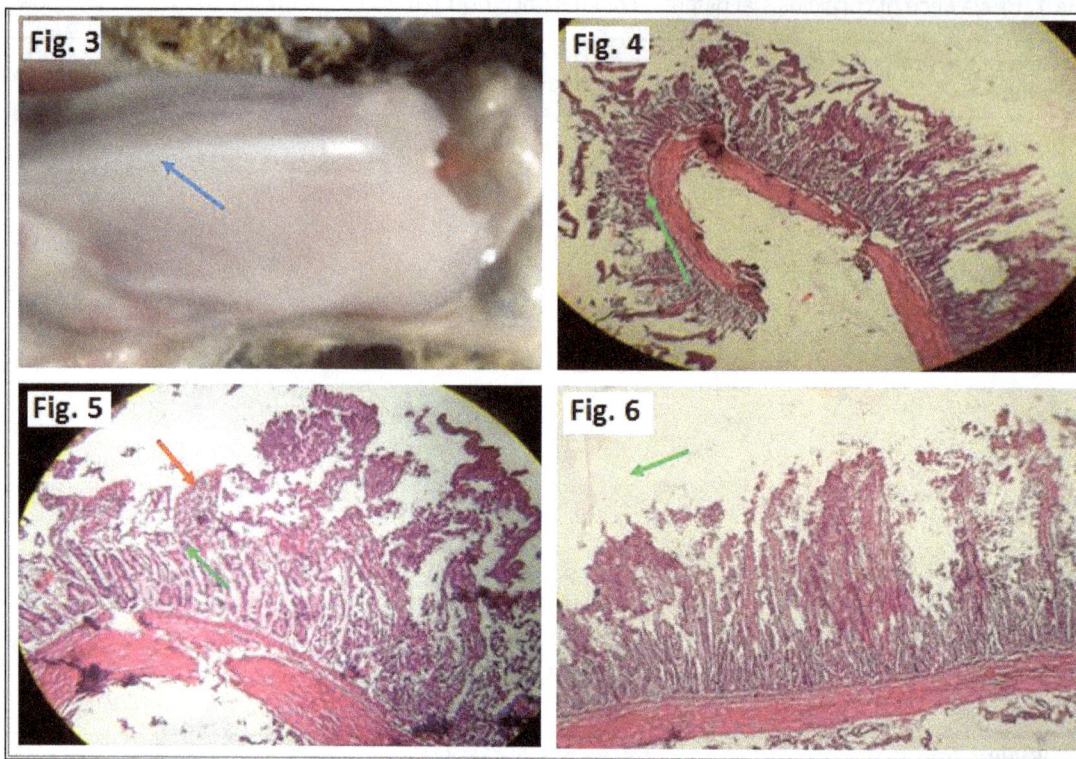

Figure 3. Anaemic carcass (blue arrow) of coccidia affected fowl; **Figure 4.** Distortion of normal architecture and desquamation of lining epithelia (green arrow) of intestine (cecum) (4 x); **Figure 5.** Formation of tissue debris on the intestinal (cecum) mucosa (red arrow) and necrotic cells infiltration in the lamina propria and submucosa (green arrow) (10x); **Figure 6.** Degeneration of epithelial cells and glands (green arrow) on intestinal villi (cecum) (10x)

DISCUSSION

The present study was conducted mainly to explore a pathological investigation of avian coccidiosis based on clinical, parasitological, gross and histopathological lesion.

Prevalence

Total 234 diseased and dead birds were examined out of which 20 birds were found to be positive for coccidiosis i.e the prevalence of coccidiosis was recorded 8.54% in relation to age and breed. This observation is similar to those of reported in other authors where the prevalence of coccidiosis was recorded as 9.40% by Bhattachrjee *et al.*, 1996. In West Bengal the cases of coccidiosis was recorded as 10.91% by Bhattacharya (1987).

In this study we found that the young birds are more susceptible to and more readily display signs of disease, whereas older chickens are relatively resistant as a result of prior infection. Typically, the disease is seen in birds of 3-6 weeks old, before they have acquired immunity. The proportional mortality rate of coccidiosis in different age group were 12.8%, 38.4%, 30.4% and 18.3% in 0-4 weeks, 5-6 weeks, 7-8 weeks and above 8 week respectively which is similar to the observation by Kamath, 1955; Rose, 1967; Humphrey, 1973 and Kogut *et al.*, 1993.

Clinical examination

During this investigation the common clinical manifestations in the chicks suffering from coccidiosis were found as bloody diarrhea, anemic carcass, and attachment of faeces around vent, blood in faeces, depression and ruffled feather. These findings are also consistent with Reid and Pitoais, 1965 and Williams, 1996.

Weight loss, reduction in egg production, damp litter and death occurs mostly on 5th or 6th day after infection were also found in this observation. This report is agreeable with the findings of Tyzzer, 1929; Waxler, 1941; Ruff *et al.*, 1976 and Levine, 1983.

Necropsy examination

Gross pathological changes of the various organs of the affected chickens were studied. At necropsy, the major pathological lesions were enlargement and discoloration of caecum with numerous hemorrhagic spots, blood mixed intestinal contents in the intestinal lumen which is vary from reddish to brown, pin point hemorrhage on the intestinal mucosa, profuse hemorrhage on intestinal wall and mucosa were recorded. These gross lesions are also reported by Bertke, 1955; Becker, 1959 and Reid, 1972.

Thickening of intestinal wall than normal, hemorrhage and extravasations of blood within the intestinal lumen, profuse congestion, hemorrhagic enteritis, and blood-tinged exudates were also found. Our observation is same to those were reported by Poul, 1967; Jagadeesh *et al.*, 1976; Arakawa *et al.*, 1981 and. Levine, 1983.

Histopathological study

The histopathological change founded in the present study were listed as severely distortion of normal architecture of intestine and desquamation of lining epithelia, formation of tissue debris on the intestinal mucosa, necrotic cells infiltration in the lamina propria and submucosa, degeneration of epithelial cells, glands, intestinal villi. The villi of the mucosa were destroyed and disorganized and there was no continuation in the lining epithelial cells of villi. This observation is similar to those reported by Noyilla *et al.*, 1972 and Jagadeesh *et al.*, 1976.

CONCLUSION

Prevalence of coccidiosis was recorded as 8.54 % in relation to age and breed. Highest mortality in 5-6 weeks (38.4%) and lowest in 0-4 weeks (12.8%) were recorded. Clinical signs including bloody diarrhea, anemia, depression, ruffled feather, reduction of feed and water intake, drooping of wings. At necropsy, enlargement and discoloration of caecum with numerous hemorrhage spots, blood mixed intestinal contents in the intestinal lumen vary from reddish to brown; pin point and profuse hemorrhage on intestinal mucosa were found. Histopathologically, distortion of normal architecture of intestine and desquamation of lining epithelia, formation of tissue debris on the intestinal mucosa and necrotic cells infiltration in the lamina propria and submucosa, degeneration of epithelial cells, glands, and intestinal villi were also present. From the above facts and findings, it could be concluded that outbreaks of coccidiosis in the commercial poultry flocks is lower due to farmers are intensely aware of coccidiosis now and they usually use coccidiostats routinely.

REFERENCES

1. Arakawa A, Baba E and Fukata T, 1981. *Eimeria tenella* infection enhances *Salmonella typhimurium* infections in chickens. Poultry Science, 60: 2203-2209.
2. Becker RF, 1959. In Diseases of Poultry. IOWA State University Press, Ames, Iowa, USA, p. 828-858.
3. Bertke EM, 1955.   Pathological effect of coccidiosis caused by the *Eimeria tenella* in chicken (Unpublished Thesis). University of Wisconsin, USA. 67: 193-199
4. Bhattacharjee D, Pan A, Dhara S, Kumar and Das SK, 1996. Evaluation of Economic Losses due to Coccidiosis in Poultry Industry in India. Agricultural Economics Research Review, 23: 91-96.
5. Bhattacharya HM and Framanik AK, 1987. Diseases of poultry in three districts of West Bengal affecting rural economy. Indian Veterinary Journal, 7: 63-65.
6. Bradfield R, 2010. Increasing and diversifying food production in Bangladesh" cited by Ahmed, U.K. (1998) Gardiner's book of production and nutrition vol.1.
7. Davies SFM and Joyner LP, 1955. Observation on the parasitology of deep litter of poultry houses. Veterinary Research, 67: 193-199.
8. Edgar SA, 1955. Sporulation of oocysts at specific temperature and notes on the prepatent period of several species of avian coccidia. Journal of Parasitology, 41: 214-216.

9. FAO, 2008. Bangladesh Bureau of Statistics (BBS), June 2006, 172-198.

10. FAO, WHO and IOE, 1976. Animal Health Year Book. Animal Health Services, Animal Health and Production Division. Food and Agricultural Organization of the United Nations (FAO), World Health Organization (WHO), International Office of Epizooties (OTE), 108-148.

11. FAO, WHO and IOE 1986. Animal Health Year Book, Animal Health Services, Animal Health and Production Division. Food and Agricultural Organization of the United Nations (FAO), World Health Organization (WHO), International Office of Epizooties (OIE), 79-144.

12. Humphrey CD, 1973. A comparison of uninfected and protozoan parasitized chick intestinal epithelium by light and electron microscopy. Veterinary Bulletin, 44: 37-98.

13. Jagadeesh KS, Seshardi SJ and Mohiuddin S, 1976. Studies on pathology of field cases of coccidiosis in poultry. Indian Veterinary Journal, 53: 47-54.

14. Kamath MG, 1955. Coccidiosis in chicken. Indian Veterinary Journal, 5: 19.

15. Karim MJ and Trees AJ, 1990. Isolation of five species of Eimeria from chicken in Bangladesh. Tropical Animal and Production, 22: 153-159

16. Kogut MH and Powell KC, 1993. Preliminary findings of alterations in serum alkaline phosphatase activity in chickens during coccidial infection. Journal of Comparative Pathology, 108: 113–119.

17. Lerine DN, 1961. Protozoan Parasites of domestic animals and man. Burgess Publishing Company, 15: 775-780

18. Levine, D.N. P. 1983. The Biology of Coccidia. University Park Press, Baltimore, USA, 28.

19. Long PL, Boorman KN and Freeman BM, 1978. Avian Coccidiosis. Proceedings of the 13th Poultry Science Symposium, 14-16th September, 1977. British Poultry Science Limited.

20. Luna LG, 1968. Mannual of Histologic staining methods of the Armed Forces Institute of Pathology (3rd edition). Mc Graw Hill Book Co. New York.

21. Macpherson, I. 1978. Avian Coccidiosis. Btritish Poultry Science Ltd. Edinburgh, 465-494.

22. Mondal AN and Ahmed, 1978. Coccidiosis; Laboratory confirmation of clinical disease. Experimental Parasitology, 28: 137–146

23. Noyilla MN and Medin CS, 1972. Pathology of experimental Eimeria mivati infection in young chicken. Veterinary Bulletin, 4: 3157.

24. PDEU, 2011. Population development and evaluation unit, Bangladesh population data sheet, planning c6ommission People's Republic of Bangladesh

25. Poul DD, 1967. Villous atrophy and coccidiosis. Nature, London, 213: 306-307.

26. Reid WM, 1972. Diseases of poultry. The Iowa State University Press. Ames, USA, 944-975.

27. Reid WM and Pitosis M, 1965. The influence of coccidiosis on feed and water intake of chicken. Avian diseases, 9: 343-348.

28. Rose ME, 1967. The influence of age of host on infection with Eimeria tenella. Journal of Parasitology, 53: 924-929.

29. Senevlranta P, 1969. Diseases of poultry (2nd Edn). John Wright and Sons Ltd Bristol.

30. Tyzzer EE, 1929. Coccidiosis in gallinaeeous birds. American Journal of Hygiene, 10: 269-283.

31. Waxler SH, 1941. Changes occuring in the blood and tissue of chickens during coccidiosis and artificial haemorrhage. American Journal of Physiology, 134: 25-26.

32. Williams RB, 1996. The ratio of the water and food consumption of chickens and its significance in the chemotherapy of coccidiosis. Veterinary Research communications, 20: 437-447.

EMPOWERMENT OF RESOURCE POOR WOMEN THROUGH INCOME GENERATION ACTIVITIES (IGAs): A CASE OF SLUM AREA IN DHAKA CITY CORPORATION OF BANGLADESH

Sadia Jahan Moon[1*] and Md. Abdul Momen Miah[2]

[1]Masters in the International Environment and Development Studies, Norwegian University of Life Sciences, Norway; [2]Department of Agricultural Extension Education, Bangladesh Agricultural University, Mymensingh-2202, Bangladesh

*Corresponding author: Sadia Jahan Moon; E-mail: sadia_4000@yahoo.com

---

ARTICLE INFO

Key words

Participation
IGAs
Empowerment
mobility
Decision making

ABSTRACT

The main purpose of the study was to determine the level of empowerment of women working at garments industry in Bangladesh. The factors influencing the achievement of empowerment of women were also analyzed. A total of 50 women serving in garment industry at Pallabi Thana in Dhaka were selected randomly from a total of 240 women. Data were collected by using structured questionnaire during January to February 2016. All the women opined for having moderate to higher level of empowerment in terms of freedom of mobility and participation in household decision making process. Even none of the women fell in low empowerment category. Before involved in IGAs it was difficult for them to go out even in relatives' house and were not consulted in making household decisions. After getting job, they contributed approximately 40 to 58 percent of their income to meet family expenses. The age, level of education, family income and job experience were related to their empowerment. The women identified that social superstition, existing value system are still hinder empowerment. The adult education programme and awareness building campaign by the development agencies could mitigate the existing limitations of empowerment of women.

INTRODUCTION

Bangladesh is a developing country with 160 million population whereas half of its population is women (BBS, 2015). The socio-economic condition of women in Bangladesh is very low. Here participation of men and women in development activities is not equal. Many indoor and outdoor activities of women are not treated as work. The devaluation of women labour in labour market also hinders women rights. But in market economy women of Bangladesh are always facing new challenges. Though women are working in different GO and NGOs still are not treated as equal to male. In the society of gender disparity it is difficult for women to stay parallel and enjoy the right of equal to men.It was found that mortality of women occurs due to discrimination and resulted imbalance sex ratio in the population where 106 men for 100 women (BBS, 2015). The empowerment of women in developing countries like Bangladesh has been considered as a potential working force of development. The national and international donors are also paying emphasize on inclusion of women in development activities. Although empowerment is often conceptualized as a process (Hanny, 2006; Naila, 2001; Anju and Ruth, 2005), most quantitative studies have been cross-sectional, comparing individual women with others in their communities or societies (Anju and Ruth, 2005). It is assumed that previously the women of Bangladesh have limited exposure in socio-econo-political arena, negligible freedom of mobility, influence, autonomy and power of making even household decisions.

With the passage of time the women are coming out of their house and involved in both indoor and outdoor activities for their survival. In the recent past women empowerment has been studied by making comparison between the past and present socio-economic and political status in the society. Some of the researchers argue that the measurement methods of empowerment are changing over time as well as mode of interventions to foster empowerment of women (UNDP 1995; Solava and Akter, 2007).The present study was undertaken to examine the status of empowerment of resource-poor women working in garments industry with the following specific objectives:

- To find out the level of empowerment achieved by the resource-poor women working in garments industry.
- To explore the factors influencing possession of empowerment of resource-poor women.

METHODOLOGY

The study was conducted with the women serving in three different garment industries in Dhaka. The garments workers living in a slum area of Mirpur, Section 11 (Baoniabad) under PallabiThanna, Dhaka form where 50 women workers were selected randomly from a total of about 240 workers. A structured questionnaire was used to collect data through interviewing during the months of January to February 2016. It was found that the sampled women workers were from different districts of Bangladesh which had the cultural diversity among the respondents.

The characteristics of women such as age, level of education, family size, family annual income, job experience, and aspiration were considered as independent variables while empowerment achieved due to involvement in IGAs was the dependent variable of the study. The age was measured in terms of years from birth to the time of interview. The year of schooling has completed by a women from formal educational institutions was considered as her level of education. The family members share their income and ate in a same pot is family size. The total income earned by the family members in a year is termed as family annual income. Job experience meant the experience which was gained by a woman from actively doing a job and it was expressed in year. The future plan of a women about her family members in terms of well-being and prosperity. Women Empowerment refers to the creation of an environment for women where they can make their own decisions by themselves and gaining benefit out of it. It also refers to increasing and improving the social, economic, political and legal strength so that they can enjoy equal-right with their male partners. They can live with self-respect, dignity access and control over resources like male members in the society. In addition, they have complete control of their life both within and outside of their home and workplace.

According to Syed and Ruth (1993), Syed et al., (1996) and Ruth et al., (1996) the empowerment of women was measured considering their choice of mobility, economic status and decision making power in household issues. In this study empowerment was measured considering their freedom of mobility and access to decision making in household activities. Twenty two items, 11 related to mobility of women in various places and 11 related to participation of women in various aspects of decision making process were considered. A 4 point rating scale such as regularly, frequently, occasionally and not at all was used. Appropriate weights such as 3 for regularly, 2 frequently, 1 for occasionally and 0 for not at all were assigned. Adding all score together the empowerment score was calculated. However, the possible empowerment score could range from 0 to 66 where, 0 indicating no empowerment and 66 for highest level of empowerment of women. The data after were compiled, coded, tabulated and analyzed. Descriptive statistical measures such as range, mean, percentage, and standard deviation etc. and coefficient of correlation test were computed to explore the relationship between the dependent and independent variables.

RESULTS AND DISCUSSION

The empowerment was the dependent variable and selected characteristics such as age, level of education, family size, family annual income, job experience, and aspiration of the women were the independent variables of the study. The description or characteristics of the respondents is presented below:

Age
The age of women ranged from 19 to 52 years with an average of 30.06 years. The standard deviation was 6.80. Based on age of the women, they were classified into 3 categories, such as young, middle aged and old aged and presented in Table 1.

Table 1. Distribution of the women according to their age

Categories	Women		Mean	Standard deviation
	Number	Percent		
Young (Up to 35 years)	8	16		
Middle age (36- 50 years)	40	80	30.06	6.80
Old age (51 years and above)	2	4		
Total	50	100		

The information of Table 1 show that large majority (96 percent) of the women were young to middle aged while a negligible number were old aged. Usually the young and middle aged women are found working in garments industry. The old aged people are not capable to work there. This may be due to the reason that comparatively the young and middle aged women are more energetic, industrious, enthusiastic, motivated and capable to perform hard work in garments industry. This is how they earn money for giving financial support to their family. It is a general assumption that young and middle aged individuals have more courage to participate IGAs for curving economic problems of the family.

Level of Education:
The level of education of the women respondents ranged from 5 to 16 years of schooling. The mean value and standard deviation were found 10.12 and 2.56 respectively. According to level of education, the women were classified into 6 categories and presented in Table 2.

The information of Table 2 indicate that all the respondent women working in the garments Industry had varying extent of education and none was found illiterate. This means all the respondents were literate. Education increases knowledge and understanding, adjustment capability in any situation and improves mental horizon. Education is also a process of development of mind of an individual which increases his/her power of observation, integration, understanding, decision making and adjustment in new situations. It is therefore, assumed that the literate women might have good sense about enjoying rights and privileges from the society and possessing empowerment as well.

Table 2. Distribution of the women according to their level of education

Categories	Women		Mean	Standard deviation
	Number	Percent		
Illiterate (0)	0	0		
Can sign only (0.5)	0	0		
Primary education(1 - 5)	5	10		
Secondary education (6 - 10)	26	52	10.12	2.56
Higher secondary education (11 - 12)	13	26		
Higher education (Above 12 years)	6	12		
Total	50	100		

Family Size

The family size of women ranged from 2 to 11 with a mean value of 4.88 and standard deviation being 1.70. The average family size of the respondents was found lower against the national average of 5.5 (BBS, 2012). The women were classified into 3 classes based on their family size and presented in Table 3.

Table 3. Distribution of the women according to their family size

Categories	Women		Mean	Standard deviation
	Number	Percent		
Small (Up to 4)	28	56		
Medium (5 – 6)	18	36	4.88	1.70
Large (7 Above)	4	8		
Total	50	100		

The data of Table 3 indicates that large majority (92 percent) of the respondents had small to medium family size while only 8 percent having large family size. This may be due to campaign against population pressure by the various GO and NGOs. However, the reasonable level of literacy of the respondents along with massive awareness activities undertaken by the development agencies the people becomes conscious about population pressure.

Family Annual Income

The family annual income of the respondents ranged from TK. 80 thousand to TK. 500 thousand. The average and standard deviation were TK. 250 and 140.19 respectively. According to family annual income the respondents were classified into 3 categories and presented in Table 4.

Table 4. Distribution of the women according to their annual family income

Categories	Women		Mean	Standard deviation
	Number	Percent		
Low (Up to TK. 100 thousand)	9	18		
Medium (TK.101 – TK. 300 thousand)	20	40	250.10	140.19
High (TK. 301 thousand and Above)	21	42		
Total	50	100		

The information of Table 4 depicted that large majority (82 percent) of the respondents fell in medium to high income category while 18 percent in low income category. This may be due to reason that the working women earning money and supporting their family by giving a lion share of their income. It may be mentioned here that the women are contributing 40 percent to 58 percent of their earnings in supporting their family expenses. This may be a reason for higher family annual income of the respondents. The economic contribution of women is acted as a driving force towards empowerment of women in Bangladesh (Syed and Ruth, 1993).

Job Experience

The job experience of the respondents ranged from 1 to 16 years with the mean and standard deviation of 5.0 and 3.37 respectively. The women based on their service length classified into 3 categories and shown in Table 5.

Table 5. Distribution of the women according to their job experience

Categories	Women		Mean	Standard deviation
	Number	Percent		
Short (1 – 5 years)	40	80		
Medium (6 – 10 years)	6	12	5.0	3.37
Long (11 – 16 years)	4	8		
Total	50	100		

The information of Table 5 depicted that 4 out of 5 respondents (80%) had short service length while only 20 percent having medium to long service length. The information seems logical because large majority (96 percent) of the respondents were young to middle aged. The job experience plays a vital role in achieving empowerment in the society. Because empowerment is highly associated with income and which eventually enhance social prestige, dignity and honour of women in the society.

Aspiration of Women

The level of aspiration score about future life of the respondents ranged from 1 to 28 against the possible range of 1 to 36. The mean value and standard deviation were 16.37 and 7.81 respectively. Based on aspiration score, the women were classified into 3 categories and presented in Table 6.

Table 6. Distribution of the women according to their aspiration

Categories	Women		Mean	Standard deviation
	Number	Percent		
Low (1 - 12)	12	24		
Medium (13 - 24)	28	56	16.37	7.81
High (25 - 36)	10	20		
Total	50	100		

The information of Table 6 depicted that about three-fourths (76 percent) of the respondents had medium to high aspiration and less than one fourth having lower aspiration about future life. This means higher the level of education and family annual income higher is the aspiration of life. The moderate to higher aspiration of the women indicates that they like to see their children having higher education, better income, and better living. The existing social empowerment might play a positive role in forming a positive aspiration for their future generations.

Empowerment of Women

The empowerment score of the women ranged from 29 to 66 against the possible range of 0 to 66. The mean value and standard deviation were 51.60 and 10.16 respectively. Based on empowerment score, the women were classified in to 3 categories and presented in Table 7.

Table 7. Distribution of the women according to their level of empowerment

Categories	Women		Mean	Standard deviation
	Number	Percent		
Low (Up to 22)	0	0		
Medium (23 - 44)	13	26	51.61	10.16
High (45 - 66)	37	74		
Total	50	100		

The information of Table 7 show that all the women respondents fell in medium to high empowerment category. None was found having low empowerment. This means the respondent women are enjoying enough privileges in respect to mobility and involvement in household decision making process. It was explored that the respondents in respect of financial support contributed 40 percent to 58 percent of their income to the family expenses. Nevertheless, all the respondents were literate along with earning through service which made them a privileged section in the community. Furthermore, functional literacy of rural people, media coverage and massive development activities of the development agencies, the mentality of people has been changing gradually towards accepting women as a power for social development. The researchers like Syed et. al., (1996) and Ruth et al. (1996) also found similar findings where women had the significant effect on different dimensions of the social program in Bangladesh.

Relationship between the variables

To explore the relationships between the variables coefficient of correlation test was employed and presented in Table 8.The information of Table 8 show that age, level of education, family annual income and job experience of the women were found significantly correlated with their empowerment. This means these characteristics of the women exerted a reasonable extent of influence in achieving empowerment by the garments workers.

Table 8. Relationship between the dependent and independent variables

Dependent variable	Personal characteristics of women	Co-efficient of correlation
Empowerment of women	Age	0.320*
	Level of education	0.599**
	Family size	0.179
	Family annual income	0.316*
	Job experience	0.338*
	Aspiration	0.183

**1% level of probability, * 5% level of probability

The women were asked to mention the problems they have been encountered in possessing empowerment. They opined that superstitions of the people especially older people, existing value system are hindering empowerment of the women. Arrangement of adult literacy programme along with awareness building campaign by the development agencies could improve the situation.

CONCLUSION AND RECOMMENDATIONS

Involvement of women in income generation activities (IGAs) is found to be a potential factor in enhancing empowerment of women in the male dominated society like Bangladesh. The findings of the study revealed that the existing value system of the society is being changing. This might be due to the reason that level of literacy of general mass is increasing, the role of mass media, communication exposure of people, and development activities of service providing agencies etc. influencing change process of value system. This is why the people are considering women as a potential force for development. The change process needs to be accelerated through adult education and media campaign towards development by collaborative efforts for men and women.

REFERENCES

1. Anju M and SS Ruth, 2005. Women's empowerment as a variable in international development, In: Narayan Deepa., editor. Measuring Empowerment: Cross-Disciplinary Perspectives. Washington, DC: World Bank.
2. BBS, 2012. Bangladesh Bureau of Statistics, Ministry of Planning, Government of the Peoples' Republic of Bangladesh, Sher-e-Banglanagar, Dhaka.

3. BBS, 2015. Bangladesh Bureau of Statistics, Ministry of Planning, Government of the Peoples' Republic of Bangladesh, Sher-e-Banglanagar, Dhaka.

4. Hanny, BC, 2006. What is missing in measures of women's empowerment? Journal Human Development, 7: 221–41.

5. Naila K, 2001. Discussing Women's Empowerment, Theory and Practice, Sida Studies. Vol. 3. Stockholm: Swedish International Development Cooperation Agency; Reflections on the measurement of women's empowerment.

6. Ruth SS, MH Syed, Riley AP and S Akter, 1996. Credit programmes, particularly men's violence against women in rural Bangladesh. Social Science and Medicine, 43: 1729 – 1742.

7. Solava I and S Akter, 2007. Agency and Empowerment: A Proposal for Internationally Comparable Indicators paper prepared for the workshop 'Missing Dimensions of Poverty Data'; 29–30 May. University of Oxford.

8. Syed MH and SS Ruth, 1993. JSI Working Paper No. 3. Arlington, VA: JSI; Defining and Studying Empowerment of Women: A Research Note from Bangladesh.

9. Syed MH, SS Ruth and AP Riley, 1996. Rural credit programmes and women's empowerment in Bangladesh. World Development, 24: 635 – 653.

10. UNDP. 1995. Human Development Report: The World's Women 1995: Trends and Statistics. New York, NY: Oxford University Press and UNDP.

EFFECT OF DIFFERENT DOSES OF IPIL-IPIL (*Leucaena leucocephala*) (LAM.) DE WIT. TREE GREEN LEAF BIOMASS ON RICE YIELD AND SOIL CHEMICAL PROPERTIES

Niloy Paul, Mohammad Kamrul Hasan[*] and Md. Nasir Uddin Khan

Department of Agroforestry, Faculty of Agriculture, Bangladesh Agricultural University, Mymensingh-2202, Bangladesh

*Corresponding author: Mohammad Kamrul Hasan, E-mail: mkhasanaf@gmail.com

---

ARTICLE INFO

Key words

Ipil-Ipil
Leaf biomass
Fertilizer
Rice
Yield

ABSTRACT

A field experiment was conducted to find out the effect of different doses of ipil-ipil (*Leucaena leucocephala*) (Lam.) de Wit. tree green leaf biomass on rice yield and soil chemical properties. Four different treatments such as T_0: Recommended fertilizer dose (Urea 195 kg/ha, TSP 50 kg/ha, MOP 142 kg/ha, Gypsum 75 kg/ha and Zinc Sulphate 4 kg/ha), T_1: 5 t/ha, T_2: 7.5 t/ha, and T_3: 10 t/ha ipil-ipil tree green leaf was used in this study in a Randomized complete block design with three replications. The results showed that the treatment T_3 was performed better than recommended fertilizer dose in case all yield contributing characters of rice except grain yield. The highest (5.29 t/ha) rice grain yield was obtained in recommended fertilizer dose followed by 10 t/ha, 7.5 t/ha and 5 t/ha ipil-ipil tree green leaf biomass amendment having 4.80, 3.16 and 2.36 t/ha respectively. The highest grain yield that was obtained from recommended fertilizer dose was 10.21% higher compared to the highest dose (10 t/ha) of ipil-ipil tree green leaf biomass. It was mentioned that among the different doses of ipil-ipil tree green leaf biomass 10 t/ha performed the best over others. The ipil-ipil tree green leaf biomass was also significantly influenced on some essential nutrient status which is very important for rice production. The highest amount of total N, available P, exchangeable K and available S were found in the treatment T_3 and the lowest in the treatment T_1. Therefore, it can be concluded that the ipil-ipil tree leaf has beneficial effects and could be combined with inorganic fertilizer for sustainable crop yield and maintaining soil fertility.

INTRODUCTION

Agriculture is the most important sector of the economy of Bangladesh. Rice (*Oryza sativa* L.) is the leading cereal crop in the world and staple food crop in Bangladesh. Agriculture in Bangladesh is dominated by intensive rice cultivation. Generally, crop production is a combined impact of soil factors, management and environmental factors. But soil varies considerably in their inherent capacity to supply nutrients, which are gradually declining over time due to intensive cropping with high yielding rice varieties. Nitrogen, phosphorus and potassium are the primary macronutrients and can play a key role to increase the production of rice to a great extent. For crop production, presence of organic matter in soil is important. Plant residues in any agricultural system are an important source of carbon and nutrient for the growth of crops. Organic resources play an essential role in soil fertility management in the tropics by their short-term effects on nutrient supply and longer-term contribution to soil organic matter formation (Palm et al., 2001). A good soil should have organic matter content of more than 3.5%. But in Bangladesh, most of the soils have less than 1.7% and some soils have even less than 1% organic matter (Sattar, 2002). In Bangladesh about 60% of arable soils have below 1.5% organic matter whereas a productive mineral soil should have at least 2.5% organic matter (Rijpma and Jahiruddin, 2004) and the decreasing trend of soil organic matter is continuing. Organic matter depletion is observed in 7.5 million ha of land and declining soil fertility affects 5.6-8.7 million hectares of land (BARC, 2000). However, the traditional approach to soil fertility restoration through spontaneous regeneration of vegetation is coming under pressure due to the expansion of cropland. Soil fertility can be improved by employing agroforestry principles of incorporating organic inputs into the soil.

Agroforestry practices, especially evergreen agriculture and conservation agriculture with trees have emerged as sustainable measures of addressing land degradation and loss of soil fertility (Mwase et al., 2015). In increasing crop productivity for a sustainable crop production, soil quality has to be restored through a sustainable and intensified form of soil fertility amendments such as biomass transfer technology. In the biomass transfer technology, green manure is mulched or incorporated into agricultural soils. The advantage of this technology is that it allows for continuous cultivation as the incorporated green manure provides sustained soil nutrient replenishment (Place et al., 2003). Typically, *Tithonia diversifolia, Leucaena leucocephala, Senna spectabilis, Gliricidia sepium,* and *Tephrosia vogelii* are the most prominent tree species used in biomass transfer systems (Place et al., 2003). Green leaf biomass of leguminous tree such as Ipil-Ipil (*Leucaena leucocephala*) can help to restore soil fertility. Ipil-ipil tree leaf contains nitrogen 20.9 kg to 43.0 kg, magnesium 3.9 kg to 10 kg, calcium 7.5 kg to 20.3 kg, phosphorus 1.5 kg to 4.0 kg and potassium 13.4 kg to 40.0 kg in per ton of dry leaves (Gordon and Wheeler, 1983). Ipil-ipil tree green leaf biomass can add these nutrients to the soil after decomposition process.

In Bangladesh the total dependence on inorganic fertilizers for crop cultivation is a common phenomenon. That is why the fertility of soils is declining day by day creating a major problem for agriculture. But there are little efforts to solve the problem. Therefore, it is necessary to examine the effect of different sources of organic fertilizers like tree leaf biomasses over inorganic fertilizers on crop yield for sustainable crop production and soil conservation. Several studies have been investigated on various aspects of tree leaf biomasses and their effects on crop yield and subsequent soil health throughout the world and in Indian sub-continent by many more researchers. Some studies on various tree leaf biomasses and their effects on crop growth and soil properties have been carried out by Haque et al. (2001), Uddin (2004), Ansari (2006), Hossain et al. (2007), Hasan et al. (2007), Dil Atia Parvin et al. (2007), Mondol et al. (2007), Mojumder et al. (2008), Khan (2009), Sarker et al. (2010), Tanzi et al. (2012), Arifin et al. (2012), Hasan (2014) in Bangladesh. However, the effects of ipil-ipil tree leaf biomass on rice yield and soil health is still in small pockets. Therefore, it thought necessary to determine the effect of ipil-ipil tree green leaf biomass on rice yield and soil chemical properties. Keeping in view of the above aspects the study was undertaken to examine the effect of different doses of ipil-ipil tree green leaf biomass added to the soil at pre-cultivation stage on the yield and yield contributing characters of transplant aman rice, and also on soil chemical properties.

MATERIALS AND METHODS

The experiment was carried out at the Agroforestry Farm, Department of Agroforestry, Bangladesh Agricultural University, Mymensingh during aman season from August-December 2014. The experimental area experiences a sub-tropical climate and belonged to Old Brahmaputra Floodplain (Agro-Ecological Zone No. 9) (UNDP and FAO, 1988). It is characterized by non-calcareous dark grey flood plain soil having pH value from 6.98 to 7.14 and the soil texture is silty loam. The relief of land is flat and above flood level and sufficient sunshine is available throughout the experiment period. The transplant aman rice cv. BRRI dhan49 a modern variety of rice was used as a test crop in this experiment. It is photosensitive and resistant to blast disease. The life cycle of this variety ranges from 100 to 135 days (BRRI, 2010). The seedlings were collected from Bangladesh Agricultural University Central Farm, Bangladesh Agricultural University, Mymensingh. It is usually produces a grain yield of 5-5.5 tha^{-1} under proper management (DAE, 2013). The experiment was conducted in a Randomized complete block design (RCBD) with three replications. There were four different treatments viz. T_0: Recommended fertilizer dose, T_1: 5 t/ha ipil-ipil tree green leaf biomass (500 g/plot), T_2: 7.5 t/ha ipil-ipil tree green leaf biomass (750 g/plot) and T_3: 10 t/ha ipil-ipil tree green leaf biomass (1000 g/plot). The treatments were randomly distributed to the unit plots in each block. The total number of plots was 12. The unit plot size was 1 m × 1 m. The spacing between blocks was 100 cm and the plots were separated from each other by 40 cm. The selected ipil-ipil (*Leucaena leucocephala*) tree green leaf biomass was collected from Bangladesh Agricultural University campus, Mymensingh and tree leaf biomass was chopped by hand and incorporated to the experimental plots soil after 15 days of transplanting of rice seedlings. Thirty three days old seedlings were transplanted with 25 × 15 cm spacing. Two or three healthy seedlings per hill were transplanted in all the plots. After transplanting, necessary intercultural operations were done as required. The rice plants were harvested on 9 December, 2014 at its full maturity when 90% of the grains became golden yellow in colour.

The data collection for studied plant parameters, five hills were randomly selected and carefully uprooted from each unit plot to record yield and yield contributing characters like plant height, total tillers/hill, effective tillers/hill, non-effective tillers/hill, panicle length, number of grains/panicle, 1000-grain weight, grain and straw yield, biological yield and harvest index (%). After plant sampling, the whole plots were harvested and necessary information recorded accordingly. Grain and straw yields were recorded plot wise and expressed as tha^{-1}. For studied soil chemical properties, an initial soil sample was collected before ipil-ipil tree green leaf biomass incorporation. Another soil samples were collected from the experimental plots at 30 days after incorporation of ipil-ipil leaf biomass or 15 days after transplanting of rice and after harvesting of rice. The soil samples were analyzed to determine the soil chemical properties like organic matter, total N, available P, exchangeable K and available S before and after incorporation of ipil-ipil tree green leaf biomass in experimental plots. Organic matter content of soil samples were estimated following the method developed by Walkey and Black (1934). The total N, available P, exchangeable K and available S were determined by following semi-micro Kjeldhal method, modified Olsen method, ammonium acetate (NH$_4$OAc) extraction method and turbidimetric method, respectively described by Page et al. (1989). The collected data both plant and soils were tabulated and analyzed through a standard computer package statistical procedure WASP. The means were ranked by Duncan's New Multiple Range Test (DMRT) (Gomez and Gomez, 1984).

RESULTS

Effect of different doses of ipil-ipil tree green leaf biomass on yield and yield contributing characters of rice cv. BRRI dhan49

Plant height

The tallest (117.79 cm) plant height was produced from the treatment T_3 which was 10 t/ha ipil-ipil tree green leaf biomass and the lowest (104.48 cm) plant height was observed in the treatment of 5 t/ha ipil-ipil tree green leaf biomass. The recommended fertilizer dose produced 114.78 cm plant height which was second highest (Table 1).

Total tillers hill^{-1}

Application of different doses of ipil-ipil leaf biomass had significant effect on the production of number of total tillers hill^{-1}. The highest number (13.07) of total tillers per hill was produced in the treatment T_3 and the lowest (9.20) was produced by the treatment T_1. The recommended fertilizer dose produced second highest (12.27) total tillers hill^{-1} which was statistically similar to the treatment T_3. From the results it is appeared that the increasing dose of ipil-ipil tree green leaf biomass produced highest number of total tillers per hill compared to the recommended fertilizer dose (Table 1).

Effective tillers hill^{-1}

The number of effective tillers per hill increased consistently and significantly with the incorporation of different doses of ipil-ipil leaf biomass. Among the different treatments of ipil-ipil leaf biomass the treatment of 10 t/ha ipil-ipil tree green leaf biomass gave the highest (12.27) number of effective tillers per hill which was statistically similar to the treatment of recommended fertilizer dose. On the other hand, the lowest (8.73) number of effective tillers per hill was produced from the treatment of 5 t/ha ipil-ipil tree green leaf biomass (Table 1).

Non-effective tillers hill^{-1}

The different doses of ipil-ipil tree green leaf biomass were not significantly affected on the production of non-effective tillers per hill. The highest (0.67) number of non-effective tillers per hill was observed from the treatment T_3 and the lowest (0.27) number of non-effective tillers per hill was observed from the treatment T_2 (Table 1).

Panicle length

Different doses of ipil-ipil tree green leaf biomass had significant effect on panicle length. However, the longest (28.95 cm) panicle was obtained from the treatment T_3 which was 10 t/ha ipil-ipil tree green leaf biomass and the shortest (26.15 cm) panicle length was obtained in the treatment T_1 which was 5 t/ha ipil-ipil tree green leaf biomass. The result revealed that the increasing dose of ipil-ipil tree green leaf biomass was gradually increased the highest panicle length (Table 1).

Number of grains panicle^{-1}

The number of grains panicle^{-1} of rice cv. BRRI dhan49 was significantly affected by the different treatments of ipil-ipil tree green leaf biomass. The number of grains panicle^{-1} ranged from 109.31 to 77.05 where the highest (109.31) number was obtained from the treatment of recommended fertilizer dose which was statistically similar with the treatment T_3. On the other side, the treatment T_1 was produced the lowest (77.05) grains per panicle which was statistically similar to the treatment T_2 (Table 1).

1000-grain weight

The results revealed that there was a significant relationship between the 1000-grain weight and different treatments of ipil-ipil leaf biomass. The highest (15.23 g) 1000-grain weight was obtained from the treatment of recommended fertilizer dose. Among the different treatments of ipil-ipil tree green leaf biomass the highest (13.38 g) 1000-grain weight obtained from the treatment of 10 t/ha and the lowest (12.84 g) was in the treatment of 5 t/ha ipil-ipil tree green leaf biomass (Table 1).

Grain yield

The effect of different doses of ipil-ipil tree green leaf biomass on transplant aman rice cv. BRRI dhan49 revealed a significant variation among the various treatments. The result showed that the highest (5.29 t/ha) grain yield was obtained from the recommended fertilizer dose treatment. Among the different doses of ipil-ipil tree green leaf biomass it was observed that the highest dose (10 t/ha) performed the better (4.80 t/ha) grain yield. Other treatments such as 7.5 t/ha and 5 t/ha of ipil-ipil tree green leaf biomass were produced 3.16 t/ha and 2.36 t/ha grain yield, respectively (Table 1).

Table 1. Effect of different doses of ipil-ipil tree green leaf biomass on vegetative and reproductive characters of rice cv. BRRI dhan49

Treatments	Plant height (cm)	Total tillers hill^{-1}	Effective tillers hill^{-1}	Non-effective tillers hill^{-1} (no.)	Panicle length (cm)	Grains panicle^{-1} (no.)	1000-grain weight (g)	Grain yield (t/ha)	Straw yield (t/ha)	Biological yield (t/ha)	Harvest Index (%)
T_0	114.78b	12.27ab	11.93ab	0.33	25.90c	109.31a	15.23a	5.29a	5.10a	10.39a	50.90c
T_1	104.48d	9.20c	8.73c	0.47	26.15c	77.05b	12.84d	2.36d	2.21d	4.57d	51.60b
T_2	111.75c	10.53bc	10.27bc	0.27	27.56b	84.41b	13.24c	3.16c	2.94c	6.11c	51.80a
T_3	117.79a	13.07a	12.27a	0.67	28.95a	106.76a	13.38b	4.80b	4.62b	9.42b	50.94c
Level of significance	**	**	*	NS	**	*	**	**	**	**	**

Note: T_0= RFD (Recommended Fertilizer Dose), T_1= 5 t/ha Ipil-ipil tree green leaf biomass, T_2= 7.5 t/ha Ipil-ipil tree green leaf biomass, T_3= 10 t/ha Ipil-ipil tree green leaf biomass. *&** Figures followed by the same letter(s) in the same column do not differ significantly at ($P \leq 0.05$ & $P \leq 0.01$) according to DMRT.

Straw yield

The straw yield of rice was significantly influenced by the addition of ipil-ipil tree green leaf biomass (Table 1). The result revealed that the highest (5.10 t/ha) straw yield was recorded from the treatment T_3 while the lowest (2.21 t/ha) was produced by the treatment T_1. It was also noted that the next highest (4.62 t/ha) straw yield was obtained from the recommended fertilizer dose treatment. The results of the ANOVA revealed that there was significant difference ($p \leq 0.01$) in the straw yield of rice across the different doses of ipil-ipil tree leaf biomasses (Table 1).

Biological yield

The ANOVA results revealed that the biological yield of rice was significantly ($p \leq 0.05$) influenced by the incorporation of various amount of ipil-ipil tree leaf biomass (Table 1). The highest (10.39 t/ha) biological yield was found from the treatment 10 t/ha ipil-ipil tree green leaf biomass while the lowest (4.57 t/ha) was found in the treatment 5 t/ha ipil-ipil tree green leaf biomass (Table 1).

Harvest index

Harvest index was significantly influenced by different doses of ipil-ipil tree green leaf biomass (Table 1). The result showed that the highest (51.80%) harvest index was obtained from the treatment 7.5 t/ha ipil-ipil tree green leaf biomass while the lowest was found in the treatment T_1 which was statistically similar to the treatment T_3 (Table 1). The results of the ANOVA revealed that there was significant difference ($p \leq 0.01$) in the harvest index of rice across the different doses of ipil-ipil tree green leaf biomasses (Table 1).

Effect of ipil-ipil tree green leaf biomass on some important essential nutrients at 30 days after incorporation and post harvest soils

Total N

Ipil-ipil tree green leaf biomass released an influential amount of nitrogen in soil at 30 days after incorporation (Table 2). The result revealed that the release of total N from ipil-ipil leaf biomass at 30 days after incorporation varied from 0.191% to 0.163% due to the different treatments. The treatment T_3 which was 10 t/ha ipil-ipil tree green leaf biomass released highest (0.191%) amount of nitrogen in the rice field soils while the lowest (0.163%) amount of nitrogen released from recommended fertilizer dose. In the initial soil sample it was found that total nitrogen content is 0.140% (Table 2). On the other hand, the incorporation of ipil-ipil tree green leaf biomass in the rice field also significantly influenced on total N content in the post harvest soils (Table 2). After harvesting of rice, total N content varied from 0.178% to 0.153% in post harvest soils. The highest total N content was 0.178% found in the treatment T_3 which was statistically similar to the treatment T_2 and T_1. The lowest total N content was 0.153% found in recommended fertilizer dose (Table 2).

Available P

The incorporation of different doses of ipil-ipil tree green leaf biomass in the experimental field was significantly influenced on the release of available P (Table 2). After 30 days of incorporation of ipil-ipil tree green leaf biomass, the treatment T_3 which was 10 t/ha ipil-ipil tree green leaf biomass released the highest (30.71 ppm) amount of available P and the lowest (19.13 ppm) amount of available P released from the treatment T_1 which was 5 t/ha ipil-ipil tree green leaf biomass. In the initial soil sample the amount of available P was 26.18 ppm. The recommended fertilizer dose released 30.18 ppm of available P in the experimental soils (Table 2).The available P was also significantly influenced due to the different treatments of ipil-ipil tree green leaf biomass in post harvest soils (Table 2). The highest amount of available P was 29.14 ppm found in the recommended fertilizer dose treatment. Among the different doses of ipil-ipil tree leaf biomass, the amount of available phosphorus was gradually increased with the increase of doses. According to the result it was found that post harvest soils contained 28.86 ppm, 20.92 ppm and 17.44 ppm available P from the treatment of 10 t/ha, 7.5 t/ha and 5 t/ha ipil-ipil tree green leaf biomass, respectively (Table 2).

Exchangeable K

Exchangeable K content in soil was greatly influenced with the application of ipil-ipil tree green leaf biomass at different growth stages of rice (Table 2). After 30 days of incorporation of ipil-ipil tree green leaf biomass, the treatment T_3 released the highest (83.52 ppm) amount of exchangeable K and the lowest (69.24 ppm) in the treatment T_1. The amount of exchangeable K in the initial soil sample was 77.40 ppm (Table 2). Exchangeable K content was also significantly influenced by the different treatments of ipil-ipil tree green leaf biomass in post harvest soils (Table 2). The highest (77.42 ppm) amount of available P of post harvest soils found in treatment T_3 and the lowest in the treatment T_1 which was 67.56 ppm. The second highest and third highest content of exchangeable K of post harvest soils was recorded in the treatment T_3 and recommended fertilizer dose which were 75.36 ppm and 73.32 ppm, respectively (Table 2).

Available Sulphur (S)

Before the application of ipil-ipil tree green leaf biomass in the experimental plots the amount of available S was 8.92 ppm in the initial soil sample (Table 2). After 30 days of incorporation of ipil-ipil tree green leaf biomass a significant increased in the amount of S was observed after soil analysis. The highest amount of available S was 26.05 ppm found in the treatment T_3 and the lowest amount of available S released from the treatment T_1 which was 8.92 ppm (Table 2). In post harvest soils, the amount of S significantly increased with the increase of doses of ipil-ipil tree green leaf biomass. The amount of available S in post harvest soils was 7.83 ppm, 7.50 ppm, 5.35 ppm and 6.43 ppm from the treatment of 10 t/ha, 7.5 t/ha, 5 t/ha ipil-ipil tree green leaf biomass and recommended fertilizer dose, respectively (Table 2).

Table 2. Chemical properties of soils influenced by ipil-ipil tree green leaf biomass at 30 days after incorporation and post harvest soils

Treatments	At 30 days after incorporation				At post harvest soils			
	Total N (%)	Available P (ppm)	Exchangeable K (ppm)	Available S (ppm)	Total N (%)	Available P (ppm)	Exchangeable K (ppm)	Available S (ppm)
T_0	0.163d	30.18b	77.45c	9.64c	0.153b	29.14a	73.52c	6.43c
T_1	0.177c	19.13d	69.24d	8.92d	0.173a	17.44d	67.56d	5.35d
T_2	0.187b	23.14cc	79.41b	12.48b	0.173a	20.92c	75.36b	7.50b
T_3	0.191a	30.71a	83.52a	26.05a	0.178a	28.86b	77.42a	7.83a
Initial Soil Status	0.140	26.18	77.40	8.92	0.140	26.18	77.40	8.92
Level of significance	**	**	**	**	**	**	**	**

Note: T_0= Recommended Fertilizer Dose, T_1= 5 t/ha Ipil-ipil tree green leaf biomass, T_2= 7.5 t/ha Ipil-ipil tree green leaf biomass, T_3= 10 t/ha Ipil-ipil tree green leaf biomass. **Figures followed by the same letter(s) in the same column do not differ significantly at ($P \le 0.01$) according to DMRT.

DISCUSSION

The present study showed that the different doses of ipil-ipil tree green leaf biomass were positively influenced on yield and yield contributing characters of rice cv. BRRI dhan49. From the results, it was found that the treatments had significant influence on plant height, total tillers hill^{-1}, effective tillers hill^{-1}, panicle length, grains panicle^{-1}, 1000-grain weight, grain yield, straw yield, biological yield and harvest index. However, the treatments had no significant effect on non-effective tillers hill^{-1}. The highest (5.29 t/ha) grain yield was obtained from recommended fertilizer dose treatment and the lowest (2.36 t/ha) grain yield was recorded in the treatment T_1. Among the different doses of ipil-ipil tree green leaf biomass, the treatment T_3 produced the highest and overall second highest (4.80 t/ha) grain yield. The above results were supported by previous findings of Apostol (1989), Zoysa et al. (1990), Akter et al. (1993) and Nahar et al. (1996). Apostol (1989) stated that organic and inorganic fertilizer produced higher length of panicle, productive tillers per hill and grain index of rice. Zoysa et al. (1990) reported that incorporation of Leucaena green manure increase N uptake throughout the vegetative period and increased grain yield of rice significantly. Nahar et al. (1996) cited that green manure with Leucaena leucocephala produced highest grain yield which was 4.36 t/ha whereas fertilized (100 kg N/ha) plots produced 4.12 t/ha grain yield of rice. Akter et al. (1993) cited that application of green manure plus chemical fertilizer produced significantly higher yield parameters of rice. Hossain et al. (2007) reported that tree litter application had a significant positive effect on the yield parameters of rice such as plant height, panicle length, tillers per hill, filled grain, 1000-grain weight, and the addition of tree litter to inorganic fertilizer produced significantly higher yield than inorganic fertilizers solely which was fully supported the results of this experiment.

Here, the results also showed that the nutrient status of soils such as organic matter, total N, available P, exchangeable K and available Sulphur were positively affected by incorporation of ipil-ipil tree green leaf biomass. The amount of organic matter, total N, available P, exchangeable K and available Sulphur was higher in the ipil-ipil tree green leaf biomass treatments than recommended fertilizer dose treatment. Among the different doses of ipil-ipil leaf biomass the treatment of 10 t/ha ipil-ipil tree green leaf biomass showed better performance in releasing nutrients than others. The uptake of N, P, K and S was higher in the treatment of 10 t/ha ipil-ipil tree green leaf biomass over the treatments of 7.5 t/ha and 5 t/ha ipil-ipil tree green leaf biomasses. Guan (1989) found that the available N and P content in soil sample taken from plots with the application of organic materials were significantly higher than control which was strongly supportive to the present findings. These results also in agreement with that of Haque et al. (2001) reported that a positive improvement in soil fertility from the application of tree pruning's. This result is also similar to the Maharudrappa et al. (2000) where they reported that the application of tree litter enhanced nutrient availability in soils. These result also in agreement with that of Das and Chaturvedi (2003) who reported that the highest decomposition rate was observed in leaf litter Leucaena leucocephala followed by Sesbania grandiflora, Dalbergia sissoo and Eucalyptus tereticornis. The release of nutrients like N, P, K was higher in Leucaena leucocephala. The species having highest N percentage showed the fastest decomposition rate. Hossain et al. (2007) found the similar type of results in their study where they reported that Ipil-Ipil (Leucaena leucocephala) and Mander (Erythrina orientalis) was the best in building organic matter and total N content over the control plots which treated by recommended fertilizer doses. Mojumder et al. (2008) reported that the addition of leaf biomass slightly increase the amount of total N, available P, exchangeable K and S in soil over recommended fertilizer dose which also supports the results of this experiment.

CONCLUSION

From the experiment it was revealed that the recommended fertilizer dose gave only the highest grain yield of rice over ipil-ipil tree green leaf biomass treatments but soil nutrient contents, growth parameters values were lowest in this treatment. Among the different doses of ipil-ipil tree green leaf biomasses, it was found that the increased dose of ipil-ipil leaf biomass were positively increased the studied parameters of rice. However, it seems that the ipil-ipil tree green leaf biomass i.e. higher dose 10 t/ha has significant impacts on rice production and would be possible to substitute of or apply in combination with inorganic fertilizer although there was some yield loss (10.21%) which was less significant compare to recommended fertilizer. Tree leaf biomass is an important source of essential nutrients which enhances the soil fertility. It is environmentally friendly because the rapid decomposition and the highest residual effect of tree leaf biomasses. Therefore, it can be suggested that for rice production we can use tree leaf biomass as a source of organic matter which is available in agroforestry system, significantly reduces the considerable amount of chemical fertilizer.

ACKNOWLEDGEMENTS

The authors would like to acknowledge giving the financial support from the authority of National Science and Technology (NST) to conduct this research work.

REFERENCES

1.   Akter MS, MKS Hasan, RC Adhikery and MK Chowdhury, 1993. Integrated management of *Sesbania rostrata* and urea-nitrogen in rice under a rice-rice cropping system. Annual Bangladesh Agriculture, 3: 189-114.

2.   Ansari HA, 2006. Effect of different tree biomass and their time of incorporation on the fertility of soil and productivity of BR11 rice. M.S. Thesis, Department of Agroforestry, Bangladesh Agricultural University, Mymensingh.

3.   Apostol EDF, 1989. Influence of mirasoil organic and X-rice liquid fertilizer in combination with inorganic fertilizer on IR66 and IR12 rice varieties. Malabeu, Metro Manila, Philippines.

4.   Arifin MSA, BN Tanzi, MA Habib, MA Mondol and MA Wadud, 2012. Effect of green leaf biomass application of different trees on the yield of rice. Journal of Agroforestry and Environment, 6: 27-31.

5.   BARC (Bangladesh Agricultural Research Council), 2000. Fertilizer Recommendation Guide-2000. Farmgate, Dhaka.

6.   BRRI (Bangladesh Rice Research Institute), 2010. Manual-Adhunik Dhner Chash.

7.   DAE (Department of Agriculture Extension), 2013. BRRI dhan49. Available at: http:// www. ViewPage_Crop_Production_Technology.aspx.htm.

8.   Das DK and OP Chaturvedi, 2003. Litter quality effects on decomposition rates on forestry plantations. Tropical Ecology, 44: 259-262.

9.   Dil Atia Parvin, TM Zakaria, MK Hasan and GMM Rahman, 2007. Effect of leaf biomass of different agroforest trees on the prevalence of insects and yield of rice cv. BR11. Journal of Agroforestry and Environment, 1: 59-62.

10. Gomez KA and AA Gomez, 1984. Duncan's Multiple Range Test. Statistical Procedure for Agricultural Research. 2nd edition, A Wiley Inter-Science Publication, Johan and Sons, New York, pp: 202-215.

11. Gordon JC and CT Wheeler, 1983. Biological nitrogen fixation in forest ecosystems: foundations and applications. Martinus Nijhoff/Dr. W. Junk Publishers, The Hague, Netherlands, p. 312.

12. Guan SY, 1989. Effect of organic manures on soil enzyme activities and N and P transformation. Acta Pedolotica Sinica, 26: 72-78.

13. Haque MA, MI Ali and MK Khan, 2001. Effect of tree pruning on soil fertility system. Pakistan Journal of Biological Science, 4: 647-650.

14. Hasan MK, GMM Rahman and MM Rahman, 2007. Effect of tree leaf biomass on the performance of rice cv. BR11 and subsequent soil health. Journal of Agroforestry and Environment, 1: 7-10.

15. Hasan MM, 2014. Performance of Kangkong and Indian spinach in ipil-ipil based alley cropping system. M.S. Thesis, Department of Agroforestry, Bangladesh Agricultural University, Mymensingh.

16. Hossain KL, MA Wadud and E Santosa, 2007. Effect of Tree Litter Application on Lowland Rice Yield in Bangladesh. Bulletin: Agronomy, 35: 149-153.

17. Khan TA, 2009. Effect of tree leaf litter on the yield and yield contributing characters of T. aman rice cv. BR 11. M.S. Thesis, Department of Agroforestry, Bangladesh Agricultural University, Mymensingh.

18. Maharudrappa A, CA Srinivasamurthy, MS Nagaraj, R Siddaramappa and HS Anand, 2000. Decomposition rate of litter and nutrient release pattern in a tropical soil. Journal of the Indian Society of Soil Science, 48: 92-97.

19. Mojumder MO, GMM Rahman, M Begum, MM Rahman and MSI Majumder, 2008. Effect of different forms of Teak leaf biomass on the yield of rice and nutrient release in the soil. Journal of Agroforestry and Environment, 2: 1-6.

20. Mondol MA, KK Islam, MH Rashid, O Farruk and GMM Rahman, 2007. Residual effect of plant biomass on the performance of mustard. Journal of Agroforestry and Environment, 1: 63-66.

21. Mwase W, A Sefasi, J Njoloma, BI Nyoka, D Manduwa and J Nyaika, 2015. Factors Affecting Adoption of Agroforestry and Evergreen Agriculture in Southern Africa. Environment and Natural Resources Research, 5: 148.

22. Nahar K, J Haider and AJMS Karim, 1996. Effect of organic and inorganic nitrogen sources on rice performance and soil properties. Bangladesh Journal of Botany, 25: 73-78.

23. Page AL, RH Miller and DR Keeny, 1989. Methods of soil analysis. Part-2, 2nd edition. American Society of Agronomy, Inc. Publication, Madison, Wisconsin, USA.

24. Palm CA, CN Gachengo, RJ Delve, G Cadisch and KE Giller, 2001. Organic inputs for soil fertility management in tropical agro ecosystems: application of an organic resource database. Agriculture Ecosystems and Environment, 83: 27-42.

25. Place F, M Adato, P Hebinck and M Omosa, 2003. The Impact of Agroforestry-Based Soil Fertility Replenishment Practices on the Poor in Western Kenya. FCND Discussion Paper No. 160. International Food and Policy Research Institute, Washington, D.C.

26. Rijpma J and M Jahiruddin, 2004. National strategy and plan for use of soil nutrient balance in Bangladesh. A consultancy report SFFP, Khamarbari, Dhaka. Sustainable environment development. Asia Pacific Journal of Environmental Development, 1: 48-67.

27. Sarkar UK, BK Saha, C Goswami and MAH Chowdhury, 2010. Leaf litter amendment in forest soil and their effect on the yield quality of red amaranth. Journal of Bangladesh Agricultural University, 8: 221–226.

28. Satter MA, 2002. Build up organic matter in soil for sustainable agriculture. In: Panorama, The Independent, October 25. p. 10.

29. Tanzi BN, MSA Arifin, MA Mondol, AK Hasan and MA Wadud, 2012. Effect of tree leaf biomass on soil fertility and yield of rice. Journal of Agroforestry and Environment, 6: 129-133.

30. Uddin MR, 2004. Effect of tree litter as green manure on foliar diseases and yield of rice. M.S. Thesis, Department of Agroforestry, Bangladesh Agricultural University, Mymensingh.

31. UNDP and FAO, 1988. Plant Resources Appraisal of Bangladesh for agricultural development. Report-2. Agro-ecological regions of Bangladesh. UNDP, FAO, Rome, p. 116.

32. Walkley A and IA Black, 1934. An examination of Degtjareff method for determining soil organic matter and proposed modification of the chromic acid titration method. Soil Science, 37: 29-38.

33. Zoysa AKN, G Keerthisinghe and SH Upasena, 1990. Effect of Leucaena leucocephala (Lam.) de Wit. as green manure on nitrogen uptake on nitrogen uptake and yield of rice. Biological Fertility Soils, 9: 68-70.

21

POSTHARVEST BEHAVIOR AND KEEPING QUALITY OF POTTED POINSETTIA

M. Ashraful Islam[1*] and Daryl C. Joyce[2]

[1]Department of Horticulture, Faculty of Agriculture, Bangladesh Agricultural University, Mymensingh 2202, Bangladesh; [2]School of Agriculture and Food Sciences, Faculty of Science, University of Queensland, Gatton, Australia.

*Corresponding author: M. Ashraful Islam, E-mail: ashrafulmi@bau.edu.bd

ARTICLE INFO	ABSTRACT
Key words *Botrytis* Bract Calcium Ethylene Poinsettia Preharvest Postharvest Quality	Poinsettia is one of the most important potted plants in the ornamentals industry. The most attractive parts of the poinsettia are its bracts and cyathia. Stem breakage, bract fading, bract edge burn (BEB), bract bruising and bract discoloration reduce its ornamental value. The ornamental value of poinsettia is affected by both pre- and postharvest factors. During cultivation as well as after harvest, sub-optimal or improper supply and / or management of temperature, light, relative humidity, nutrition and hygiene affect the quality of potted poinsettia. A low K : Ca ratio can reduce the incidence of bract necrosis and stem breakage. These disorders are negatively affected by high K and NH_4-N fertilizer due to an antagonistic relationship with Ca. Spraying either $CaCl_2$ or $Ca(NO_3)_2$ (400mg/l of Ca per week) or a silicon spray (7.12 mM Na_2SiO_3) and increasing the transpiration rate by reducing the relative humidity can reduce the incidence and severity of bract necrosis. Appropriate variety selection can lessen the risks of leaf drop and BEB. Propagating with thick stemmed cuttings (> 7.5 mm diameter) and carefully watering of plants can also help to prevent stem breakage. The optimum temperature range to achieve expected bract color, size and bud formation during the growth period is 16 to 22°C. Inappropriate light intensity and quality can delay flowering and promote cyathia abscission by decreasing plant carbohydrate status. Ethylene is central to the leaf and cyathia abscission. 1-MCP treatment can be used during short time (2-3 day) transportation to prevent ethylene responses by blocking ethylene receptors. Avoiding rough handling and shortening transportation play an important role in maintaining the quality of harvested poinsettia. The most important pathological problem is Botrytis infection. Low temperatures of 20 to 21°C during cultivation and 10 to 13°C during transport combined with low relative humidity of 60 to 70% RH can reduce infection by *Botrytis cinera*.

INTRODUCTION

The ornamentals or floriculture industry is a very important part of the overall horticulture industry. Cut flowers and foliage, potted flowering and foliate plants, bedding plants and nursery plants for gardens and landscapes are divisions in the floriculture industry (Chandler, 2003; Chandler and Brugliera, 2011). Poinsettia (*Euphorbia pulcherrima* Willd. ex Klotzsch) is one of the most important flowering potted plant lines. They are targeted to the Christmas market in regions, including in North America, Europe, Asia and Australia. This sector represents as an industry valued at over $ 154 million per annum (Ecke et al., 2004; USDA, 2009). Poinsettia is also called 'Christmas Star'; December 12th is designated as 'national poinsettia day' in the USA. Potted poinsettias have red, pink or white coloured bracts. Poinsettia production in the USA and the EU is ~ 50 and ~ 100 million plants per year, respectively (Lütken et al., 2012). Potted poinsettia plants are exported where road transport is adequate and are moved in large numbers between countries, especially in Europe.

Poinsettia is sold for its colorful inflorescence (cyathia) and bracts (transition leaves). Important indicators of a good postharvest quality in poinsettia plants are fully formed, turgid, blemish free evenly coloured bracts and leaves along with intact, fresh-looking cyathia. The estimated postharvest loss of flowers is around 10 to 30% over all stages in the value chain (Personal communication, Sissel Torre, Norway). In the 1960's, poinsettia was sold just a week before Christmas time due to the risks of epinasty and abscission of cyathia. The epinasty problem has been largely overcome by new or modern cultivars (Personal communication, Jim Faust, USA). The most serious current postproduction disorders of poinsettia are stem breakage, bract fading, leaf yellowing, bract edge burn (BEB) or necrosis and bract bruising (Ranch, 2012; Whipker, 1999). These disorders are influenced by to varietal selection, environmental conditions and nutrition during growth and postharvest management. Ethylene plays an important role in the abscission of leaves and cyathia of poinsettia. Breeders today are focusing more on the postharvest quality of poinsettia. New cultivars with a longer postharvest life are available. Poinsettia is now-a-days in the marketplace by early November (Bævre, 1994; Odula, 2011).

As an ornamental product in an economic context, it is important to maintain poinsettia quality throughout postharvest handling and distribution from the producer to the consumer (Fig. 1). Postharvest management begins with determining the harvest maturity stage that the plants are ready for market. It continues through establishing the proper storage, transport, retail and consumer conditions. The overarching goal is to maximize crop performance and quality. Considerable research effort has been devoted to improving the keeping quality of poinsettia plants.

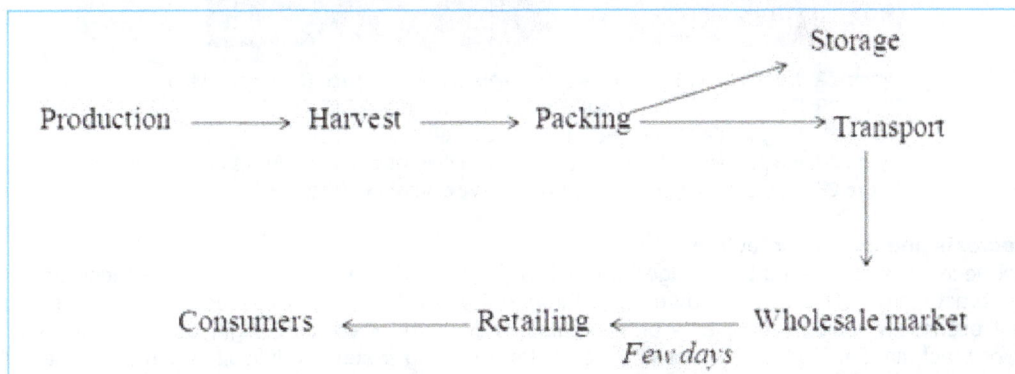

Figure 1: Flow diagram showing the serial steps of postharvest handling of ornamental plants such as poinsettia, orchid and others.

As noted above, considerable research has been carried out on poinsettia. For example, Islam et al. (Islam et al., 2012; 2013; 2014) recently worked on controlling poinsettia shoot elongation investigating light quality, hormonal physiology, genetic engineering and postharvest quality. However, there is no recent review on the postharvest quality of poinsettia. In this regard, recent research on keeping quality and postharvest behavior of poinsettia is overviewed herein. This review considers genetics, climatic factors, production techniques, nutritional disorders, transport and storage.

POSTHARVEST PROBLEMS OF POINSETTIA

Stem breakage

Plant pinching promotes the growth of lateral shoots resulting compact plants. However, fewer shoots may develop in a horizontal orientation leading to pressure at the branch junction resulting in stem breakage (Kuehny et al., 2000). Low light levels during the vegetative growth phase after pinching can contribute to poor structural development in the crotch area and / or lateral weak branch growth (Fig. 2A). Small diameter cuttings, improper plant spacing and inappropriate fertilization can also influence lateral stem breakage. Stem breakage is problematic during production and / or after sleeving. Such breakage markedly reduces market value and, thus, economic returns.

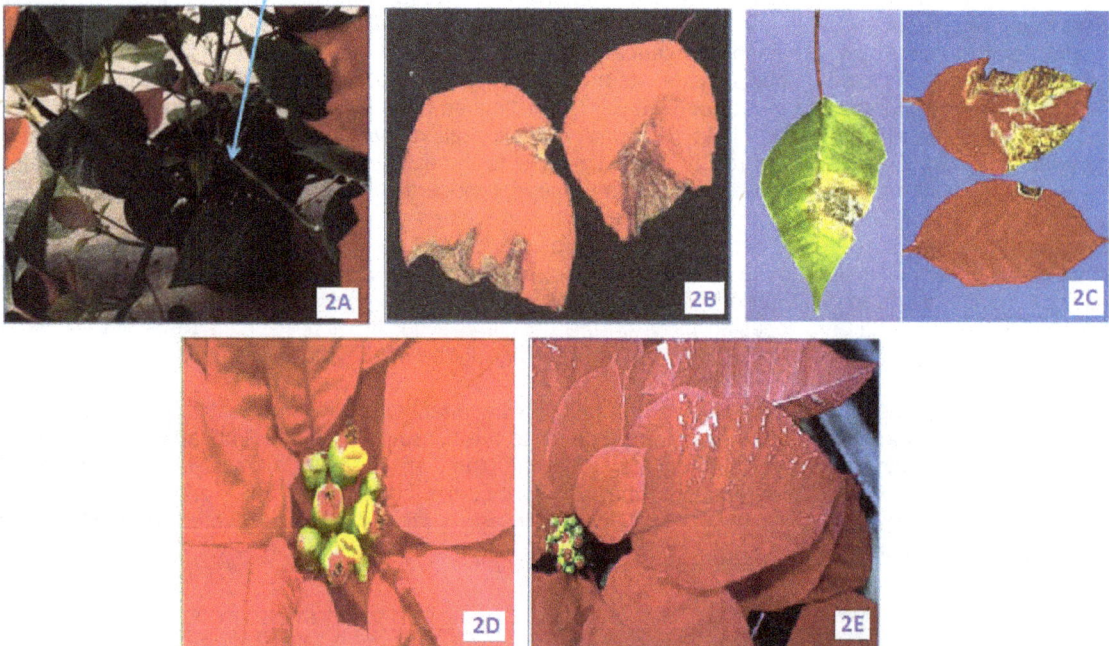

Figure 2A: Weak lateral stems (arrows) that tend to stem breakage (Hammer, 1999).
Figure 2B: Bract necrosis/bract edge burn (BEB) (Tayma and Roll, 1990)
Figure 2C: *Botrytis* infected leaf (left) and bracts (right) (Jones, 1999)
Figure 2D: Bracts (transition leaves) and cyathia of poinsettia (Photo: Md. Ashraful Islam)
Figure 2E: Bract abrasion (arrow) due to excessive contact (Whipker, 1999)

Bract necrosis and Botrytis infection

Bract necrosis is also called bract edge burn (BEB) (Fig. 2B). BEB starts as small brown necrotic lesions along the bract margin. The lesions may expand through the develop *Botrytis* infection to damage the entire bract that essentially renders the plants unmarketable. Low level of calcium might be the reasons of the disorder or bract margin. On the other hand, calcium translocating system is difficult in some varieties which are essential elements for the cell expansion. It will not be possible to solve using more application of calcium which are cultivar dependant (Ranch, 2011). Botrytis blight or gray mold (Fig. 2C) is caused by the fungal pathogen *Botrytis cinerea*. In poinsettia, Botrytis blight can occur at all stages during production and postharvest. Regardless of the tissue affected, the rot begins as water soaking developing into tan to brown lesions.

Excess fertilizer application, particularly ammonium sources of nitrogen at the late stage production, can lead to bract burn and then *B. cinerea* infection. It is considered that contributory factors to bract necrosis include excess soluble salt levels in the growing medium that causes root injury, reduced water absorption stress from excessive or insufficient irrigation, damage from pesticides, pollutants, and high relative humidity.

Leaf drop and yellowing

A lack of irrigation water can cause leaf drop that affects older leaves. Leaf fall under stress conditions occurs at an abscission layer at the junction of petiole and the stem. It is believed that leaf drop is mediated by a loss of auxin export across the abscission zone from the leaf blade under stress condition. Irregular irrigation, low light intensity, warm temperature and low relative humidity are thought to influence leaf drop at any level in the supply chain up to and including where the poinsettia place is placed by the consumer. In particular, the lower leaves yellow and drop if the plants are kept under very low light intensity for several days.

Cyathia abscission

Premature abscission of one or more of the cyathia or true flowers on poinsettia also reduces the ornamental value (Fig. 2D). Abscission *per se* results from degradation pectin in the mid-lamella and of pectin, cellulose and arabinose in cell walls of the abscission zone (AZ). The levels of xylan and xyloglucan in the AZ zone is shown to increase prior to abscission, rendering the cell walls more loosely organized and susceptible to degradation (Lee et al., 2008).

Bract discoloration / abrasion

Poinsettia bracts are fragile and easily discolor. They can be severely damaged by temperatures below 10°C (Love, 1999). Expansion of discoloration depends on the duration of exposure to such cold temperature. Red bracts develop a blue to silver white color (Fig. 2E). During shipping, bract abrasion occurs due to rubbing against the sleeve or against each other. Sleeves and boxes which are taller than the plants can help to prevent bract abrasion as well as confer a degree of protection against chilling injury.

IMPROVING THE POSTHARVEST QUALITY OF POINSETTIA

Genotype

Genotype or cultivar selection is one of the most critical decisions for growers. Poinsettia breeding has attracted considerable attention resulting in increased demand for better quality cultivars. Thus, cultivar selection is important to the commercial grower. New poinsettia cultivars are introduced by poinsettia breeding companies each year. A national poinsettia trial programme has been evaluating poinsettia cultivars with the collaboration of poinsettia breeding companies for over 19 years since started in 1993. Barrett and Dole (2012) surveyed 21 new poinsettia cultivars of varying color and type in three locations in a comparison with established cultivars. Cultivar 'Ice Punch' was selected as the best cultivar with strong (46%) consumers support. Dunn et al. (2011) studied 40 cultivars at different temperatures and determined differences in bract numbers, size and lateral shoots among the cultivars. BEB emerged as a major problem in the late 1970s with the introduction of 'Gutiber V-14 Glory'. Now-a-days it is considered that BEB is cultivar dependent (Ranch, 2011).

Self-branching cultivars, such as 'Annette Hegg Dark Red' and 'Gutibar' havethin stems that are prone to bend and break during handling or shipping (Larson et al., 1978). Some cultivars like 'Success' and 'Red Splendor' are relatively more resistant to stem breakage (Whipker, 1999). More recently introduced cultivars show better postproduction quality as compared to older ones. Some newer varieties were evaluated by Ranch (2012) for their relative postproduction problems. All varieties are ranked into the categories of 'excellent', 'good' or 'below average' after 4 weeks considering postproduction performances like bract edge burn (BEB), bract fading, leaf yellowing, leaf drop, bract bruising etc.

Environmental factors

Light

Light intensity during the growth period can affect the post-production keeping quality of potted ornamental plants (Fjeld, 1990). In poinsettia, low light intensity was shown to reduce stem strength, delay flowering and increase cyathia abscission (Faust et al., 1998; Fjeld, 1992; Moe et al., 1992a; 2006). A stem breakage problem was observed by Kuehny et al. (2000) in cv. 'Freedom Red'. Due to pinching, a large number of lower laterals grew out and upwards towards light. This growth habit caused weaker lateral stems and wider lateral branch angles, especially in lower stems. In general, after pinching the competition for light during the vegetative growth causes weak lateral branching (Smith and Cox, 2009).

An increase in irradiation from 12 to 73 $\mu mol . m^{-2} . s^{-1}$ during production increased cyathia retention in the 1st and 2nd weeks of postharvest testing under indoor climate conditions (Moe et al., 1992b). Extending the photoperiod from 10 h to 24 h the last 3 weeks of production resulted in higher postharvest cyathia abscission rate and discoloration. Moe et al. (2006) tested postharvest performance of cvs. 'Lilo' and 'Millenium' and established differences. Bract discoloration occurred on 'Lilo' under production in a 24 h photoperiod as compared to the 10 h control treatment. One hundred (100) % bract discoloration occurred in 'Lilo' during the last 3 weeks of cultivation under the 24 h photoperiod. However, 'Millenium' showed no discoloration in red bract under control as well as with the supplemental light of 24 h photoperiod. Carbohydrate depletion might be the reason for cyathia abscission. It is not clear about the retention capacity of cyathia in different cultivars

(Moe et al., 2006). Bract discoloration may be a result of accumulation of anthocyanin or other phenolic compounds under high light intensity or long photoperiods. Petal blackening of roses occurred due to this phenomena at low night temperature exposure (Zieslin and Halevy, 1969). Overall, premature abscission is interlinked with long photoperiod and low light intensity (Scott et al., 1983; 1984a; Staby and Kofranek, 1979).

Carbohydrate status and sugar supply to floral organs likely play a role in cyathia abscission (van Doorn, 2001). In *Christmas begonia*, Fjeld (1992) measured sucrose content and abscission rate during the marketing stage for plants grown at different irradiance level. The sucrose content was a higher contributor to the dry matter of inflorescences and leaves (10%) and the abscission rate was lower for flowers and buds (19%) on plants grown under 60 μmol . m^{-2} . s^{-1} as compared to 15 μmol . m^{-2} . s^{-1} (2.1% sucrose and 36% abscission).

Towards saving energy and manipulating plant growth in an environmentally sounder manner that also improves plant health, the use of specific light qualities provided by light emitting diodes (LED) is attracting attention. It is well known for a variety of long day plants that blue (B) and red (R) light promote flowering while far-red (FR) delays flowering (Simpson and Dean, 2002). No difference in the keeping qualities of poinsettia cvs. 'Christmas Spirit' and 'Christmas Eve' was found after exposure to LED lamps providing 80% R and 20% B light irradiance at 100 μmol . m^{-2} . s^{-1} (Islam et al., 2012). Light quality manipulation resulting in an increased R : FR ratio at the EOD (end of day) for 30 min used to avoid unwanted extension growth. It is well known that both phytochrome and cryptochrome are involved in the content of hormone gibberellin (Hisamatsu et al., 2005; Zhao et al., 2007). However, this treatment did not have any effect on the flowering of the two cultivars of poinsettia (Islam et al., 2014). Nevertheless, postharvest behavior is cultivar dependent. It has been recorded that the senescence of poinsettia was faster under fluorescent lamps as compared to under incandescent tungsten lamps (Scott et al., 1984a).

Temperature

Stem breakage, flower bud formation, flower development, bract color, bract size, bract discoloration, bract necrosis and *Botrytis* infection are all influenced by temperature during the plant growth period. Pritchard et al. (1996) observed that the bract and foliage infected area increased during the postproduction, if the temperature is increased both at day and night time from 16 to 22°C during the poinsettia production. Higher temperature was considered to increase turgor pressure that resulted in the extracellular leakage of nutrients, salts and sugars in the bract and to the leaf surface which influence the germination of *Botrytis cinera* conidia (Blakeman, 1975; Salinas et al., 1989; Van Meeteren, 1980). As per above, common symptoms caused by *Botrytis* infection are leaf spots and blighting which may affect leaves, stems and petioles (Hausbeck and Moorman, 1996). Schnelle and Barrett (2011) recorded chroma values for bract fading or discoloration at 27°C as compared to at 21 and 24°C. Bract discoloration at the higher temperature of 27°C gave a chroma value of 39 as being significantly lower than 55 at 21°C and 54 at 24°C. Also, it has been observed that ~ 34% lateral shoots were broken at a relatively lower temperature and 4% at a higher temperature while the night temperature was maintained at 20°C with the day temperature was 30 or 37°C (Faust et al., 1998). Too low average daily temperatures cause slowed plant development, delayed color development, reduced bract size and elicited lower leaf loss from poinsettia plants (Whipker, 1999). Runcle and Faust (2008) reported that poinsettias growth stopped at < 13°C.

Moe et al. (1992a) tested the keeping quality of poinsettia plants grown under different day and night temperatures (DIF). A negative temperature difference between day and night (DT < NT) was found to delay flowering (Moe et al., 1992a; 1995). In these plants, bract necrosis and cyathia abscission were higher during the 1st and 2nd week postharvest as compared to plants that experienced positive (DT > NT) and zero (DT = NT) DIF. After 4 weeks, there was no difference in bract necrosis between the plants grown under the different temperature treatments. Although there is limited information on the effect of temperature on carbohydrate status in poinsettia, premature cyathia abscission is probably linked to carbohydrate depletion (Miller and Heins, 1986). Photosynthetic rates are correlated with temperature and a lower photosynthetic rate was found under negative DIF as compared to positive and zero DIFs (Berghage et al., 1990). Moe et al. (1992a) mainly applied negative DIF to get compact plants, but the postharvest quality was not good. However, such a treatment given early in poinsettia production with an ADT (average daily temperature) 23 to 26°C from visible bud formation to flowering can result in compact plants without serious reduction in postharvest quality (Berghage, 1989). Temperature after flowering appears to be important for the keeping quality of poinsettia. In general, the optimum temperature for bract development is 22 to 23°C during the day and 19 to 20 °C at night. Before shipping for 2 weeks, the temperature should be maintained at 21°C and 20 °C day and night, respectively, during production (Whipker, 1999). The lowest postharvest quality rating was scored at <17°C in regard of leaf yellowing, cyathia drop and bract edge burn (Syngenta (2009).

Water and air humidity

Water supply and air humidity conditions affect the morphological and physiological condition of ornamental plants, including their postharvest life. It is claimed that excess water stress or water deficiency stress both interfere with the longevity of plant organs. Despite this general understanding, the effects of water supply on the longevity of ornamental plants have not been comprehensively examined. Water supply and air humidity both have correlative influences on plant transpiration s. For instance, the water consumption per unit leaf area is lower at higher humidity and Mortensen (2000) mentioned that water consumption per unit area of leaf is decreased 39% in poinsettia during the water vapour pressure deficits (VPD) decreased from 660 to 155 Pa.

The effect of air humidity on keeping quality is unclear for poinsettia and other greenhouse grown flowering plants (Grange and Hand, 1987). However, deficit irrigation and low air humidity can both stimulate tissue dehydration to cause lower stomatal conductance values (gs) reflecting efficient stomatal regulation (Sánchez-Blanco et al., 2004). Drought stress to water potentials of -1.3 to -1.1 MPa can elicit leaf drop in poinsettia during the reproductive growth stage (Gilbertz et al., 1982; Whipker, 1999). Drought stress may induce endogenous ethylene resulting premature loss of organs.

In contrast, high humidity can reduce the rate of transpiration (Ehret and Ho, 1986). Mortensen (2000) observed transpiration rates for poinsettia during the growth at air humidities of 70%, 81% and 93%. The transpiration rate was higher at 31.9 ± 2.9 g . day^{-1} under 70% RH as compared to 23.7 ± 2.2 g . day^{-1} under 93% RH. It has been shown that Ca plays an important role for postharvest life of ornamental plants like cut roses and *Antirrhinum*. Ca uptake and transport are affected by transpiration rate (Higaki et al., 1980; Marschner, 1995; Michalczuk et al., 1989; Nielsen and Starkey, 1999; Starkey and Pedersen, 1997). As previously described, Ca is integrally related to bract necrosis and stem breakage of poinsettia. Tissues having low xylem flow due to low transpiration rate have low translocation of Ca into the bracts leading to bract necrosis. Thus, the Ca content becomes lower in bracts and leaves of poinsettia grown at high RH as compared to for those grown at moderate RH. Similar results were found for flowers and leaves of cut roses grown under high RH, resulting in their comparatively shorter postharvest life (Baas et al., 1998; Strømme et al., 1994; Torre et al., 2001). There also appears to be a difference in transpiration rates between the bracts and leaves of poinsettia. Consequently, the Ca content is commonly lower in the bract (Robichaux, 2008; Strømme et al., 1994). Insufficient Ca in bracts may be due to low stomatal density in this relatively low transpiring organ (Nell and Barrett, 1986a).

Both high air humidity and water stress favor *B. cinerea* in greenhouses (Hausbeck and Moorman, 1996; Whipker, 1999). During the growing phase, roots are primarily affected. Infection may later cause stem rot and plant death. Stressed or wounded leaf tissues and true flowers of poinsettia are particularly affected by *B. cinerea*. Low Ca levels favour this pathological disorder (Ranch, 2011). Lowering the air humidity, maintaining good air circulation, reducing moisture in the plant canopy, cleaning away plant debris, and application of approved fungicides like benzimidazole and dicarboximide can prevent *B. cinerea*. During production, small physical breaks sometimes occur at branch and main stem junctions under to force of hand watering. Watering gently can reduce this type of stem breakage.

Management practices

Cuttings and space
Poinsettias are vegetatively propagated by cuttings. Stem breakage, which is the most important production problem in poinsettia, is affected by cutting diameter. Two times more stem breakage was observed for 4.5 mm or less cutting diameters as compared to 7.5 mm or greater diameters (Ranch, 2011). Poinsettia plant development is weak from thin diameter cuttings. Consequently, resulting inflorescences cannot be kept erect without the staking. Plants growing in pots from cuttings should be placed close together at the beginning of production cycle in order to encourage upright growing. Otherwise, angled to horizontal lateral shoots will be produced in the open space. In addition to thicker stem cuttings, support rings can help prevent stem breakage.

Nutrition
Fertilization influences the postproduction performance of poinsettia. The greatest difference in Ca contents in the interior part of leaf blades was between healthy leaves and those suffering from leaf edge burn (LEB). At the end of the postharvest period, Ca content was analyzed in the bracts of poinsettia (Strømme et al., 1994). The Ca content was lower at the margin as compared to middle or interior bract sections, indicating localized Ca deficiency. McAvoy et al. (1998) observed condensed tannins in bract tissues with necrotic lesions as compared to healthy tissue.

Application of a high NH_4-N fertilizer level indirectly resulted in necrotic spots by reducing Ca absorption (Nell and Barrett, 1985). On the other hand, high NH_4-N treatment in solution 'A' was associated with the lowest incidence with no influence on Ca uptake (Strømme et al., 1994). It has been reported that BEB increased by two fold when the nitrogen (N) supply concentration was increased from 200 ppm to 400 ppm (Ranch, 2011). The reason may be that NH_4 is converted to NO_3 quickly such that very little NH_4 is left to compete with Ca uptake. Ammonia-nitrogen promotes vegetative growth that may lead to also vegetative growth as well as weaker stem resulting stem breakage (Whipker, 1999). After application of different ratios of NH_4-N : NO_3-N (0.1, 1.2 and 2.1) during production, a difference was detected in Ca contents between the outer margins and the interior section of leaf blades (Biermann et al., 1990).

Although the mechanism is not entirely clear, spraying with $CaCl_2$ and Si can suppress bract necrosis development, (McAvoy and Bible, 1996a; 1996b; McAvoy et al., 1998; Woltz and Harbaugh, 1985). Si spray reduces the incidence of bract necrosis / LEB without effects on macro or micro nutrient contents in the leaf / bract margin tissues (McAvoy and Bible, 1996b). Si is taken up by plant roots in the form of silicic acid and transported to the shoot. Si treatment can reduces the plant transpiration rate (Ma and Takahashi, 2002). Termination of Si fertilization 1 to 4 weeks prior to harvest appears to reduce leaf drop (Prince and Cunningham, 1988; Ranch, 2011). Ca and Si treated poinsettia plants typically show better stem strength (Kuehny et al., 2000; Robichaux, 2008). Ca plays a key role in cell wall structure. It is involved in cross-linking pectin molecules (Ferguson and Drøbak, 1988; Hepler, 2005). As above, poinsettia stem breakage, leaf edge burn (LEB) and bract necrosis are reportedly linked to calcium (Ca) supply and can be improved by spraying with Ca and silicon (Si) (McAvoy and Bible, 1995; Robichaux, 2008; Woltz and Harbaugh, 1985).

There is an apparent antagonistic relationship between potassium (K) and Ca (Tsutsui and Aoki, 1981). Ca is translocated from the root to the different parts of the shoot through the transpiration stream in the xylem and Ca mobility through the phloem is limited (Strømme et al., 1994). Substrate pH is important for poinsettia production. Leaf edge burn occurs and mature leaves in the middle of the plants become yellow due to molybdenum (Mo) deficiency (Whipker et al., 2002). At pH 6.5, new leaves become yellow due to iron (Fe) deficiency.

Storage and transportation

In addition to production factors, the keeping quality of poinsettia is dependent on postproduction factors. Deterioration occurs quickly such that the ornamental value is reduced due to poor treatment, handling, transportation and storage. It is important to understand the plants physiological requirements in respect to postharvest environmental factors (e.g. light, temperature, air humidity) from harvest to the consumer (Fig. 1). In general, growers, distributors and wholesalers use preservative chemicals, hygiene and cold chain management to maintain product quality (Eason, 2006). In poinsettia, problems such as bract discoloration, leaf yellowing, and shrinkage in terms of plant losses occur during the postharvest period. These issues mean lower returns and ultimately to cost cutting on the input side. Thus, it is vitally important to give due care and attention to proper handling, transportation, and storage to retain harvest quality.

In Norway by way of example, poinsettia plants are placed either in open framed Danish containers or into sleeves and boxed before leaving the growers premises for transportation. As poinsettia is a chilling sensitive species, the paper sleeves may protect the plants from low temperature injury. Leaf drop and bract discoloration can occur due to non-optimal temperatures and long durations of storage. These disorders are associated with ethylene and auxin physiology (Sacalis, 1978; Saltviet et al., 1979; Staby et al., 1980; 1978). Active auxin transport inhibits the formation of abscission layers by reducing abscission zone sensitivity to ethylene. On the other hand, ethylene inhibits the transport of both indole-3-acetic acid (IAA) and α-naphthalene acetic acid (NAA) in midrib sections, resulting in reduced auxin transport through the abscission zone (Riov and Goren, 1979). Ethylene is considered the primary regulator of abscission process and auxin acts as a suppressor of the ethylene effect (Taiz and Zeiger, 2010).

Faust and Enfield (2010) suggested that 10 to 15°C is the optimum temperature range for shipping of un-rooted poinsettia. Higher temperatures can increase ethylene concentrations. For example, ethylene increased over 48 h from 0.0 to 0.5 ppm with increasing temperature from 10 to 26°C in plastic bags packaged un-rooted poinsettia cuttings while ethylene was measured at 12, 24, 48 and 72 h (Faust and Lewis, 2005). Following 72 h storage, 4 days in propagation, the leaf abscission was increased at the temperature from 10 to 26°C.

Light is also known to influence auxin content and thus leaf drop. Decreased cyathia retention occurred during storage for 3 to 6 d under darkness (Scott et al., 1982,1984b; Shanks et al., 1970). In retail shops, poinsettia plants are typically packed tightly on shelves tightly in low light. Their deterioration under such conditions reduces their ornamental value.

Ethylene can stimulate abscission of whole flowers and flower buds from ornamental plants (Woltering, 1987). Among other chemical compounds, 1-methylcyclopropene (1-MCP) can effectively block ethylene responses at the receptors (Serek et al., 2006). 1-MCP has been shown to protect many potted plants as well as other horticultural produce against ethylene action (Faust and Lewis, 2005; Serek et al.,1994; Serek et al., 2006). It can be used in shipping boxes or containers. Preventing gas leakage during treatment makes for effective delivery. Faust and Lewis (2005) used 1-MCP sachets during shipment to prevent the ethylene action and reduce leaf yellowing and abscission in un-rooted poinsettia cuttings. Similar findings were found in Pothos (*Epipremnum pinnatum*) (Muller et al., 1997). Faust and Lewis (2005) showed a potential benefit of placing 1-MCP sachets in with un-rooted poinsettia cuttings shipped at the relatively warm temperature of 18 to 26°C. At low temperature (5 to 10°C), 1-MCP is not necessarily needed due to low ethylene production at low temperature (Jiang et al., 2002).

Botrytis disease is favored at room temperature due to injured or succulent tissue (Ecke et al., 1990). Infection can occur during transportation, particulary if the plants become too wet with moisture on their foliage. Temperature fluctuations during transportation and / or storage commonly result in high humidity and condensation of water onto bracts or leaves. Thus, maintaining a stable temperature is important. The optimum temperature appears to be about 12 to 13°C. A low temperature of < 4°C results in chilling injury and higher temperatures of 16 or 24°C are associated with leaf abscission (Nell and Barrett, 1986b; Scott et al., 1983).

In general terms, it can be difficult to manage poinsettia plants during shipping or on retail display in respect to providing proper light and water management (Barrett, 2011). Darkness is a particularly poor condition for the plant. Sleeving plants can avoid damage during handling by customers.

RECOMMENDATIONS

The keeping quality of poinsettia plants is influenced by cultivar selection, optimum pre-harvest cultural and pre- and postharvest environmental conditions (viz., temperature, light, air humidity) as well as general management practices, including during transportation and storage. To maintain high-quality poinsettia plants, the following guidelines may be followed.

- Selection of 'new' poinsettia cultivars which are less susceptible to BEB and leaf drop as compared to 'old' cultivars.
- Avoidance of high temperature after flowering and maintenance of an optimum growing temperature of 20 to 21°C.
- Provision of at least100 $\mu mol . m^{-2} . s^{-1}$ B (20%) and R (80%) LED light for growing poinsettia for no negative effect on postharvest behaviour. Under indoor conditions, a minimum irradiation of 10 $\mu mol. m^{-2} . s^{-1}$ should be provided by fluorescent tubes.
- Maintenance of 60 to 70% air humidity for poinsettia production and around 40 to 60% RH for postharvest and / or indoor conditions.
- Achievement of an optimum temperature range at between 10 to 13°C during transportation and avoidance of excess water during storage with a view to minimise BEB and *Botrytis* infection. Long distance transport for > 3 d should be avoided and 1-MCP sachets can be used.
- At retail level, realising a minimum light level (10 $\mu mol . m^{-2} . s^{-1}$) and a temperature of about 18 to 20°C in the shop. Plants should ideally be kept on shelves or racks with spacing that ensures optimum light to them all. A few plants might be separated display with the aim of avoiding rough handling of the others by consumers.
- Optimum fertilization should be practised with thought to K, Ca and N due to potential antagonistic effects. EC and pH levels should be considered. Towards the end of the production phase, applications of K and NH_4-N should be reduced. Conversely, Ca should be applied during the last weeks before harvest.
- Spraying of leaves and bracts with calcium chloride or calcium nitrate (400 mg/l of Ca per week) or with silica (7.12 mM Na_2SiO_3) to reduce the incidence and severity of necrosis. Also, maintenance of low air humidity with a view to increasing plant organ Ca contents and minimising *Botrytis* blight.

ACKNOWLEDGEMENTS

First author would like to give thanks to Professor J.E. Olsen, Dr Sissel Torre and Dr J.L. Clarke, Norway who encouraged to write this paper. Also, thanks to Dr Jim Faust, USA who updated on some information on poinsettia. First author is grateful to Professor Daryl C. Joyce, Australia for continuous support as well as for the critical reviewing of this manuscript. Finally, both authors are decided to dedicate this manuscript to Professor Adel A Kader who was very devoted to science especially in postharvest technology in his life.

REFERENCES

1. Baas R, N Marissen and A Dik, 1998. Cut rose quality as affected by calcium supply and translocation. Acta Horticulturae, 518: 45-54.
2. Bævre O, T Fjeld and R Moe, 1994. Poinsettia production in the north. In the scientific basis of poinsettia production, The AgriculturalUniversity of Norway, Ås, Norway.
3. Barrett J, 2011. The effect of shipping and handling on poinsettia marketability. GPN Magazine. (http://www.gpnmag.com/effect-shipping-handling-poinsettia-marketability-0).

4. Barrett J and J Dole, 2012. The consumers response to poinsettia cultivars. North American poinsettia trials. GPN Magazine in March. (http://www.gpnmag.com/sites/default/files/06_Poinsettia_GPN0312%20FINAL.pdf).
5. Berghage RD, 1989. Modeling stem elongation in the poinsettia. PhD Dissertation, Michigan State University, USA.
6. Berghage RD, JA Flore, RD Heins and JE Erwin, 1990. The relationship between day and night temperature influences photosynthesis but not light compensation point or flower longevity of Easter lily, *Lilium longiflorum* Thunb. Acta Horticulturae, 272: 91-96.
7. Biermann MR, J Dole, C Rosen and H Wilkins, 1990. Leaf edge burn and axillary shoot growth of vegetative poinsettia plants. Journal of the American Society for Horticultural Science, 115: 73-78.
8. Blakeman J, 1975. Germination of *Botrytis cinerea* conidia *in vitro* in relation to nutrient conditions on leaf surfaces. Transactions of the British Mycology Society, 65: 239-247.
9. Chandler S, 2003. Commercialization of genetically modified ornamental plants. Journal of Plant Biotechnology, 5: 69-78.
10. Chandler SF and F Brugliera, 2011. Genetic modification in floriculture. Biotechnology letters, 33: 207-214.
11. Dunn BL, C Goad and S Stanphill, 2011. Performance of 40 poinsettia cultivars grown under two different temperatures. Journal of Horticulture and Forestry, 3: 72-77.
12. Eason JR, 2006. Molecular and genetic aspects of flower senescence. Stewart Postharvest Review, 2: 1-7.
13. Ecke IP, JE Faust, A Higgins and J Williams, 2004. The Ecke poinsettia manual. Ball Publishing, Illinois, USA.
14. Ecke P, O Matkin and DE Hartley, 1990. The Poinsettia Manual (3rd edition). Paul Ecke Poinsettias, Encinitas, California, USA.
15. Ehret DL and LC Ho, 1986. Translocation of calcium in relation to tomato fruit growth. Annals of Botany, 58: 679-688.
16. Faust J and A Enfield, 2010. Effect of temperature and storage duration on quality and rooting performance of poinsettia (*Euphorbia pulcherrima*'Prestige Red') cuttings. Acta Horticulturae, 877: 1799-1807.
17. Faust J and K Lewis, 2005. Effect of 1-MCP on the postharvest performance of un-rooted poinsettia cuttings. Acta Horticulturae, 877: 807-812.
18. Faust JE, E Will and R Klein, 1998. Stock Plant environment impacts lateral stem strength of finished poinsettias. HortScience, 33: 448
19. Ferguson I and B Drøbak, 1988. Calcium and the regulation of plant growth and senescence. HortScience, 23: 262-266.
20. Fjeld T, 1990. Effects of temperature and irradiance level on plant quality at marketing stage and the subsequent keeping quality of Christmas begonia (*Begonia x cheimantha* Everett). Norwegian Journal of Agricultural Sciences, 4: 217-223.
21. Fjeld T, 1992. Effects of temperature and irradiance level on carbohydrate content and keeping quality of Christmas begonia (*Begonia cheimantha* everett). Scientia Horticulturae, 50: 219-228.
22. Gilbertz DA, JE Barrett and TA Nell, 1982. Effects of water stress on flowering in poinsettia. HortScience, 17: 516.
23. Grange R and D Hand, 1987. A review of the effects of atmospheric humidity on the growth of horticultural crops. Journal Horticultural Science, 62: 125-134.
24. Hammer PA, 1999. Poinsettia problem, diagnostic key: physiological disorders of poinsettias. (http://www.ces.ncsu.edu/depts/hort/poinsettia/corrective/a11.html).
25. Hausbeck M and G Moorman, 1996. Managing *Botrytis* in greenhouse-grown flower crops. Plant Disease, 80: 1212-1219.
26. Hepler PK, 2005. Calcium: a central regulator of plant growth and development. The Plant Cell Online, 17: 2142-2155.
27. Higaki T, H Rasmussen and W Carpenter, 1980. Color breakdown in anthurium (*Anthurium andreanum* Lind.) spathes caused by calcium deficiency. Journal of American Society for Horticultural Science, 105: 441-444.
28. Hisamatsu T, RW King, CA Helliwell and M Koshioka, 2005. The involvement of *gibberellin 20-oxidase* genes in phytochrome-regulated petiole elongation of Arabidopsis. Plant Physiology, 138: 1106-1116.
29. Islam MA, D Tarkowská, JL Clarke, D-R Blystad, HR Gislerød, S Torre and JE Olsen, 2014. Impact of end-of-day red and far-red light on plant morphology and hormone physiology of poinsettia. Scientia Horticulturae, 174: 77-86.

30. Islam MA, G Kuwar, JL Clarke, D-R Blystad, HR Gislerød, JE Olsen and S Torre, 2012. Artificial light provided by light emitting diodes (LEDs) with a high portion of blue light results in more compact poinsettia compared to the traditional high pressure sodium (HPS) lamps. Scientia Horticulturae, 147: 136-143.

31. Islam MA, H Lütken, S Haugslien, D-R Blystad, S Torre, J Rolcik, SK Rasmussen, JE Olsen and JL Clarke, 2013. Overexpression of the *AtSHI* gene in poinsettia, *Euphorbia pulcherrima*, results in compact plants PLoS ONE 8, doi:10.1371/journal.pone.0053377.

32. Jiang W, Q Sheng, XJ Zhou, MJ Zhang and XJ Liu, 2002. Regulation of detached coriander leaf senescence by 1-methylcyclopropene and ethylene. Postharvest Biology and Technology, 26: 339-345.

33. Jones R, 1999. Poinsettia problem, diagnostic key: *Botrytis* blight or gray mold. The North Carolina State University, USA. (http://www.ces.ncsu.edu/depts/hort/poinsettia/corrective/c13.html).

34. Kuehny J, P Branch and P Adams, 2000. Stem strength of poinsettia. Acta Horticulturae, 515: 257-264.

35. Larson RA, JW Love, DL Strider, RK Jones, JR Baker and KR Horn, 1978. Commercial poinsettia production. North Carolina Agricultural Extension Service. AG-108.

36. Lee YK, P Derbyshire, JP Knox and AK Hvoslef-Eide, 2008. Sequential cell wall transformations in response to the induction of a pedicel abscission event in *Euphorbia pulcherrima* (poinsettia). The Plant Journal , 54: 993-1003.

37. Love J, 1999. Poinsettia problem-diagnostic key: physiological disorders of poinsettias. North Carolina State University, USA. (http://www.ces.ncsu.edu/depts/hort/poinsettia/corrective/a11.html).

38. Lütken H, JL Clarke and R Müller, 2012. Genetic engineering and sustainable production of ornamentals: current status and future directions. Plant Cell Reports, 31: 1141-1157.

39. Marschner H, 1995. Mineral nutrition of higher plants. Academic Press, Berlin.

40. McAvoy R and B Bible, 1995. Bract necrosis on poinsettia. Connecticut Greenhouse Newsletter, 188: 17-23.

41. McAvoy R and B Bible, 1996a. Silicate sprays as effective as calcium sprays at suppressing bract necrosis in poinsettia. HortScience, 31: 654

42. McAvoy R and B Bible, 1996b. Silica sprays reduce the incidence and severity of bract necrosis in poinsettia. HortScience, 31: 1146-1149.

43. McAvoy R, B Bible and MR Evans, 1998. Localized accumulation of condensed tannins associated with poinsettia bract necrosis. Journal of the American Society for Horticultural Science, 123: 916-920.

44. Michalczuk B, D Goszczynska D, R Rudnicki and A Halevy, 1989. Calcium promotes longevity and bud opening in cut rose flowers. Israel Journal of Botany, 38: 209-215.

45. Miller S and R Heins, 1986. Factors influencing premature cyathia abscission in poinsettia 'Annette Hegg Dark Red'. Journal of American Society for Horticultural Science, 111: 114-121.

46. Moe R, T Fjeld and LM Mortensen, 1992a. Stem elongation and keeping quality in poinsettia (*Euphorbia pulcherrima* Willd.) as affected by temperature and supplementary lighting. Scientia Horticulturae, 50: 127-136.

47. Moe R, E Fløistad and D-R Blystad, 2006. Impact of light on cyathia abscission and bract discoloration in poinsettia (*Euphorbia pulcherrima*). Acta Horticulturae, 711: 285-290.

48. Moe R, N Glomsrud, I Bratberg and S Valsø, 1992b. Control of plant height in poinsettia by temperature drop and graphical tracking. Acta Horticulturae, 327: 41-48.

49. Moe R, K Willumsen, I Ihlebekk, A Stupa, N Glomsrud and L Mortensen, 1995. DIF and temperature drop responses in SDP and LDP, a comparison. Acta Horticulturae, 378: 27-33.

50. Mortensen LM, 2000. Effects of air humidity on growth, flowering, keeping quality and water relations of four short-day greenhouse species. Scientia Horticulturae, 86: 299-310.

51. Muller R, M Serek, E Sisler and AS Andersen, 1997. Poststorage quality and rooting ability of *Epipremnum pinnatum* cuttings after treatment with ethylene action inhibitors. Journal of Horticultural Science, 72: 445-452.

52. Nell T and J Barrett, 1985. Nitrate-ammonium nitrogen ratio and fertilizer application method influence bract necrosis and growth of poinsettia. HortScience, 20: 1130-1131.

53. Nell T and J Barrett, 1986a. Growth and incidence of bract necrosis in 'Gutbier V-14 Glory'poinsettia. Journal of American Society for Horticultural Science, 111: 266-269.

54. Nell T and J Barrett, 1986b. Influence of simulated shipping on the interior performance of poinsettias. HortScience 21: 310-312.

55. Nielsen B and KR Starkey, 1999. Influence of production factors on postharvest life of potted roses. Postharvest Biology and Technology, 16: 157-167.

56. Odula FO, 2011. Effects of temperature, light and plant growth regulators on fuel-efficient poinsettia production. Clemson University, M.S thesis in Plant and Environmental Sciences.

57. Prince T and M Cunningham, 1988. Leaf abscission of poinsettias affected by preharvest fertilization termination and sleeving stress. HortScience, 23: 138-139.
58. Pritchard P, M Hausbeck and R Heins, 1996. The influence of diurnal temperatures on the postharvest susceptibility of poinsettia to *Botrytis cinerea*. Plant Disease, 80: 1011-1014.
59. Ranch PE, 2011. Postproduction issue, commercial floriculture. University of Florida, Environmental Horticulture, Gainesville, USA. (http://hort.ifas.ufl.edu/floriculture/poinsettia/beb.shtml).
60. Ranch PE, 2012. Commercial floriculture. University of Florida. Gainesville, USA. (http://hort.ifas.ufl.edu/floriculture/poinsettia.shtml).
61. Riov J and R Goren, 1979. Effect of ethylene on auxin transport and metabolism in midrib sections in relation to leaf abscission of woody plants. Plant Cell Environment, 2: 83-89.
62. Robichaux MB, 2008. The effect of calcium or silicon on potted miniature roses or poinsettias. Master's Thesis, University of Louisiana at Lafayette, USA.
63. Runcle E and J Faust, 2008. Energy-efficient poinsettia production. GPN Magazine.(http://www.gpnmag.com/energy-efficient-poinsettia-production).
64. Sacalis J, 1978. Ethylene evolution by petioles of sleeved poinsettia plants [*Euphorbia pulcherrima*]. HortScience 13: 594-596.
65. Salinas J, DCM Glandorf, F Picavet and K Verhoeff, 1989. Effects of temperature, relative humidity and age of conidia on the incidence of spotting on gerbera flowers caused by *Botrytis cinerea*. European Journal of Plant athology, 95: 51-64.
66. Saltviet MEJ, DM Pharr and RA Larson, 1979. Mechanical stress induces ethylene production and epinasty inpoinsettia cultivars. Journal of American Society for Horticultural Science, 104: 452-455.
67. Sánchez-Blanco J, T Ferrández, A Navarro, S Bañon and J Alarcón, 2004. Effects of irrigation and air humidity preconditioning on water relations, growth and survival of *Rosmarinus officinalis* plants during and after transplanting. Journal of Plant Physiology, 161: 1133-1142.
68. Schnelle R and J Barrett, 2011. High temperatures and marginally inductive photoperiods impact the floral display of poinsettias'freedom red', 'early red splendor', and 'prestige early red'. Acta Horticulturae, 893: 873-878.
69. Scott L, T Blessington and J Price, 1983. Postharvest effects of temperature, dark storage duration, and sleeving on quality retention of 'Gutbier V-14 Glory' poinsettia. HortScience, 18: 749-750.
70. Scott LF, TA Blessington and JA Price, 1982. Postharvest performance of poinsettia as affected by micronutrient source, storage, and cultivar. HortScience, 17: 901-902.
71. Scott LF, TA Blessington and JA Price, 1984a. Influence of controlled-release fertilizers, storage duration, and light source on postharvest quality of poinsettia. HortScience, 19: 111-112.
72. Scott LF, TA Blessington and JA Price, 1984b. Postharvest effects of storage method and duration on quality retention of poinsettias. HortScience, 19: 290-291.
73. Serek M, EC Sisler and MS Reid, 1994. Novel gaseous ethylene binding inhibitor prevents ethylene effects in potted flowering plants. Journal of the American Society for Horticultural Science, 119: 1230-1233.
74. Serek M, E Woltering, E Sisler, S Frello, S Sriskandarajah, 2006. Controlling ethylene responses in flowers at the receptor level. Biotechnogy Advances, 24: 368-381.
75. Shanks J, W Noble and W Witte, 1970. Influence of light and temperature upon leaf and bract abscission in poinsettia. Journal of the American Society for Horticultural Science, 95: 446-449.
76. Simpson GG and C Dean, 2002. Arabidopsis, the rosetta stone of flowering time? Science, 296: 285-289.
77. Smith T and D Cox, 2009. Lateral stem breakage on poinsettias. University of Massachusetts, USA. (http://www.negreenhouseupdate.info/index.php/september/201-lateral-stem-breakage-on-poinsettias).
78. Staby G, B Eisenberg, J Kelly, M Bridgen and M Cunningham, 1980. Leaf petiole epinasty in poinsettias. HortScience, 15: 635-636.
79. Staby G and A Kofranek, 1979. Production conditions as they affect harvest and postharvest characteristics of poinsettias. Journal of the American Society for Horticultural Science, 104: 88-92.
80. Staby GL, J Thompson and A Kofranek, 1978. Postharvest characteristics of poinsettias as influenced by handling and storage procedures. Journal of the American Society for Horticultural Science, 103: 712-115.
81. Starkey KR and AR Pedersen, 1997. Increased levels of calcium in the nutrient solution improves the postharvest life of potted roses. Journal of the American Society for Horticultural Science, 122: 863-868.
82. Strømme E, A Selmer-Olsen, HR Gislerød and R Moe, 1994. Cultivar differences in nutrient absorption and susceptibility to bract necrosis in poinsettia (*Euphorbia pulcherrima* Willd. ex Klotzsch). Gartenbauwissenschaft 59: 6-12.

83. Syngenta, 2009. Energy efficient poinsettia production.
(http://www.syngentaflowersinc.com/pdf/cultural/FINALPoinsettiaColdGrowing.pdf).

84. Taiz L and E Zeiger, 2010. Plant Physiology (5th edition), chapter: ethylene. Sinauer Assoc. Inc., Publishers, Sunderland, Massachusetts USA. pp: 649-668.

85. Tayma H and T Roll, 1990. Tips on growing poinsettia: (Second edition). The Ohio State University, USA. (http://melissa560.tripod.com/braconbrac1.html).

86. Torre S, T Fjeld and HR Gislerød, 2001. Effects of air humidity and K/Ca ratio in the nutrient supply on growth and postharvest characteristics of cut roses. Scientia Horticulturae, 90: 291-304.

87. Tsutsui K and M Aoki, 1981. Response of poinsettias to major nutrient supply in relation to nutrient uptake and growth. Bull. Veg. Orn. Crops Res. Station. Series A 8: 171-207.

88. USDA [NASS] 2009. Floriculture crops - 2008 summary. Sp Cr 6-1. (http://usda.mannlib.cornell.edu/usda/current/FlorCrop/FlorCrop-04-23-2009.pdf).

89. van Doorn WG, 2001. Role of soluble carbohydrates in flower senescence: a survey. Acta Horticulturae, 503: 179-183.

90. Van Meeteren U, 1980. Role of pressure potential in keeping quality of cut gerbera inflorescences, Acta Horticultuare, 113: 143-150.

91. Whipker B, 1999. Poinsettia problems, diagnostic key: physiological disorders of poinsettia. North Carolina State University, USA.
(http://www.ces.ncsu.edu/depts/hort/poinsettia/corrective/a11.html#a11-01).

92. Whipker BE, C Warfield, R Cloyd, JL Gibson and TJ Cavins, 2002. Avoiding top problems of poinsettias. Green house production news (GPN) in July. North Carolina State University. 24-29. (http://www.gpnmag.com/sites/default/files/Whipker_Warfield.pdf).

93. Woltering E, 1987. Effects of ethylene on ornamental pot plants: a classification. Scientia Horticulturae, 31: 283-294.

94. Woltz S and B Harbaugh, 1985. Effect of nutritional balance on bract and foliar necroses of poinsettia. Proceeding of Florida State Horticulture Society, 98: 122-123.

95. Zhao X, X Yu, E Foo, GM Symons, J Lopez, KT Bendehakkalu, J Xiang, JL Weller, X Liu and JB Reid, 2007. A study of gibberellin homeostasis and cryptochrome-mediated blue light inhibition of hypocotyl elongation. Plant Physiology, 145: 106-118.

96. Zieslin N and A Halevy, 1969. Petal blackening in 'Baccara'roses. Journal of the American Society for Horticultural Science, 94: 629-631.

ABUNDANCE AND COMPOSITION OF ZOOPLANKTON AT SITAKUNDA COAST OF CHITTAGONG, BANGLADESH

Md. Shahzad Kuli Khan[1*], Sheikh Aftab Uddin[2] and Mohammed Ashraful Haque[1]

[1]Bangladesh Fisheries Research Institute, Marine Fisheries and Technology Station, Cox's Bazar-4700, Bangladesh
[2]Institute of Marine Sciences and Fisheries, University of Chittagong, Chittagong-4331, Bangladesh

*Corresponding author: Md. Shahzad Kuli Khan, E-mail: khanbfri@gmail.com

ARTICLE INFO

Key words
Mangrove area
Zooplankton
Copepoda
Fish larvae

ABSTRACT

Eight groups of zooplankton were found at Sitakunda coast, Chittagong, northeastern part of the Bay of Bengal during January to June 2007. The identified groups were Appendicularia (2.46%), Chaetognatha (2.45%), Cladocera (2.31%), Copepoda (26.05%), Ctenophora (5.86%), Crustacean zooplankton (21.64%), Ichthyoplankton (17.77%) and Meroplankton (21.45%). Abundance of zooplankton varied from 413 to 7730 individuals/m^3.Mangrove vegetate area (station- VI) has the highest abundant possibly due to the organic and inorganic matters dissolved in the water while ship breaking area (station- IV) has the lowest abundant. Zooplankton population was significantly ($p>0.05$) higher in the mangrove vegetate area than the fishermen community area and ship breaking area. The mangrove vegetate area has the highest composition (57.06%) of zooplankton than the fishers community area and ship breaking area (29.77% and 13.16%, respectively). *Calanus* sp. (12.29%) belonging to Copepods and fish eggs (9.25%) belonging to Ichthyoplankton were the most abundant and *Oikopleura albicans* (0.66%) from Appendicularia, *Metapenaeus brevicornis* (0.71%) and *Metapenaeus monoceros* (0.90%) belonging to Crustacean larvae were the lowest abundant species found at three major investigated area.

INTRODUCTION

Zooplankton is an aquatic animal community that has limited swimming capacity against the ambient currents. Even with their quite limited swimming capacity, they carry out day-night periodic movements of hundreds of meters. They prefer to feed at night on the water surface and effectively graze the phytoplankton, and hence they referred to as living machines. They habitually represent a vital link between the microbial portion and the large grazers (Laval-Peuto et al., 1986; Pierce and Turner, 1994).The zooplankton, secondary consumer plays a key role in the food chain of aquatic ecosystem by transferring energy from phytoplankton to higher tropic levels leading to the production of fisheries to human exploitation. The health of marine ecosystems inherently linked to the abundance of zooplankton and their biodiversity. The potentiality of marine pelagic fishes directly or indirectly depends on the availability of zooplankton. In the aquatic ecosystem zooplankton are being used as the indicator species for the physical, chemical and biological processes due to their universal distribution, small size, and rapid metabolic and growth rates (Heinbokel, 1978; Fenchel, 1987), huge density, tinier life span, drifting nature, great species diversity and diverse tolerance to the stress (Gajbhiye, 2002).

A survey report of FAO (1985) stated that the tidal areas of Bangladesh are relatively rich in zooplankton. The abundance of zooplankton and their ecology in the coastal and estuarine environment of Bangladesh is little studied. Islam and Aziz (1975) studied on zooplankton of the northeastern part of the Bangladesh coastal area and identified a total of 18 genera and 18 species. Bhuyain et al. (1982) made an observation on the macro-zooplankter of the continental shelf of the Bay of Bengal and reported the occurrence and distribution of 18 calanoid copepods. Ali et al. (1985) recorded a periodic variation of zooplankton in the coastal estuarine water in the southeastern part of Bangladesh. The major groups of zooplankton are copepoda, decapoda, chaetognatha, cladocera and fish and shellfish larvae. Zooplankton diversity of salt marsh habitat in the Bakkhali river estuary, Cox's Bazar, Bangladesh has also studied by Ali (2006).

Coastal zone contains critical terrestrial and aquatic habitats, such as mangrove forests, wetlands and tidal flats. Sitakunda coast under the Chittagong district, northeastern part of the Bay of Bengal is adjacent to the Sandwip Chanel, having tidal mangrove, ship breaking yard and fishermen community area and an important source of fisheries resources. The purpose of this study is to provide more information on the abundance and composition of the zooplankton community on the Sitakunda Upazila coastal water, north of the Chittagong city, which is currently affected by ship-breaking activity on the shore.

MATERIALS AND METHODS

Sitakunda coast, which is the northeastern part of the Bay of Bengal, located in between 22°22′ and 22°42′ northern latitudes and in between 91°34′ and 91°48′ east longitudes. For the present investigation this coastal area was divided into three pre define activities community with six sampling stations (Fig. 1).Station-I (Salimpur) and station-II (Saidpur) was considered as a fishermen community area, station-III (Grisubedar Ship yard) and station-IV (PHP Ship yard) located in Bhatiari area was considered as ship breaking yard and station-V (Barabkunda) and station-VI (Muradpur) was considered as a tidal mangrove vegetate area.

Zooplankton sampling and isolation

The sampling was conducted during January to June 2007 by using a wooden boat. Zooplanktons were collected using a net (Hydrobios model 55 μm mesh size) ending with a cod end to retain the organisms which was towed horizontally. A flow meter (FMC 0.3) was attached within the aperture of the net to measure the amount of water displaced. At each station, the net was slanted three times for 45 minutes each while the boat was moving slowly. The sampling was taken place in the sub-surface layer (0.2m-0.5m) of the water column. Abundance of organisms was calculated from the volume of water displaced through the plankton net and expressed as numbers of individuals per cubic meter. Immediately after collection, the samples were preserved in 4% formalin (45% formaldehyde) in 250 ml plastic bottles and labeled. Then the samples brought to the laboratory of Institute of Marine Sciences and Fisheries, University of Chittagong for qualitative and quantitative analysis. For efficient sorting, a vital stain "Rose Bengal" was added and the sample left for overnight. Zooplanktons were sorted out with the help of fine brushes, needle, forceps and an inverted microscope (Model-Axiovert 25, CFL) and Sedgwick-Rafter chamber was used for counting.

Major groups were identified by the works of Patel (1975), Kasturirangan (1963), Koga (1984), Zafar and Mahmud (1989) for Copepoda; Wickstead (1965) and Smirnov (1996) for Cladocera; Srinivasan (1988), Andreu et al.(1989) and Bieri (1991) for Chaetognatha; Haq and Hasan (1975), Muthu et al. (1978), Amin and Mahmud (1979), Paulinose (1982), Deshmukh and Kagwade (1987), Rothlisberg (1983, 1987), Tirmizi et al. (1987) and Zafar (2000) for Crustacean zooplankton; Peter (1969), Newell and Newell (1979), Omori and Ikeda (1984), Zafar and Mahmud (1989), Olivar and Fortuno (1991) and Goswami & Padmavati (1996) for Meroplankton and Ichthyoplankton.

Data analysis

The zooplankton abundance was calculated using the following formula:

a. Total number of zooplankton specimens =Total counts of the specimens (say x)/ Volume of water filtered (V).

No. /m3 = x/v (No. can also be expressed/ 100 m^{-3} or 1000 m^{-3}).

b. Total number of specimens of a particular zooplankton taxon

= Total counts (x)/Volume of water filtered (Y)

No. /m3 = x / y.

SAS (2003) was used to analyze the data for analysis of variance (ANOVA).

RESULTS

Eight groups of zooplankton were identified, i.e. Appendicularia, Chaetognatha, Cladocera, Copepoda, Ctenophora, Crustacean zooplankton, Ichthyoplankton and Meroplankton at six different stations on Sitakunda coast, Chittagong, Bangladesh. In total 10 known species of Crustacean, 7 known and unidentified species of Meroplankton, 6 species of Copepoda, each 2 species of Appendicularia, Ctenophora and Ichthyoplankton and each one species of Cladocera and Chaetognatha were identified during the investigation. Abundance of zooplankton varied from 413 to 7730 individuals/m^3. Figure 2 (A and B) shows the composition of the various zooplankton group on Sitakunda coast and the contribution of those groups in each station.

Appendicularia

This class includes *Oikopleura albicans* and *O. dioica*, comprising together 2.46 % of the total zooplankton population. They live in the pelagic zone, especially in the upper sunlight portion of the ocean. These zooplanktons were found in all stations, but in low number (63 indi/m^3) was observed in the Bhatiyari area near the ship breaking yard and large number (1150 indi/m^3) were observed in the mangrove vegetate area. Among them, a few *O. albicans* (12 indi/m^3) was found in ship breaking area.

Cladocera

Cladocera the lowermost group made only 2.32 % of the total zooplankton population and *Evadue sp.* was the only identified zooplankton, which was very common in all stations. The abundance of *Evadue sp.* was 56 indi/m^3 to 503 indi/m^3.

Ctenophora

The ctenophores designed 5.86 % of the total zooplankton population. This group composed of *Bolinopsis vitrea* and *Pleurobrachia* sp. and the percentage occurred 2.81 % and 3.05 % respectively.

Chaetognatha

Chaetognatha were the second lowermost group, forming 2.45 % of total zooplankton. In mangrove vegetate and fishermen community area, they found great number compare to ship breaking area near Bhatiyari. The highest abundance was 507 indi/m^3 and the lowest was 64 indi/m^3.

Copepoda

Copepods were the most abundant group encompassing 26.05 % of the total zooplankton population. This group consisted with *Calanus* sp., *Microsetella* sp., *Oncaea* sp., *Calanopia* sp., *Coryeacus* sp. and *Oithona* sp. During the study highest abundance 1,937 indi/m^3 was found in mangrove vegetate area (station VI) due to the high number of *Calanus* sp. while station-III & IV (Ship breaking area) was the lowest abundance 2 indi/m^3 and 7indi/m^3 respectively owing to *Oncaea* sp. *Calanus* sp. was the most abundant and found at all stations, comprising 12.29 % of the total zooplankton population.

Crustacean zooplankton

Crustaceans were the second most plentiful group of zooplankton, founding 21.64% of the total population. This group was composed of *Acetes* larvae (7.87%), *Lucifer* larvae (4.14%), *Penaeid* larvae (7.41%) and *Sergestes* larvae (2.22%). The *Acetes* larvae were very common in this study. *Assets erythraeus* (2.67%), *Acetes indicus* (2.69%) and *Acetes japonicas* (2.51%) accounted for the majority of the crustacean zooplankton. The highest number (819 indi/m^3) of *Acetes erythraeus* occurred in the mangrove vegetate area (i.e. st. VI) and the lowest number (19 indi/m^3) in ship breaking area.

Lucifer

Lucifer sp. was very common and made only 4.14 % of the total zooplankton population. The amount of *Lucifer sp.* was quite high in mangrove vegetate and fishermen community water while the number was lower in the ship breaking area. The average abundance showed substantial differences in those places.

Shrimp larvae

Penaeus and *Metapenaeus* larvae were regular component in the Penaeid zooplankton, constituting 5.81% and 1.61% of the total zooplankton population respectively. Among them *P. indicus* was most dominant species (1473 indi/m^3) occurred in mangrove vegetate area (station VI). The abundance was very low at ship breaking area for all species, i.e. *P. monodon* (19 indi/m^3), *P. indicus* (100 indi/m^3), *P. merguiensis* (14 indi/m^3), *Metapenaeus monoceros* (12 indi/m^3) and *Metapenaeus brevicornis* (15 indi/m^3).

In all stations, *Sergestes similis* also found in worthy number. In mangrove vegetation (926 indi/m^3) and fishermen community area (395 indi/m^3) the number was high, but in the ship breaking area (74 indi/m^3) the number was very low.

Meroplankton

Meroplankton consisted of Polychaete larvae (5.28%), Snail veliger (2.77%), Snail larvae (2.34%), Barnacle nauplius (2.18%), Barnacle cyprid (2.36%), Crab megalopa (3.12%) and Crab zoea (3.41%). Polychaete larvae were very common zooplankton and high in number. The average abundance in all stations showed no significant differences. Snail veligers and Snail larvae were also available at all stations. The amount of Barnacle nauplius and Barnacle cyprid was high in mangrove and fishermen community area rather than ship breaking area.

Ichthyoplankton

Fish eggs and larvae were very common and high in number, creating 9.25 % and 8.52% of total zooplankton respectively. Fish eggs and larvae found available in all investigated areas. The average abundance showed no significant differences between mangroves vegetate area and ship breaking area.

DISCUSSION

A sensible variation was observed in the zooplankton abundance in all stations. Mangrove vegetate area have the highest abundant 35,755 individuals/m^3 and fishermen community area and ship breaking yard has 18,825 individuals/m^3 and 8,321 individuals/m3, respectively around the sampling period. Statistical analysis showed that the abundance of zooplankton population in the mangrove vegetate area was significantly higher ($p > 0.05$) than the fishers community area and ship breaking area. The mangrove vegetate area has the highest composition (57.06%) of zooplankton then the fishers community area and ship breaking area (29.77% and 13.16%, respectively).

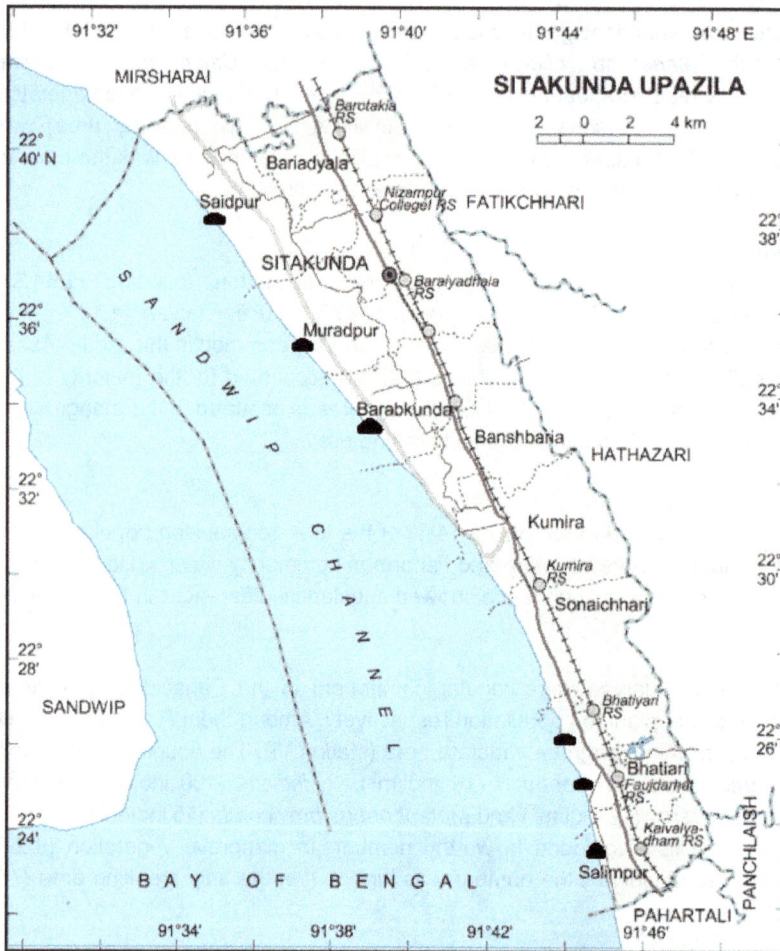

Figure 1. Map of study area (Sitakunda coast) with the location of sampling stations.

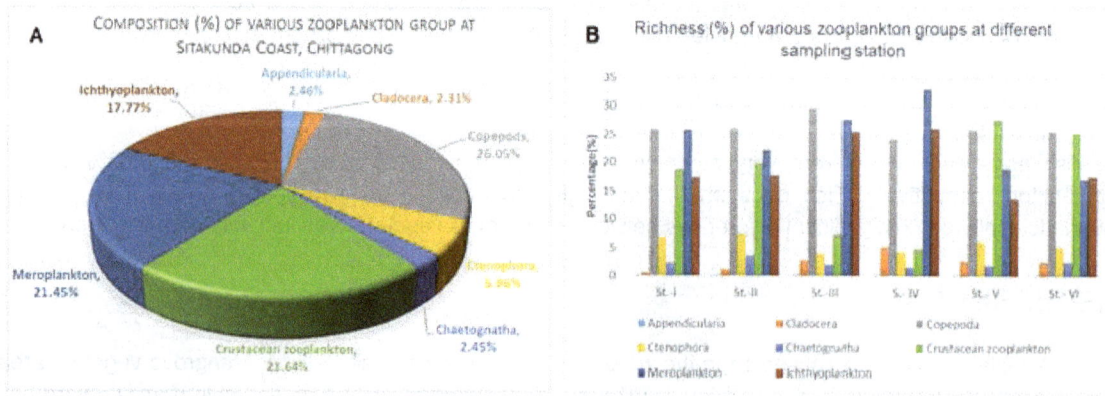

Figure 2. Percent composition of various zooplanktons (A) and their richness at different sampling station (B).

Table 1. List of major groups and species of zooplankton identified and their number and percentage at Sitakunda coast, Chittagong

Group	Species	Total No.	Percentage (%) within group	Overall (%)
Appendicularia	*Oikopleuradioica*	1135	73.32	1.80
	Oikopleuraalbicans	413	26.68	0.66
Copepods	*Calanus* sp.	7730	47.18	12.29
	*Microsetella*sp.	4993	30.48	7.94
	*Oncaea*sp.	852	5.20	1.35
	*Calanopia*sp.	639	3.90	1.01
	*Coryeacus*sp.	1312	8.01	2.08
	*Oithona*sp.	857	5.23	1.36
Cladocera	*Evadue*sp.	1457	43.17	2.32
	Pleurobrachia sp.	1918	56.83	3.05
Ctenophores	*Bolinopsisvitrea*	1767	100	2.81
Chaetognatha	*Sagitta*sp.	1544	100	2.45
Crustacean	*Lucifer* sp.	2602	19.11	4.14
	Sergestessimilis	1395	10.25	2.22
	Penaeus monodon	1361	10.00	2.16
	Penaeus merguiensis	818	6.00	1.30
	Metapenaeus monoceros	566	4.16	0.90
	Metapenaeus brevicornis	447	3.28	0.71
	Penaeus indicus	1473	10.82	2.34
	Aceteserythraeus	1694	12.44	2.67
	Acetesindicus	1681	12.35	2.69
	Acetesjaponicus	1577	11.58	2.51
Meroplankton	Polychaete larvae	3320	24.61	5.28
	Snail veliger	1740	12.90	2.77
	Snail larvae	1472	10.90	2.34
	Barnacle nauplius	1369	10.15	2.18
	Barnacle cyprid	1483	10.99	2.36
	Crab megalopa	1960	14.53	3.12
	Crab zoea	2148	15.92	3.41
Ichthyoplankton	Fish eggs	5820	52.07	9.25
	Fish larvae	5358	47.93	8.52
Total		**62901**		**100**

Table 2. Zooplankton abundance (individual/m^3) and their averages in fishers community area (St.-I and II), ship breaking area (St.-III and IV) and mangrove vegetate area (St.-V and VI) at Sitakunda coast, Chittagong.

Species	St-I	St-II	St- III	St- IV	St- V	St- VI	Average	Total	Overall %
Aceteserythraeus	109	98	32	19	372	487	186.17	1117	1.78
Acetesindicus	211	47	9	3	112	179	93.5	561	0.89
Acetesjaponicus	149	1236	867	693	1884	1937	1127.67	6766	10.76
Barnacle nauplius	357	913	118	107	1353	1749	766.17	4597	7.31
Barnacle cyprid	297	112	2	7	286	366	178.33	1070	1.70
Bolinopsisvitrea	218	49	98	24	177	189	125.83	755	1.20
Calanopiasp.	102	118	79	81	298	419	182.83	1097	1.74
Calanus sp.	1113	94	98	72	253	311	323.5	1941	3.09
Coryeacussp.	317	119	118	214	447	503	286.33	1718	2.73
Crab megalopa	277	399	66	91	533	427	298.83	1793	2.85
Crab zoea	319	307	107	87	472	576	311.33	1868	2.97
Evaduesp.	56	354	87	64	321	507	231.5	1389	2.21
Fish eggs	719	565	27	49	703	754	469.5	2817	4.48
Fish larvae	892	188	72	2	521	405	346.67	2080	3.31
Lucifer sp.	504	201	12	7	425	532	280.17	1681	2.67
M. brevicornis	37	127	2	12	254	327	126.5	759	1.21
M. monoceros	56	83	7	5	211	204	94.333	566	0.90
Microsetellasp.	753	49	12	3	142	204	193.83	1163	1.85
Oikopleuraalbicans	63	201	58	42	477	518	226.5	1359	2.16
Oikopleuradioica	127	147	23	19	577	819	285.33	1712	2.72
Oithonasp.	29	181	54	34	689	512	249.83	1499	2.38
Oncaeasp.	79	173	36	27	554	638	251.17	1507	2.40
Penaeus indicus	177	649	451	521	553	589	490	2940	4.67
Penaeus merguiensis	96	247	117	98	537	429	254	1524	2.42
Penaeus monodon	184	171	68	59	414	501	232.83	1397	2.22
Pleurobrachia sp.	402	209	17	21	346	419	235.67	1414	2.25
Polychaete larvae	557	119	9	54	483	521	290.5	1743	2.77
Sagittasp.	211	321	236	361	354	411	315.67	1894	3.00
Sergestessimilis	207	431	276	234	427	461	339.33	2036	3.24
Snail larvae	259	1031	612	503	1236	1719	893.33	5360	8.52
Snail veliger	312	697	473	565	1019	1712	796.33	4778	7.60
Total kind	31	31	31	31	31	31	31	31	
Total individual	9189	9636	4243	4078	16430	19325	10483.5	62901	100%

Large carnivorous zooplankters namely, the Ctenophora and Chaetognatha are planktonic predators of fish larvae. The correlation between fish larvae and their predators, i.e. Chaetognatha, and Ctenophora was 0.8611, 0.8083 respectively, at 95 % confidence. The correlation of fish larvae and Copepoda, which their prey species was 0.9100 at 95 % confidence.

At all stations, the dominant species in the Sitakunda coast were as *Calanus* sp., *Microsetella* sp. belonging to Copepods, fish eggs and fish larvae belonging to Ichthyoplankton, Polychaete larvae and Crab zoea belonging to Meroplankton and *Lucifer* sp. belonging to Crustacean larvae. All most all species were lower at station III and IV, which was denoted as the ship breaking area probably due to oil pollution and other human activities. Copepods were the main contributors in the present investigation. Wimpenny (1966) and Omori and Ikeda (1976) reported that copepods are the most abundant zooplankton communities sampled in the world ocean. Houde and Lovdal (1982) showed that copepods are important components of larval fish food. The present investigation on Crustacean zooplankton found five commercially important species such as *Penaeus monodon*, *Penaeus merguiensis*, *Metapenaeus monoceros*, *Metapenaeus brevicornis* and *Penaeus indicus*. *Penaeus* and *Metapenaeus* have worldwide commercial importance in fisheries and aquaculture, and the larvae of many species have been reared in the respected shrimp hatchery.

In general, particularly in coastal waters, the composition and abundance of zooplankton varied remarkably due to the seasonal variations and their sheltered systems like coastal and mangrove waters. On the Sitakunda coast, in the mangrove vegetate area, total abundance of zooplankton was higher than the fishermen community area and ship breaking area. This is because of organic and inorganic matters dissolved in the water, which is ultimately support directly or indirectly to the zooplankton growth. Similar results have also been reported in the coastal waters of Bangladesh by Bhuiyan et al. (1982), Ali et al. (1985) and Zafar (2000).

Fraser (1969) and Suwanrunpha (1983) reported that big carnivorous zooplankters namely Ctenophora, Chaetognatha, Medusae and Siphonophora are planktonic predators of fish larvae. In this study, a high correlation between fish larvae and their predator, especially chaetognatha was observed. Thus, their presence in numbers of zooplankton could have a serious effect on the recruitment of larval fish and could be very significant for the fish stocks and for the fishing industry. Houde and Lovdal (1982), Balbontin et al. (1986) and Anderson (1994) presented that small zooplankton e.g. Copepods, Tintinnids, Cladocerans, larval molluscs etc. are important components of larval fish food. The present study found a high correlation between fish larvae and their prey, especially copepods. Positive correlations indicated that fish tend to aggregate where the standing stock of copepods is highest. However, Sameoto (1972) found no significant correlation between standing stock of copepods and the valued abundance of herring larvae. Manyauthors point out that zooplankton was influencing on fisheries. Krisshnapillai and Bhat (1981) found that the fish-catching rate was maximum in while the zooplankton productive rate was high. Jacob et al. (1981) reported that the peak times in the zooplankton biomass coincided with the peak periods of pelagic fisheries.

Unfortunately, information about the fisheries in the present studied areas was not available, so that correlation of fish catch and zooplankton abundance was not measured.

CONCLUSION

The zooplankton abundance in the three locations showed a much different from each other. The zooplankton abundance in mangrove vegetate area was higher than the fishermen community and ship breaking area. The abundance and composition of the zooplankton can be used as an indicator of marine productivity.

REFERENCE

1. Ali A, S Sukanta and N Mahmood, 1985. Seasonal abundance of plankton in Moheskhali channel, Bay of Bengal. In: Proceedings of SAARC Seminar on Protection of Environmental from Degradation, Dhaka, Bangladesh, p 128-140.

2. Ali, M. 2006. Zooplankton diversity of salt marsh habitat in the Bakkhali river estuary, Cox'Bazar, Bangladesh.4th Year Project Paper, Institute of Marine Sciences and Fisheries (IMSF), University of Chittagong. 56 p.

3. Amin MN and N Mahmood, 1979. On identification of post larvae of penaeid shrimp *Metapenaeus brevicornis* (H. Milne Edwards), Bangladesh Journal of Scientific and Industrial Research,14: 97-100.

4. Anderson JT, 1994. Feeding ecology and condition of larval and pelagic juvenile redfish, *Sebastes*spp. Marine Ecology Progress Series, 104: 211-226.

5. Andreu P, C Marrase and E Berdalet, 1989. Distribution of epiplanktonicChaetognatha along a transect in the Indian Ocean. Journal of Plankton Research, 11 (2): 185-192.

6. Balbontin F, M Garreton and J Neuling, 1986. Stomach content and prey size of the fish larvae from Bransfield Strait (SIBEX-Phase 2, Chile). SerieCientifica. InstitutoAntarticoChileno, 35: 125-144.

7. Bhuiyan AL, SA Mohi, SA Khair and NG Das, 1982. Macro-zooplanktons of the continental shelf of the Bay of Bengal. Chittagong University Studies, 6: 51-59.

8. Bieri R, 1991. Systematics of the Chaetognatha. In: The biology of chaetognaths: Eds,. Bone, Q. H. Kapp and A.C. Pierrot-Builts. Oxford University Press.p 122-136.

9. Deshmukh VD and PV Kagwade, 1987. Larval abundance of non-penaeid prawns in the Bombay Harbor. Journal of Marine Biological Association India, 29(1&2): 291-296.

10. FAO 1985. Reported on tidal area study Bangladesh.Fisheries Resources Survey System FAO/UNDP BGD/79/015, 32 pp.

11. Fenchel T, 1987, Ecology of Protozoa - The Biology of Free Living PhagotrophicProtists Springer-Verlag, Berlin, p. 197.

12. Fraser JH, 1969. Experimental feeding of some Medusae and Chaetognatha. Journal of the Fisheries Research Board of Canada, 26: 1743-1762.

13. Gajbhiye SN, 2002. Zooplankton - Study methods, importance and significant observations.In: Quardros G, (Ed.) The National Seminar on Creeks, Estuaries and Mangroves - Pollution and Conservation, 28- 30th November, 2002, Thane, p. 21-27

14. Goswami SC, and G Padmavati, 1996. Zooplankton production, composition and diversity in the coastal waters of Goa. Indian Journal of Marine Sciences, 25: 91-97.

15. Haq SM and H Hassan, 1975. Larvae of shrimps of the genera *Penaeus, Parapenaeopsis,* and *Metapenaeus*from the coast of Pakistan. Pakistan Journal of Zoology, 7: 145-159.

16. Heinbokel JF, 1978.Studies on the functional role of tintinnids in the Southern California Bight: 1. Grazing and growth rates in laboratory cultures. Marine Biology, 47: 177-189.

17. Houde E and JD Lovdal, 1982. Variability in Ichthyoplankton and microzooplankton abundances and feeding of fish larvae in Biscayne Bay, Florida.Estuarine, Coastal and Shelf Science, 18: 403-419.

18. Islam AKMN and A Aziz, 1975.A preliminary study on the zooplankton of the North-eastern Bay of Bengal. Bangladesh Journal of Zoology, 3: 125-138.

19. Jacob RM, NK Ramachandram and KR Vasantha, 1981. Zooplankton in relation to hydrography andpelagic fisheries in the inshore waters of Virhinjam, Trivandrum. Journal of the Marine Biological Association of India, 23: 62-76.

20. Kasturirangan LR, 1963. A key for the identification of the more common planktonic Copepoda of Indian coastal waters. Indian National Committee on Oceanic Research, Publication No. 2: 1-87. New Delhi.

21. Krisshnapillai S and GJ Subramonia Bhat, 1981. Note on the abundance of zooplankton and trawler catchduring the post monsoon months along the northwest coast of India. Journal of the Marine Biological Association of India, 23: 208-21.

22. Koga F, 1984 .Morphology, ecology, classification and specialization of copepods nauplius. Bulletin of Nansei Regional Fisheries Research Laboratory, 16: 95-229.

23. Laval-Peuto M, JF Heinbokel, OR Anderson, F Rassoulzadegan and BFSherr, 1986. Role of micro- and nanozooplankton in marine food webs. Insect Science and its Application, 7: 387-395.

24. Muthu MS, NN Pillai and KV George, 1978.Larval development - Pattern of penaeid larval development and generic characters of the larvae of the genera *Penaeus, Metapenaeus*and *Parapenaeopsis.* Central Marine Fisheries Research Institute, Cochin, 28: 75-86.

25. Newell GE and RC Newell, 1979. Marine Plankton, a practical guide.Hutchinson of London. 244p.

26. Olivar MP and JM Fortuno, 1991. Guide to Ichthyoplankton of southeast Atlantic (Benguela Current region). Scientia Marina, 55: 1-383.

27. Omori M and T Ikeda, 1976. Methods in marine zooplankton ecology. A wiley-Interscience Publication, 332p.

28. Omori M and T Ikeda, 1984.Methods in Marine Zooplankton Ecology. John Wiley & Sons, New York, 332 pp.

29. Patel MI, 1975, Pelagic copepods from the inshore waters off Saurashtra coast. Journal of the Marine Biological Association of India, 17: 658-663.

30. Paulinose VT, 1982. Key to the identification of larvae and post larvae of the penaeid prawns (Decapoda: Penaeidae) of the Indian Ocean. Mahasagar–bulletin of the National Institute of Oceanography, 15: 223-229.

31. Peter KJ, 1969. Preliminary report on the density of fish eggs and larvae in the Indian Ocean. Bulletin of National Institute of Sciences of India, 38: 854-863.

32. Pierce RW and JT Turner, 1994. Plankton studies in Buzzards Bay, Massachusetts, USA: IV. Tintinnids, 1987 to 1988. Marine Ecology Progress Series, 112: 235-240.

33. Rothlisberg PC, CJ Jackson and RC Pendrey, 1983. Specific identification and assessment of distribution and abundance of early penaeid shrimp larvae. The Biological Bulletin (Woods Hole) 164: 279-298.

34. Rothlisberg PC, CJ Jackson and RC Pendrey, 1987.Larval ecology of penaeids in the Gulf of Carpentaria, Australia.I. Assessing the reproductive activity of five species of *Penaeus*from the distribution and abundance of the zoeal stages. Australian Journal of Marine and Freshwater Research, 38: 1-17.

35. Sameoto DD, 1972. Distribution of Herring (*Clupleaharengus*) larvae along the southern coast of NovaScotia with observations on their growth and condition factor.Journal of the Fisheries Research Board of Canada, 29: 507-515.

36. Statistical Analysis System (SAS) 2003. User's Guide SAS/STA-t version.8th Edition. SAS, Institute, Inc. Cary, N. C., US.

37. Smirnov NN, 1996. Guides to the identification of the micro invertebrates of the continental waters of the world. cladocera: the chydorinae and sayciinae (chydoridae) of the world. SPB Academic Publishing.The Netherlands.11, 197p.

38. Srinivasan M, 1988.Species associations in Chaetognatha from the Arabian Sea.Journal of the Marine Biological Association of India, 30: 206-209.

39. Suwanrumpa W, 1983. Zooplankton in the western Gulf of Thailand. III. Relation between the distributionof zooplankton predators and fish larvae collected during January to October, 1981.Tech.Paper No.25/13. Marine Fisheries Division, Dept. of Fisheries, 18 p.

40. Tirmizi NM, N Aziz and WM Qureshi, 1987. Distribution of planktonic shrimp SergestessemissisBurkenroad, 1940 (Decapoda, Sergestidae) in the Indian Ocean with notes on juveniles. Crustaceana, 53: 15-28.

41. Wickstead JH, 1965. An introduction to the study of tropical plankton. Hutchinson and Co. Ltd. London. 155pp.

42. Wimpenny RS, 1966. The plankton of the sea. Faber and Faber LTD, London, 426p.

43. Zafar M, and N Mahmood, 1989. Studies on the distribution of zooplankton communities in the Satkhira estuarine system, Chittagong University Studies Science, 13: 115- 122.

44. Zafar M, 2000. Study on Sergestid shrimp Acetes in the vicinity of Matamuhuri river confluence, Bangladesh", *Ph. D. Thesis*, University of Chittagong, Bangladesh, 320p.

PERMISSIONS

All chapters in this book were first published in RALF, by AgroAid Foundation; hereby published with permission under the Creative Commons Attribution License or equivalent. Every chapter published in this book has been scrutinized by our experts. Their significance has been extensively debated. The topics covered herein carry significant findings which will fuel the growth of the discipline. They may even be implemented as practical applications or may be referred to as a beginning point for another development.

The contributors of this book come from diverse backgrounds, making this book a truly international effort. This book will bring forth new frontiers with its revolutionizing research information and detailed analysis of the nascent developments around the world.

We would like to thank all the contributing authors for lending their expertise to make the book truly unique. They have played a crucial role in the development of this book. Without their invaluable contributions this book wouldn't have been possible. They have made vital efforts to compile up to date information on the varied aspects of this subject to make this book a valuable addition to the collection of many professionals and students.

This book was conceptualized with the vision of imparting up-to-date information and advanced data in this field. To ensure the same, a matchless editorial board was set up. Every individual on the board went through rigorous rounds of assessment to prove their worth. After which they invested a large part of their time researching and compiling the most relevant data for our readers.

The editorial board has been involved in producing this book since its inception. They have spent rigorous hours researching and exploring the diverse topics which have resulted in the successful publishing of this book. They have passed on their knowledge of decades through this book. To expedite this challenging task, the publisher supported the team at every step. A small team of assistant editors was also appointed to further simplify the editing procedure and attain best results for the readers.

Apart from the editorial board, the designing team has also invested a significant amount of their time in understanding the subject and creating the most relevant covers. They scrutinized every image to scout for the most suitable representation of the subject and create an appropriate cover for the book.

The publishing team has been an ardent support to the editorial, designing and production team. Their endless efforts to recruit the best for this project, has resulted in the accomplishment of this book. They are a veteran in the field of academics and their pool of knowledge is as vast as their experience in printing. Their expertise and guidance has proved useful at every step. Their uncompromising quality standards have made this book an exceptional effort. Their encouragement from time to time has been an inspiration for everyone.

The publisher and the editorial board hope that this book will prove to be a valuable piece of knowledge for researchers, students, practitioners and scholars across the globe.

LIST OF CONTRIBUTORS

Moonmoon Nahar Asha, Atiqur Rahman, Quazi Forhad Quadir and Md. Shahinur Islam
Department of Agricultural Chemistry, Faculty of Agriculture, Bangladesh Agricultural University, Mymensingh 2202, Bangladesh

Shah Md. Yusuf Ali, Sharmin Akhter and M Abdul Matin Biswas
Department of Agricultural Extension, Gazipur Sadar-1701, Bangladesh

Md. Ahiduzzaman, Nafis Iqbal, Jakaria Chowdhury Onik and M Hafizur Rahman
Department of Agro-processing, Faculty of Agriculture, Bangabandhu Sheikh Mujibar Rahman Agricultural University, Salna, Gazipur-1706, Bangladesh

Akida Jahan, Nushrat Jahan, Farjana Yeasmin, Mohammad Delwar Hossain and Muhammed Ali Hossain
Department of Plant Pathology, Faculty of Agriculture, Bangladesh Agricultural University, Mymensingh-2202, Bangladesh

Mohammad Monirul Hasan
Department of Economic and Technological Change, Center for Development Research, University of Bonn, Walter-Flex-Str. 3, D-53113, Germany

József Tóth
Corvinus University of Budapest, Department of Agriculture Economics and Rural Development, H-1093, Budapest, Fovam ter 8, Hungary

Md. Sarowar Alam
Scientific Officer, Regional Agricultural Research Station, BARI, Akbarpur, Moulvibazar, Bangladesh

Md. Sultan Mia
Scientific Officer, Regional Agricultural Research Station, BARI, Hathazari, Chittagong, Bangladesh & PhD fellow, School of Plant Biology , University of Western Australia, Perth, WA, Australia

Md. Salim
Scientific Officer, Hill Agricultural Research Station, BARI, Ramgarh, Khagrachari

Jubair Al Rashid
Executive, R & D, MATEX Bangladesh Ltd. Dhaka, Bangladesh

Md. Saidur Rahman
Senior Scientific Officer, RARS, BARI, Akbarpur, Moulvibazar and PhD fellow, Department of Horticulture, BAU, Mymensing, Bangladesh

Md. Rafiqul Islam and Aurunima Kanchi Suprova Shawon
Department of Soil Science, Bangladesh Agricultural University, Mymensingh-2202, Bangladesh;

Most. Lutfun Nesa Begum
Practical Skill Development Training Department, Proshika, Mirpur-2, Dhaka-1216, Bangladesh

Azmul Huda
Department of Soil Science, Bangladesh Agricultural University, Mymensingh-2202, Bangladesh; Department of Soil Science, Sylhet Agricultural University, Sylhet-3100, Bangladesh

Sushan Chowhan
Bangladesh Institute of Nuclear Agriculture (BINA), Sub-station, Khagrachari, Bangladesh

Shapla Rani Ghosh
Department of ICT, Mawlana Bhashani Science and Technology University, Tangail, Bangladesh

Tushar Chowhan
Department of Geography and Environment, University of Dhaka, Bangladesh

Md. Mahmudul Hasan
Apex Organic Soya Industries Ltd., Sapmara, Gaibandha, Bangladesh

Md. Shyduzzaman Roni
Department of Horticulture, Bangabandhu Sheikh Mujibur Rahman Agricultural University, Gazipur, Bangladesh

Monika Nasrin and Md. Abdus Salam
Department of Agronomy

Md. Akhter Hossain Chowdhury
Department of Agricultural Chemistry, Bangladesh Agricultural University, Mymensingh-2202, Bangladesh

Md. Arif Hossain Khan
Joint Director (Fertilizer), Fertilizer Management Division, BADC, Rajshahi

Md. Muzammel Hoque
Principal Scientific Officer, BANSDOC, Agargaon Dhaka, Bangladesh

Sherajum Monira, M. Abdur Rahim, MAB Khalil Rahad and M. Ashraful Islam
Department of Horticulture, Faculty of Agriculture, Bangladesh Agricultural University, Mymensingh-2202, Bangladesh

Amit Malaker
Department of Seed Science and Technology

AKM Zakir Hossain
Department of Crop Botany

Tahmina Akter
Department of Biochemistry and Molecular Biology

Md. Shariful Hasan Khan
Department of Soil Science; Faculty of Agriculture, Bangladesh Agricultural University, Mymensingh-2202, Bangladesh

Fatema Zahan, Md. Masudul Karim, Tahmina Akter1 and Md. Alamgir Hossain
Department of Crop Botany and 1Department of Biochemistry and Molecular Biology, Faculty of Agriculture, Bangladesh Agricultural University, Mymensingh-2202, Bangladesh

Masuma Habib
Graduate Training Institute Bangladesh Agricultural University (BAU), Mymensingh-2202

Abu Jafur Md. Ferdaus and Md. Shawkat Ali
Department of Poultry Science, BAU, Mymensingh-2202

Md. Touhidul Islam
Nourish Poultry and Hatchery Ltd. Uttara, Dhaka-1230

Begum Mansura Hassin
Department of Livestock Services (DLS), Bangladesh, Dhaka- 1215

Rajib Jodder, Mohammad Asadul Haque and Tapan Kumar
Department of Soil Science, Patuakhali Science and Technology University, Patuakhali-8602, Bangladesh

M Jahiruddin
Department of Soil Science

M. Zulfikar Rahman
Department of Agricultural Extension Education, Bangladesh Agricultural University, Mymensingh-2202, Bangladesh

Derek Clarke
Department of Civil and Environmental Engineering, University of Southampton, UK

Tangina Akhter and Azmira Nasrin Riza
Bangladesh Agricultural Development Corporation (BADC), Dhaka, Bangladesh

Md. Ali Ashraf and Md. Monirul Hassan
Department of Farm Structure and Environmental Engineering, Bangladesh Agricultural University, Mymensingh-2202, Bangladesh

Farzana Akhter
Bangladesh Agricultural Research Institute (BARI), Gazipur, Bangladesh

Imam Mehedi, Afia Sultana and Md. Amanut Ullah Raju
Department of Plant Pathology, Faculty of Agriculture, Bangladesh Agricultural University, Mymensingh-2202, Bangladesh

Md. Mahbubur Rahman and Mohammad Mahir Uddin
Department of Entomology, Faculty of Agriculture, Bangladesh Agricultural University, Mymensingh-2202, Bangladesh

Md. Manik Hossain, Md. Shahadat Hossain and Md. Nazrul Islam
Department of Pathology and Parasitology

Md. Tareq Mussa
Department of Anatomy and Histology, Faculty of Veterinary and Animal Science, Jhenidah Government Veterinary College, Jhenidah, Bangladesh

SM Harunur Rashid
Department of Pathology and Parasitology, Faculty of Veterinary and Animal Science, Hajee Mohammad Danesh Science and Technology University, Dinajpur, Bangladesh

Sadia Jahan Moon
Masters in the International Environment and Development Studies, Norwegian University of Life Sciences, Norway

Md. Abdul Momen Miah
Department of Agricultural Extension Education, Bangladesh Agricultural University, Mymensingh-2202, Bangladesh

Niloy Paul, Mohammad Kamrul Hasan and Md. Nasir Uddin Khan
Department of Agroforestry, Faculty of Agriculture, Bangladesh Agricultural University, Mymensingh-2202, Bangladesh

M. Ashraful Islam
Department of Horticulture, Faculty of Agriculture, Bangladesh Agricultural University, Mymensingh 2202, Bangladesh

Daryl C. Joyce
School of Agriculture and Food Sciences, Faculty of Science, University of Queensland, Gatton, Australia

Md. Shahzad Kuli Khan and Mohammed Ashraful Haque
Bangladesh Fisheries Research Institute, Marine Fisheries and Technology Station, Cox's Bazar-4700, Bangladesh

Sheikh Aftab Uddin
Institute of Marine Sciences and Fisheries, University of Chittagong, Chittagong-4331, Bangladesh

Index

A

Acrylamide, 100-101, 106-109
Adaptive Measure, 124, 131
Aloe Vera, 84-91
Anthracnose, 25-32
Asparagines, 100-101, 103, 107-108
Avian Coccidiosis, 156-158, 160, 162

B

Bau-biofungicide, 25-32
Biogas, 134-135, 139-140
Bioinoculants, 1
Bitter Gourd, 126-128, 132, 150-154
Botanical Extract, 142
Botrytis, 180, 182, 184, 187-190
Bract, 180-186, 189-191
Broiler Chickens, 110, 116-117, 123
Brri Dhan28, 51-53, 55
Brri Dhan29, 76-82

C

Calcium, 6, 8, 10, 112, 119-120, 171, 180, 182, 186-190
Chitosan, 84-87, 89
Climate Change, 49, 57-62, 64-67, 70-75, 124-125, 132-133
Cluster Analysis, 44
Coccidiostats, 156, 161
Commercial Broiler Farms, 156-159
Composting, 134, 138-140
Copepoda, 192-195, 199-200
Coping Strategies, 57, 75
Corruption Control, 33

D

Dinajpur District, 156-159

E

Economic Crisis, 33-35, 38-41
Empowerment, 163-169
Ethylene, 15, 68, 90, 180-181, 185-187, 189-191
European Agriculture, 33

F

Fish Larvae, 192, 197-201

G

Genetic Divergence, 44-45, 48, 50
Genetic Variations, 92
Glucose, 15, 100-101, 103
Govt. Effectiveness, 33
Growth Depression, 110, 117

I

Igas, 163-165, 168
Inorganic Fertilizers, 51, 171, 177
Ipil-ipil, 170-178

K

Koroch Seed Cake, 110-115, 117-122

L

Leaf Biomass, 170-179
Leaf Infestation, 150-153

M

Magic Growth Spray Solution, 76-80, 82
Mango, 24, 84-87, 89-91
Mangrove Area, 192
Meat Yield Characteristics, 110-111, 114-115, 117, 121-123
Morphology, 92, 188, 200

N

N-fixation, 1
Nitrogen Uptake, 51-53, 55, 179

O

Organic, 6, 8, 12, 51-52, 55, 57, 61, 93, 171-172, 177-179, 192, 199

P

Performance, 1, 25, 43-44, 49, 51, 83-85, 89-90, 93, 98-99, 110-111, 113, 115, 117-119, 121-123, 142, 148, 177-179, 181, 183, 185, 188-190
Ph Meter, 9
Phenology, 92
Phosphorus, 6, 10, 52-53, 109, 112, 119-120, 171, 175
Pineapple, 9-24
Plant Growth, 1-2, 5-8, 31, 43, 59, 77, 93, 184, 188-189
Poinsettia, 180-191

Postharvest, 10-11, 24, 84-85, 88-91, 143, 180-191
Preharvest, 180, 190
Productive, 42-43, 79, 98, 110-111, 117, 119, 171, 177, 199
Productive Performance, 110-111
Pummelo, 44-50

R
Reducing Sugar, 9, 15-16, 23-24, 100-103, 106-109
Refract Meter, 9, 86
Regional Disparity, 33
Rhizobacteria, 1-8
Ridge Gourd, 126-128, 150-154

S
Seed Borne Fungi, 142, 144-146, 148-149
Simulation Models, 57, 60, 75
Snake Gourd, 127-128, 150-154
Solubilization, 1-6, 8
Soybean, 25-32, 65, 72-73, 111-112, 119-120
Sstorage Period, 9

T
Technical Efficiency, 33-43
Tomato, 63-64, 68, 73-75, 92-99, 127-128, 142-149, 188
Total Soluble Sugar, 100, 106-108
Trichoderma, 25-26, 31-32

U
Urea, 45, 53, 56, 76-83, 93, 170, 178

W
Waste Management, 134-135, 140
Water Scarcity, 124
Weighing Balance, 9, 12

Y
Yield, 2, 7, 32, 45, 49, 53, 57, 62, 64-68, 71, 73, 83, 85, 91-94, 97-99, 103, 106, 108, 111, 115, 118, 121-123, 125, 132-133, 143, 148, 155, 170-175, 177-179

Z
Zooplankton, 192-195, 197-201

www.ingramcontent.com/pod-product-compliance
Lightning Source LLC
Chambersburg PA
CBHW082027190326
41458CB00010B/3294